THE NEW ASPECTS OF TIME

BOSTON STUDIES IN THE PHILOSOPHY OF SCIENCE

VOLUME 125

MILIČ ČAPEK

THE NEW ASPECTS OF TIME

Its Continuity and Novelties

Selected Papers in the Philosophy of Science

KLUWER ACADEMIC PUBLISHERS
DORDRECHT / BOSTON / LONDON

Library of Congress Cataloging-in-Publication Data

Čapek, Milič.
The new aspects of time : its continuity and novelties / Milič Čapek.
p. cm. -- (Boston studies in the philosophy of science ; v. 125)
Includes bibliographical references.
ISBN-13: 978-94-010-7455-1 e-ISBN-13: 978-94-009-2123-8
DOI: 10.1007/978-94-009-2123-8

1. Space and time. 2. Time--Philosophy. I. Title. II. Series.
Q174.B67 vol. 125
[QC173.59.S65]
001'.01 s--dc20
[530.1'1] 90-44196

ISBN-13: 978-94-010-7455-1

Published by Kluwer Academic Publishers,
P.O. Box 17, 3300 AA Dordrecht, The Netherlands.

Kluwer Academic Publishers incorporates
the publishing programmes of
D. Reidel, Martinus Nijhoff, Dr W. Junk and MTP Press.

Sold and distributed in the U.S.A. and Canada
by Kluwer Academic Publishers,
101 Philip Drive, Norwell, MA 02061, U.S.A.

In all other countries, sold and distributed
by Kluwer Academic Publishers Group,
P.O. Box 322, 3300 AH Dordrecht, The Netherlands.

Printed on acid-free paper

TABLE OF CONTENTS

ACKNOWLEDGEMENTS

All previously published papers are included in this volume with the kind permission from editors and publishers, which is hereby gratefully acknowledged. The original places of publication are as follows:

1. 'Stream of Consciousness and '*durée réelle*,' *Philosophy and Phenomenological Research*, X, No. 3 (1950), 331–353.
2. 'The Elusive Nature of the Past,' *Experience, Existence, and the Good*. Southern Illinois University Press, (1961), 126–142.
3. 'The Fiction of Instants', *Studium Generale* 24 (1971), 31–43.
4. 'Two Types of Continuity,' *Boston Studies in the Philosophy of Science 13*. D. Reidel Publishing Company, (1974), 364–375.
5. 'Process and Personality in Bergson's Thought,' *The Philosophical Forum 17* (1959), 25–42.
6. 'Russell's Hidden Bergsonism,' *Bergson and Modern Physics (Boston Studies 7*). D. Reidel Publishing Company, (1971), 333–345.
7. 'The Development of Reichenbach's Epistemology,' *The Review of Metaphysics 11* (1957), 42–67.
8. 'The Significance of Piaget's Researches on the Psychogenesis of Atomism,' *Boston Studies VIII*. D. Reidel Publishing Company, (1970), 446–455.
9. 'Toward a Widening of the Notion of Causality,' *Diogenes*, no. 28 (Winter 1959), 63–90.
10. 'Simple Location and Fragmentation of Reality,' *The Monist 46* (1964), 195–218.
11. 'Particles or Events?,' *Boston Studies 82*. D. Reidel Publishing Company, (1984), 1–28.
12. 'The End of the Laplacian Illusion,' *Philosophical Impact of Contemporary Physics*. D. van Nostrand Company, (1961), 289–332.

13. 'Eternal Recurrence – Once More,' *Trans. of C.S. Peirce Soc., 19,* no. 2 (1983), 141–153.
14. 'Note About Whitehead's Definitions of Co-Presence,' *Philosophy of Science 24* (1957), 79–86.
15. 'Bergson and Louis de Broglié,' *Bergson and Modern Physics (Boston Studies 7).* D. Reidel Publishing Company, (1971), 292–301.
16. 'What Is Living and What Is Dead in the Bergsonian Critique of Relativity,' trans. of French original in *Revue de Synthese* (1980), 313–344.
17. 'Time-Space Rather than Space-Time,' *Diogenes*, No. 123 (1983), 30–49.

EDITORIAL PREFACE

At last his students and colleagues, his friends and his friendly critics, his fellow-scientist and fellow-philosophers, have the works of Milič Čapek before them in one volume, aside from his books of course. Now the development of his interests and his thoughts, always led centrally by his concern to understand 'the philosophical impact of contemporary physics', becomes clear. In the nearly 90 essays and papers, and in his book on the philosophical impact as well as his classical restatement of process philosophy in his *Bergson and Modern Physics*, Professor Čapek establishes one of the fundamental alternatives to the comprehension of human experience, and thereby of the world.

Čapek is certainly to be seen with respect and admiration, for he has dealt with the deepest and toughest of scientific as well as metaphysical problems: his major efforts in the philosophy of mind focussed upon the time of experience, and in the philosophy of physics focussed upon continuity, causality and again the temporal, now in the world-picture. Has Čapek brought a renaissance of Bergson's thought, working as he has in the writings of Whitehead, Russell, Piaget, Einstein, de Broglie, and the others? Must the fellow-philosopher always share a deep presupposition in order to answer Čapek's question, the title of a paper here, 'Particles or Events?'

I am so glad my dear friend agreed to speak with Professor Ronald Martin about his life, his times as he lived them, and his work. The introductory essay will gently but firmly bring readers into his context, into the plausibility and the reality of Milič Čapek's 'open universe'.

September 30, 1990 R.S. Cohen

MILIČ ČAPEK WITH RONALD E. MARTIN

RM: How and at what point in your intellectual development did you get interested in the problem of time?

MC: Even before my university education I began to realize the problematic nature of time – in particular the aspect of its future. If all future events result *inevitably* from their antecedent states, they do possess *genuine reality*, even though their pre-existence is not yet perceived *now*, but will later be *inevitably* perceived. Such causal necessity seems to be equivalent to logical necessity: just as in the traditional system of logic the conclusion is *contained* in the premises, the causal effect is *contained* in its causal antecedent, even though our psychological grasp of the future is not instantaneous. In other words, the strict causal relation seems to be indistinguishable from the logical implication.

RM: Do you remember your first awareness of this problem?

MC: I do remember in 1926 reading the novel *The Time Machine*, by H.G. Wells, in a good translation. The author represents the past, present, and future alongside each other; even though they are perceived by us in succession, they truly co-exist, juxtaposed along a static dimension of time which seemed like a fourth dimension of space. Wells imagines that while one is riding on a magical time machine, he could move in either direction and *see* the allegedly dead past as well as the allegedly not-existing (for us) future. But I could see in that an inherent difficulty.

RM: And what difficulty was that?

MC: As an ordinary physicist or biologist could point out, the machine and its passenger should be affected by travelling in either direction in time. By moving into the future, the machine would rust and then disintegrate; its passenger would age and finally die and enter the grave. When the machine went back into the past it would become increasingly new and shining until it would decompose (by an inverted logic) into parts and finally into the minerals. The passenger would diminish in size, become a child, and finally reenter his mother's womb. The situation is absurd. Furthermore, it was a contradiction: while time was excluded from the universe, viewed as becomingless and without succession, the tiny piece which was the time machine retained its moving, its succession, and therefore its temporality. This is why physicists, biologists and philosophers would reject such an absurd notion; it was retained only in science fiction and 'time tunnel' phantasies.

RM: But were the physicists who were so likely to reject Wells's notion likely to negate the concept of the becomingless universe?

MC: Not quite. That concept had begun rather early, in ancient Greek philosophy, in the thought of Parmenides; it was retained with some modifications in medieval philosophy, and strengthened in modern philosophy by the development of classical physics. I began to be aware of these tendencies as a student in the King Charles University at Prague.

RM: Could you explain more specifically?

MC: A completely radical elimination of time was at first confined only to the Eleatic School, with its absolutist concept of being. Becomingless time was retained in subsequent thought, but it was moderated by the concession of at least a sort of existence to the time and change of our empirical experience. Its status was far less perfect than the primary supreme principle often endowed with such different names as Being, Supreme Idea, Ineffable One, *Deus* of Aquinas and of Calvin, Substance of Spinoza, and Eternal Energy or Matter of classical physics. In other words, time was retained as a shadowy and less perfect copy of the fundamental, eternal and changeless reality. The textbooks I encountered in the history of philosophy treated this approvingly and seldom criticized it. In my first reading of Bergson, then, especially of the concluding chapter of *Creative Evolution*, I began to doubt with him the prevailing tendencies to reduce the status of time to a mere 'appearance' or even to 'illusion.' With Bergson and with William James's *Pragmatism* in my university and post-university readings, I was getting acquainted with serious critical analyses of the attempts at degrading the status of time.

RM: Have you been fully satisfied with both Bergson and James?

MC: I was impressed by both of them, though not without some doubts. I was impressed by the first chapter of *Creative Evolution* in which Bergson insisted on the primary character and genuine reality of becoming, in opposition to the static spatialization of time, which he dismissed as a mere 'mechanistic illusion.' He insisted that there is an *emergence of novelty* at every moment of duration, especially conspicuous in our free decisions (as he explained in his first book, *Time and*

Free Will), and probably too in the creation of organic forms, as vitalists often claimed. But how could there be a convincing proof about the presence of indetermination even on the *physical* level, in inorganic nature? Could he deny the general belief in the strict causal law, generally then accepted in physics (Laplace) and biology (Huxley) of which he was aware? On the other hand, he pointed out that in mechanism "time is... deprived of efficacy, and if it *does* nothing, it *is* nothing. Radical mechanism implies a metaphysics on which the totality of the real is postulated complete in eternity, and in which the apparent duration to things expresses merely the infirmity of a mind that cannot know everything at once." (*Creative Evolution*, 1907) Physicists generally adhered to strict determinism, some even after 1927. Thus it was that I decided to study an unusual combination of fields in the university – philosophy and physics – and, unavoidably, mathematics (at least infinitesimal calculus and differential geometry).

RM: Did you encounter similar challenges in William James's thought?

MC: In reading *Pragmatism* I was impressed by some critical passages favoring pluralism based on our experience, as opposed to the extremely abstract structure of monism. But in my second university year I got acquainted with *A Pluralistic Universe* in which he used profound analysis of our concrete experience of change and diversity, producing a new metaphysics of process, incompatible with the traditional systems of static monism. It was undeniable that he was positively responding to Bergson; but some of his insights into the nature of time and the problem of determinism were anticipatory.

RM: Could you have been aware then of some of James's anticipations of Bergson's views?

MC: I was not yet aware, as I would be later, in reading his early essay 'Dilemma of Determinism' (1884) in the United States, and his *Principles of Psychology* (1890). Then I noted especially his criticism of the doctrine of 'the automaton theory of mind' and of the atomization of mind (associationism), his very subtle insights into the 'stream of thought,' consciousness of self, and the perception of duration. Also his sympathy for Renouvier's daring belief in elementary 'absolute novelties' even in nature. This I read about in Prague in 1946–47, when I returned to my native country from which I had been away during the war. Would you kindly allow me a brief outline of an intermezzo period (1936–46) of my life?

RM: Very gladly, especially if you would also tell of your finishing your studies before the political tension and the coming of war.

MC: I finished my studies of all three areas – philosophy, physics, and mathematics – in 1929–36. It required two degrees equivalent to the A.M. degree in this country (in philosophy and mathematics) and the equivalent of a Ph.D. in philosophy. My interest in physics was awakened by more than just the examinations: by reading books on the history of physics (by Mach and Duhem especially), and books offering philosophical insights on the theories of relativity and quanta (by Reichenbach and Eddington, and especially Émile Meyerson's *La deduction relativiste*, on which Einstein himself had commented so favorably in an article in a French review). I was deeply interested in the explanations and philosophical comments by de Broglie and J. Frenkel on wave mechanics and on the meaning of the principle of indeterminacy. Such views were discussed in my doctoral thesis on Bergson's daring anticipations of some modern trends in physics; the thesis was published in March 1938, and Bergson read an enlarged French résumé and wrote me expressing his deep appreciation. That reached me in July, 1938, only three months before the mutilation of my country, followed by its occupation by the Nazi armies in March, 1939.

RM: When did you leave Czechoslovakia?

MC: I was encouraged by L'Institut d'Ernest Denis at Prague, and endowed with a scholarship for study at the Sorbonne. That French institute still remained during the last days of peace, untouched by the occupying Germans who were probably hoping that another Munich appeasement would be offered by the Western powers. But the war started on September 1st, and the institute was abolished at once. I had left Prague just ten days before.

RM: Would you briefly describe your experience of the intellectual situation of wartime France?

MC: I had immediately lost my financial support from the French institute at Prague, now closed, but fortunately I soon was aided and generously endowed by the French Foreign Office (January 1st, 1940). Thus I could establish a contact with the Sorbonne. I attended the lectures of outstanding philosophers: Léon Brunschvicg, Émile Bréhier, Jean Wahl. Brunschvicg was committed to rationalism and the strict determinism of classical physics, while being interested in, as well as critical of, the coming of the new physics. Bréhier was thoroughly acquainted with the whole history of philosophy: ancient, medieval, and modern.

Wahl was interested in modern forms of metaphysics in France (Bergson), Germany (Heidegger), England (Whitehead), and the U.S.A. (William James and others). Both Bréhier and Wahl welcomed the contributions to the philosophical reviews which I had begun writing in the spring of 1940, but publication was interruptd by the German invasion of France. Thus my French papers could appear only after the war, in the early fifties.

RM: Was any contact with Bergson possible?

MC: I was deeply gratified by his kind and moving response (November 9, 1939) to my letter. He was sincerely saddened by the tragedy of my native country, but still hopeful for the defeat of Germany. His illness kept him away from Paris in his eightieth year; he died hardly more than one year later (January 5, 1941) when France, including Paris and the whole Atlantic coast, had been taken over by the enemy.

RM: When did you leave Paris and France?

MC: Allow me, before speaking about that, to mention another personel regret. I was looking forward to attending the lectures to be given by Louis de Broglie, the winner of the Nobel Prize for his work on the foundation of wave mechanics. I was interested in his articles on philosophical problems and his leaning toward the admission of indeterminacy on the microphysical level. (That concept had been daringly anticipated by Émile Boutroux in his 1874 book *De la contingence de la nature*, which was ignored in that period of classical physics; after 1927 the situation changed.) Professor de Broglie was registered for military research with the beginning of the war, and his lectures were never given. He remained in Paris and the Nazis did not dare to touch him. Leon Brunschvicg escaped to southern France where he died in 1944. Bergson bravely returned to Paris to face the racist registration by the Germans. I learned at Marseille about his death, six months after I had run from Paris with several of my Czech friends; it had been only one day ahead of the Germans coming there.

RM: How were you able to leave for the U.S.A.?

MC: It took more than fifteen months to get from Paris to New York – June 13, 1940 to September 24, 1941. That was too much time, and there were great difficulties and even some dramatic events. First, escaping from Paris by bicycle on roads crowded with refugees and retreating and disorganized French troops. The Nazi planes bombed constantly and indiscriminately, all the way to Bordeaux. Many of the French were sympathetic, but the new Vichy administrators were often suspicious.

Passage through Spain, an ally of Hitler, was impossible, and we had to wait at Marseille for the Vichy government to decide to allow us to leave. Finally, the departure on a French ship on January 15, 1941 – bound for Brazil! But the ship never reached Brazil because of the disagreement between Vichy and England, which was still at war; hence we waited, first at Dakar, and then at Morocco, where we were held at a 'settlement camp' at Kasba Tabla, in the desert. Our final wait was at Casablanca for another departure, this one by a Portuguese ship which reached New York at that date indicated above.

RM: What was your new life like in the new country?

MC: I still gratefully remember all the kind and understanding people who helped to bring me here and helped me adjust to the new country. I remember kindly all the organizations, institutions, and even persons: first, the Czechoslovak consulate at New York (in 1941 still recognized by the U.S., although the Germans occupied my country; only two European countries – England and Portugal – also did so). Also there was the National Unity of Czechoslovak Protestants in Chicago, and in particular Professor Matthew Spinka of the University of Chicago (an outstanding scholar and a specialist in Slavic history, especially the religious development of the Czech reformation). He provided me with a scholarship to study physics at his University.

I was able to finish my review course in optical physics because Professor George Monk allowed me to answer all the questions in the test examinations in French! I had hardly finished the course when the Japanese attack on Pearl Harbor brought the U.S. into the war. At that time the Attorney General changed the student visas of the refugees into visitor visas, thus allowing them to work. The Institute of International Education suggested that I be accepted by Pendle Hill, a Quaker center for study and contemplation, where discussions of the philosophical problem of time were not excluded. At the same time I was able to work in a modest job as a laboratory assistant in physics at the University of Pennsylvania (1942–43).

Before my sojourn in Pennsylvania was over I was invited by Professor René Wellek, of Czech origin and later famous for his several volumes on the theory of literary criticism (translated into sixteen languages – except Czech!). He organized the Army Special Training Program (ASTP) for teaching four different languages – German, Italian, Russian, and Czech – to the inductees, and he invited me to teach the boys of Slavic origin. That was done on the campus of the University of Iowa; like many other universities and colleges, the University of Iowa survived the extensive induction of male youth by having semi-military status and admitting the Army and Navy units. The ASTP program was shortened two months before the invasion of Normandy when the army withdrew the soldiers from the campus. I was then invited to teach physics, including analytical mechanics, at a V-12 Navy course at Doane College, and at the Air Corps program at the University of Nebraska, Lincoln (1944–46). Although the war ended in 1945, one year later I returned, for several reasons, to my liberated country for seventeen months.

RM: Had you lost your interest in philosophy during the war?

MC: Not at all, but naturally the amount of philosophical reading was reduced. But it was impossible to forget the problem of time, even in teaching physics, at least for me. The first paper that I wrote and read in English was on the relation of the concept of time with that of determinism. I read it to the class of Dr. Howard Brinton at Pendle Hill Center, and he was very appreciative. In Nebraska I got acquainted with Professor W.H. Werkmeister, then the author of *A History of the Philosophy of Science*, and I attended some of his classes. As early as 1943 I got acquainted with the development of American philosophy from its precolonial history to the twentieth century. One book helped especially. It was written by an outstanding philosopher-mathematician, Karel Vorovka of the University of Prague; it had been published in the same year as his death (1929). The book was rather rare in European philosophical literature, and this was the reason why I read it so much later, in this country. Before the war I had read two of his outstanding books, unfortunately untranslated from my native language: *Reflections on the Intuition in Mathematics* (1917) and *Kant's Philosophy and Its Relations to Exact Science*. His interest in Poincaré, Brouwer, Einstein and modern physics had stimulated my thinking in philosophy of science. In his *Polemos* (1925) he pointed out a certain affinity of the philosophy of T.G. Masaryk with William James, although they were not acquainted with each other. (James was known in Europe only superficially, as a mere 'pragmatist,' not in relation to his metaphysics of real time and pluralism.) When Vorovka died prematurely, the significance of his work was greatly appreciated by Daniel Essertier, a French sociologist, in *La nouvelle Revue de Prague* (1930). During my last years in the U.S.A. I was reading Whitehead's early writings such as *The Concept of Nature*, William James's 'The Dilemma of Determinism,' Z. Zawirski's *L'Evolution de la notion du Temps*, an introduction of Victor Lowe on the development of Whitehead's philosophy, some short books dealing with the problem of emergence of evolution and life by Charles Sherrington and W.M. Wheeler, some articles in *Philosophy and Phenomenological Research* on the dynamic nature of 'the stream of thought,'and other papers. But it was difficult to forget about political and military developments and the changing situation in Europe. Would you allow me to mention one particular political problem that especially concerned me?

RM: Why not?

MC: The re-establishment of my native country came in May 1945. One minor problem at that time concerned the boundaries of Czechoslovakia with Germany. The pre-Munich shape of Czechoslovakia was restored, but I felt that some geographical – and historical – irregularities should be corrected. I wrote *A Key to Czechoslovakia* (New York, 1946) before my return to Europe.

RM: What corrections did you propose?

MC: I proposed to re-incorporate the county of Kladsko (Glatz) into the territory of Czechoslovakia. It was carved out of the body of the kingdom of Bohemia by Frederick the Great of Prussia in 1742 for his strategic purposes. My thesis was that if Kladsko could be re-integrated into Czechoslovakia, Germany could be compensated by two tiny salients of Bohemia outside the frame of her mountains. I enlarged this book with wider documentation, and it was accepted for publication by Orbis at Prague in 1947; its title was *The History of the Czech Population of Kladsko*. Czech publication was completely forbidden, though, when I returned to the U.S. after my sojourn in my country. And the historical Prussianization of Kladsko was replaced by its Polonization by the dictate of Stalin, who favored Poland over Czechoslovakia, which was not yet completely subject to the sovietization. This happened only one month after my departure.

RM: What about your life in those seventeen months, 1946–48? Did you have any contact with philosophy or physics in Prague?

MC: I was not at Prague, since I was assigned to teach the introduction to physics at a new university established at Olomouc, Moravia. My interest in philosophy had not declined, but the risk of the concern about the genuine problems in metaphysics and philosophy of science grew. Political developments were a prevailing concern, and I had increasing anxiety about the destiny of the country and of the whole of Eastern Europe. This is why I left my country for the second time for the U.S.A.

RM: Have you then returned completely to philosophy, including philosophy of science?

MC: I faced three offers to teach; philosophy, physics, and mathematics. My choice eventually was philosophy which I taught for nearly three decades (1948–1978), first at Carleton College, then Boston University and as emeritus professor at three universities (Yale, Emory and the North Texas University). My interest was philosophy and history of science, not unrelated even to such broader fields as metaphysics and philosophy of religion. The contact with *Boston Studies in Philosophy of Science* and my participation in several congresses in this country and abroad, especially in France and Switzerland, stimulated and broadened my search for a greater insight into the nature of time and the revision of the concept of determinism. All its essence is contained in Professor Robert S. Cohen's selection of my articles and especially the relevant sections of two of my books.

RM: Do you have interest in the broader implications and impacts of different attitudes toward the problems of time and determinism which transcend the limits of contemporary physics?

MC: I pointed out in several essays the influences of different views on time before the coming of physics. The dilemma of static Being and dynamic Becoming appeared at the dawn of the Western thought; the former was generally preferred to the latter. Although the reality of time was very rarely completely denied, it was generally subordinated to the supreme timeless Being whether divine or impersonal in various systems of metaphysics. Thus the traditional theology was affected. The dogma of the timeless divine Being implied the absolute omniscience, foreknowledge and predestination; therefore the notion of spontaneity and freedom virtually or even explicitly eliminated. Modern thought, both in philosophy and modern science, including classical Newtonian physics and Darwinian biology, insisted on the strict predetermination of every event. In this way the personal God was replaced by the impersonal, Spinoza's substance (*Deus sive nature*), or by Laplace's static mathematical order. Such 'iron determinism' was applied not only to nature, but also to man and to the whole history of mankind. It led to discouraging pessimism in psychology and even in ethics. More than once it was claimed that the ontological distinction between good and evil was eliminated: they are both inevitable products such as 'the production of sugar and sulphuric acid.'

RM: You have indicated that such extreme determinism was and is challenged in the physics of the twentieth century. Does this challenge reopen the discussion on the real character of freedom and spontaneity? Does it restore the objective distinction between values and disvalues lost almost a century ago by naturalism and pessimism?

MC: There are two questions which we face. It is generally recognized (except by very conservative thinkers) that microphysical indeterminacies make free actions possible. John Dewey, with his penetrating insight, rightly formulated that microphysical contingency must be *necessary* though not sufficient for free actions; they result only from the convergent summation of microphysical events by the acts of mind. The second question is concerned about the ethical character of freedom: does it mean that *every* free choice results in the creation of positive values? A mere glance at the last decades shows depressingly many instances in which evil is deliberately and consciously created. The philosophical systems of Heidegger and Sartre show clearly that 'freedoms of existentialism' were going hand in hand with the most extreme forms of both ethical and intellectual irrationalism. But *not* every novelty can be identified with a positive value. The distinction between values and disvalues remains justified, however, in the open universe.

PART I

THE PROBLEMS OF TIME IN PSYCHOLOGY

CHAPTER 1

STREAM OF CONSCIOUSNESS AND "DURÉE RÉELLE"

I

A future historian of ideas will certainly consider James and Bergson as two founders and initiators of a new and fresh movement in human thought to which Arthur Lovejoy has given a very appropriate name—temporalism. With such few exceptions as Heraclitus, Schelling in his last period, and, to a certain extent, Hegel, there has been a persistent tendency in the philosophy of all ages to interpret reality in static terms and to consider temporal existence as a shadowy replica of a timeless and Platonic universe. Even apparently dynamic conceptions of the world did not avoid eventual fascination by the changeless and static. Nothing sounds more dynamic than Schopenhauer's term "Will"; and yet in setting "Will" beyond time he arrived at something hardly less static and immutable than the Eleatics and Spinoza. Even Hegel's notion of "Werden" (Becoming) is not dynamic in the true sense of the word, since it is conceived to be a result of the synthesis of two opposites which are equally static—Being and Non-being. Only in the transition period between the nineteenth and twentieth century did the deep meaning of the temporal character of reality begin to be understood. Although it is difficult to assign a definite date to the birth of an idea, the rediscovery of time in modern thought will be connected with a decade between 1880–1890. In 1884 James published an article in *Mind* under the title, "On Some Omissions of Introspective Psychology," where the main features of his concept of the stream of thought appeared; while on the other side of the Atlantic Henry Bergson silently meditated on his *Essai sur les données immédiates de la conscience*, whose title is strikingly similar to that of James. Thus an extremely fertile period in philosophy began which found its natural continuation in the far-reaching revolution in physics and the corresponding profound modification of basic notions including that of time. The recent generation of temporalists, marked by such names as Ch. D. Broad, Samuel Alexander, A. N. Whitehead, betrays the double influence of modern physics and Jamesian and Bergsonian thought.

It is not our purpose to follow historically the relations and mutual influences between both thinkers. An attempt to compare them has already been made by Horace M. Kallen, while R. B. Perry dealt in two chapters of his voluminous work with their personal and literary contacts. R. B. Perry also stated in a convincing way that the question of philosophical priority is meaningless; the influences between James and Bergson were mutual with the initiative passing alternately from one to another.[1] For our purpose, however, it is sufficient to state that the notion of the stream of thought, as well as that of *durée réelle*, were established independently, though both philosophers already knew each other at that time. It is interesting to investigate similarities and differences between two conceptions which, in spite of their independent origins are so strikingly similar.

Bergson himself was aware how close his concept of duration was to that of James. It is interesting to notice, however, that he was always ready to emphasize and even to exaggerate the differences between himself and James on this point. In a letter to William James on January 6, 1903, in which he expressed his admiration for James's *Varieties of Religious Experience*, he did not forget to point out the difference between "durée réelle" and "the stream of thought." In a letter to H. M. Kallen, published in *The Journal of Philosophy*, October 28, 1915, he opposed the metaphysical meaning of his notion of duration to the merely psychological meaning of the stream of consciousness. In a letter to M. Delattre on August 23, 1923, an even sharper distinction was drawn by him between both notions; or, in his own words, the analogy between them, being much more superficial than it is generally admitted, hides from us a "basic difference" ("différence fondamentale").[2]

What was considered by Bergson as a "basic difference"? He refers to the distinction made by James between "substantive" and "transitive" parts of the streams of consciousness which the author of *Psychology* calls "places of rest" and "places of flight." This abstract statement would hardly give us any precise and concrete idea of what James really meant. Only by reading and rereading directly the passages of *Psychology* can the meaning of the distinction become clear. "Like a bird's life, it (conscious life) seems to be made of an alternation of flights and perchings. The rhythm of language expresses this, where every thought is expressed in a sentence, and every sentence closed by a period. . . . The resting places are usually occupied by sensorial imaginations of some sort, whose peculiarity is that they

[1] It would require more space to show that the mutual contacts between James and Bergson passed through three distinct phases. The first one, marked by the influence of the older James on Bergson who not only quoted James's article "The Feeling of Effort," but also was stimulated by certain passages in his *Psychology* in setting up several important problems which *Matière et Mémoire* tried to answer. In the second period James expressly admitted his debt to Bergson in his last two books: *Pluralist Universe* and *Some Problems of Philosophy*. The third period, which is almost completely overlooked, concerns the posthumous influence of James as a philosopher of religion on the author of *Les deux sources de la morale et de la religion*, whose central problem of the mystical experience and divine personality was set by *Varieties of Religious Experience* as well as by *A Pluralist Universe*.

[2] *Etudes bergsoniennes*, pp. 9–10.

can be held before the mind for an independent period of time, and contemplated without changing; the places of flight are filled with thoughts of relations, static or dynamic, that for the most part obtain between the matters contemplated in the periods of comparative rest." (Let us note an important adjective used by James: "*comparative* rest.") According to James, "the main end of our thinking is at all times the attainment of some other substantive part than the one from which we have just been dislodged. And we may say that the main use of the transitive parts is to lead us from one substantive conclusion to another." But this peculiar character of the transitive parts accounts for an extreme difficulty in observing them introspectively. "Let anyone try to cut a thought across in the middle and get a look at its section, and he will see how difficult the introspective observation of the transitive tracts is. . . ." By focusing our attention on "the feeling of relation" (which is another name given by James to the transitive parts) we arrest it in flight and transform it into something fixed and crystallized which has always a character of a sensory image. Thus by the very act of introspective focusing its real dynamic nature is destroyed in the same way that the mobility of Zeno's arrow is frozen into an instantaneous static position.[3]

It is obvious that James was one of the first discoverers of so-called "imageless thought" which so powerfully stimulated the scientific investigation of the logical processes of thought as it was carried on later, especially by the Würzburg school in Germany (Külpe, Messer, Ach). To the old associationism and sensualism, logical thought was a simple succession of ideas which were conceived as mere faint replicas of sensory images. The difficulty of observing the vague and imageless "feeling of tendency" favored and apparently justified sensualism in reducing the differences between "conception" and "perception" to a simple degree of sensorial vividness. The unity of thought denied by sensualism and looked for by intellectualism in some "transcendental act" beyond concrete conscious states, was found by James in the transitive parts of the stream of thought.

Let us consider Bergson's criticism of the Jamesian distinction. In the 1903 letter he said: "I see in the 'resting places' themselves 'places of flight,' on which a fixed regard of attention confers an apparent immobility." Twenty years later the same objection was repeated by him even more strongly and the difference between his and James's notion was classified as "fundamental." Yet, it is quite obvious that Bergson imputed to James the absurd assumption that "the substantive parts" of the stream of conscious-

[3] *The Principles of Psychology*, Vol. I, pp. 243–44. How close James and Bergson were together already then is illustrated by the fact that both used Zeno's paradox to point out the insufficiencies of the conceptual treatment of dynamic psychological reality.

ness were something static and resting. Bergson's attention was caught exclusively by a metaphor "place of rest" and overlooked the all-important adjective "comparative" which James joined to the word "rest". Substantive part, "sensorial kernel," have naturally a certain temporal breadth; they are not instantaneous, that is, they are certain temporal wholes even when they are glimpsed "at once." No one was aware of this more clearly than James who in another chapter of his *Psychology* emphasized that the shortest sensation lasts at least 0.002 of a second. "At once" never amounts to a mathematical durationless instant.[4] The meaning of James's distinction is that while "substantive parts" are more easily observed and recalled, the transitive parts evaporate when the attention is focused on them. The possibility of observing and reproducing the "sensorial kernels" of thought lends them an apparent stability; they appear independent of time in the way a crest of wave traveling on the surface of a stream looks like a solid object seemingly independent of the fluid and moving water. The moving surface between two successive crests would illustrate the "feelings of tendency" or "transitive parts" of the stream of thought.

The difference between James and Bergson does not consist in the fact that for James the "substantive parts" are immobile, while for Bergson they are moving. For both of them the immobility of the "substantive tracts" is only relative. However, while for Bergson their apparent immobility is produced by the fixing of our attention, for James it has foundations in a real differentiated structure of our consciousness. It is a difference between a homogeneous, undifferentiated, even-flowing current and a stream on which a smooth surface alternates with whirls and waves. Of course, James also insisted that the "substantive parts" can be sharply separated from the "feelings of tendency" only by an artificial abstraction. He illustrated it in a suggestive way when he pointed out that we never hear "pure thunder," but "thunder-breaking-upon-silence-and-contrasting-with-it." A substantive part of the mental stream cannot be considered as a "break in the thought."[5] Nevertheless, in spite of this intimate connection with the whole stream of thought, it has its specific character different from its "halo" or "fringe," which are only different names for James's notion of the "transitive tract."

Apparently, Bergson, more fascinated than James by the continuous fluidity of psychological time, overlooked the relatively steady, though still temporal, whirl-like structures in the general fluidity of the stream of thought. But did he really overlook them? Nobody insisted more vigor-

[4] The whole chapter about the perception of time is a refutation of the concept of a "knife-edged present."

[5] *Principles*, Vol. I, p. 240.

ously on "heterogeneity of duration"; nobody protested more consistently against the supposed homogeneity and mathematical continuity of time, which, according to him, does not belong to true duration, but only to its graphic symbol.[6] One of his most valuable contributions to the analysis of the notion of time was his distinction between the concept of *mathematical continuity* and that of *dynamic continuity*. While dynamic continuity means for him the mutual penetration of the successive parts, "the intellectual representation of continuity is negative, being at bottom, only the refusal of our mind, before any actually given system of decomposition to regard it as the only possible one."[7] Mathematical continuity is nothing but discontinuity endlessly repeated or, as it is called, "infinite divisibility." In applying this concept to time we become entangled in Zeno's paradoxes which all result from the confusion of true duration with its underlying spatial symbol. Real time is not a succession of infinitely thin instants, but it grows by finite, indivisible drops.

It is curious that Bergson did not see that the admission of the "resting places" in the stream of thought is only the positive side of his refutation of the mathematical continuity or homogeneity of time. His disciple, Albert Bazaillas, saw this particular point more clearly. According to him, the psychologists were fascinated by a vertigo of succession and of restless change; the tendency to see change even inside the smallest drops of time led them to the artificial concept of infinitely divisible change considered as a succession of instantaneous (or almost instantaneous) and homogeneous sensations, whose combinations and rearrangement produce the apparently inexhaustible variety of psychological life. The individuality, structure, and specific character of the psychical wholes was, according to Bazaillas, ignored by the associationist school and was not fully appreciated even by Bergson in spite of his repeated attacks against the "atomistic" concepts in psychology. Bazaillas was more successful in finding more accurate expression for what James called "places of rest." At the end of the nineteenth century William Thomson set up a physical theory which considered the atoms as steady gyrostatic formations in the ether. The old-fashioned solidity of

[6] *Time and Free Will*, the whole chapter II, concerning the distinction between "true time" and "time-length".

[7] *Creative Evolution*, p. 154. Bergson points out the practical origin of the same notion: "No doubt, it is useful to us, in view of our ulterior manipulation, to regard each object as divisible into parts arbitrarily cut up, each part being again divisible as we like, and so on *ad infinitum*. . . . To this possibility of decomposing matter as much as we please, and in any way we please, we allude when we speak of the continuity of material extension; but this continuity, as we see it, is nothing else but our ability, an ability that matter allows to us to choose the mode of discontinuity we shall finding it." (*Ibidem*)

the atoms was thus given up and replaced by the fluidity of the ethereal whirls or "vortex-atoms"; their individuality, though preserved as far as their form was concerned, was resolved into the fluid continuity of the ether. According to Bazaillas, the psychical structures are analogous to the "vortex-atoms," being relatively steady forms in the mental stream and like the "vortex-atoms," never consisting of the same substance.[8] This metaphor, in spite of its limitations, has definitely less static connotation than James's term "place of rest."

Psychological duration is, therefore, not an amorphous fluidity, but a droplike succession of temporal wholes which can be divided into two distinct groups. The first group consists of more or less clear sensorial images, easily observed and recalled, which, being replicas of the direct sensorial impressions, borrow their apparent immobility from external objects which they represent. It is this group which attracted the exclusive attention of the associationist school. The second category is represented by the vague and fleeting "feelings of relation," "halos," "fringes," devoid of sensorial character, impossible to observe by a prolonged attention which disturbs and solidifies their fleeting mobility. Both are, however, the temporal pulsations of the stream of consciousness and the spurious immobility of the first group is rather imagined than directly experienced; only by an unconscious association with the idea of an external object do "the places of rest" appear static. The difference between both groups is not in their different relation to time, but in their different qualitative content.

Obviously, the opposition between the true duration of Bergson and stream of consciousness of James is more apparent than real. The misunderstanding was caused by an unfortunate metaphor used by James and also

[8] A. Bazaillas, *La musique et l'inconscience*, p. 264: "A force de vouloir échapper aux conditions de l'espace et aux formes immobiles, les psychologues laissent se dominer par le vertige de la succession et du changement. C'est un nouveau fantôme, le fantôme du temps qui les obsède: ils emièttent et dispersent la vie psychologique en matière impalbable, et ils ignorent qu'elle est faite précisément des reprises et des contractions de cette matière...." *P. 277*: "Ce sont de véritable atomes-tourbillons dans l'ordre mental...." Herbert Spencer in his *Principles of Psychology* and H. Taine in his *De l'Intelligence* are probably the most famous representatives of this tendency which in a serious effort to express the radical fluidity of time arrives only to its clumsy imitation by multiplicity of instants and "mental-dust" elements, both essentially static and discontinuous. The lasting conflict between a direct intuition and its spatializing symbol can hardly be better illustrated. According to Bazaillas, not even did Bergson escape entirely "the illusion of continuity" (homogeneous fluidity) of time by ignoring "the perceptions of difference" by which true duration is realized. (*La vie personnelle. Etudes sur quelques illusions de la perception intérieure*, pp. 190–2.) In view of Bergson's insistence on *duration* and *persistence of the past*, Bazaillas's reproach is not just.

by Bergson who in criticizing James misunderstood himself. That only shows how extremely complex and difficult is any discussion concerning the nature of time; a confusion between two entirely opposite meanings covered by the same term is very frequent, like, for instance, the confusion of mathematical continuity with qualitative continuity. On the other hand, two apparently different terms very often express the same fact. The terms "transitive tracts of the stream," "fringe," "halo," "feeling of relation" have nearly the same meaning in the writings of William James; his distinction between "places of rest" and "places of flight," expressing the difference between the sensorial and static thought and imageless and dynamic thought, corresponds very closely to Bergson's distinction between two aspects of the Self: "the one clear and precise, but impersonal; the other confused, ever changing and inexpressible, because language cannot get hold of it without arresting its mobility or fit it into its commonplace forms without making it into public property."[9] Let us compare this with the words of James in his *Psychology*: "As a snowflake crystal caught in the warm hand is no longer a crystal but a drop, so, instead of catching the feeling of relation moving to its term, we find we have caught some substantive thing, usually the last word we were pronouncing, statically taken, and with its function, tendency, and particular meaning in the same sentence quite evaporated."[10] The same idea was expressed later by Bergson in *L'Énergie spirituelle*: "Remember that real concrete living thought is something of which psychologists have so far told us very little, because it is very ill adapted to internal observation. What is usually studied under this head is not so much thought itself, as an artificial imitation of it obtained by putting together images and ideas. But with images, and even with ideas, you can no more reconstitute thinking, than with positions you can make movement. The idea is a *halt of thought*; it arises when the thinking, instead of continuing its train, makes a pause or is reflected back on itself. . . . You will see that it is impossible, that the thought translated by the sentence is an indivisible movement, and that the ideas corresponding to each of the words are simply the images or concepts which would arise in the mind at each moment of the thinking *if* the thinking halted; but it does not halt."[11] Not only the content, but also a metaphor used is the same: the "place of rest" of James is called the "halt of thought" ("l'arrêt de la pensée") by Bergson: both terms have an equally static connotation and Bergson in criticizing the James's term, involuntarily criticized himself. It is true, however, that he succeeded in the same essay in finding a

[9] *Time and Free-Will*, p. 129.
[10] *Principles of Psychology*, Vol. I, p. 244.
[11] *Mind-Energy*, pp. 55–56.

more accurate and less static expression for the rhythmical structure of time: "I believe that our whole psychological existence is just like this single sentence, continued since the first awakening of consciousness, *interspersed with commas, but never broken by full stops.*"[12] The full meaning of this sentence will be discussed later.

It is evident that both James and Bergson, while insisting on the continuity of time admit at the same time its discrete structure. Hardly any intellectual problem is more difficult than that involving the antinomy of temporal continuity and discontinuity. It has already been pointed out how extremely careful we have to be in dealing with such terms; for instance how mathematical continuity is the very opposite of a "sensible continuity." The complexity of the same problem is also illustrated not only by Bergson's misinterpretation of James, but also by the attitude of Charles Renouvier who misunderstood James in *exactly the opposite sense.* While to Bergson "stream of thought" with its "resting places" appeared to be too discontinuous, to Renouvier, on the contrary, it seemed too continuous and amorphous; he even reproached James for bringing τὸ ἄπειρον into the mind whose functions are essentially discrete.[13]

II

Yet there *was* a difference between the Jamesian and Bergsonian notions of duration, a very important and one might even say essential difference. It concerns different views of the *past* and its relation to the present. Both agreed in rejecting the concept of a dimensionless, punctual present, advancing "in the direction of the future" in the way a mathematical point slides along a geometrical line. Both arrived at this negative conclusion by the same method: by a direct introspection of the "immediately given" ("donnée immédiat") of the stream of thought ("durée réelle"). There was, however, one additional motive in Bergson's attitude: his distrust of all graphic symbols which by their static character naturally distort the true nature of time. Bergson insisted in all his books that time cannot be represented by a geometrical line as a kind of "fourth dimension." A line is a rigid, timeless thing, a "fait accompli," whose different parts exist simultaneously; true duration, on the other hand, is a "fait accomplissant," something not yet closed in its gradual and perpetual growth. Not much is gained, if we imagine the line as continuously extending its length in the "direction" of the future; our visual imagination will always set this growing line into an imaginary space in which the positions successively occupied

---

[12] *L'energie spirituelle*, pp. 56–7.

[13] R. B. Perry, *Thought and Character of William James*, Vol. I, p. 697. (Letter of Charles Renouvier to W. James on September 11, 1884.)

by the "present-point" virtually preexist; involuntarily, we imagine again the future as simultaneously juxtaposed to the present instant. Only by radically freeing our imagination from our visual habits and geometrical symbols, are we able to arrive at an undistorted picture of time, if we have the right to use the word "picture." It is not without interest that Bergson purposely avoided visual metaphors in describing the structure of duration; auditive and kinesthetic images (melody, elan, explosion) definitely prevail in his writings. Once we get rid of a symbolical representation of time as a line, we also escape the tendency to imagine its "parts" geometrically, that is as *points*. In other words, the moments of time are not punctual instants; time is not infinitely divisible. The homogeneity of time is thus denied; each temporal moment is heterogeneous and involves not a pure abstract present, but present tinged with "immediate recency."

Homogeneity and mathematical continuity belong, according to Bergson only to *empty time*, as for instance to the absolute and even-flowing time of Newton. But such a time, when looked at closely, appears to be only a verbal one, being at bottom, only a spatial, one-dimensional diagram, associated with the word "time." The reason why we attribute to such a static graphic symbol a temporal character, is that we associate it vaguely with our feeling of directly experienced duration; but the visual picture of a graphic symbol pushes the intuition into the background. There is no distinction between time and its qualitative content; a contentless duration is a meaningless notion, because a qualitative difference of the successive parts is the *very essence* of any temporal reality. Obviously, Bergson was right when he insisted that his analysis of true duration has deeper meaning than a purely psychological one. It aimed to explore the nature of time in general, not only that of "the stream of thought." To the Bergsonian "durée réelle" the words of Whitehead could be applied—that "the texture of observed experience, as illustrating the philosophical scheme, is such that all related experience must exhibit the same texture."[14]

This also accounts for the contrasting attitudes of these thinkers toward the theory of matter. It is obvious that Bergson by attacking in his *Essai* the concept of homogeneous time, virtually declared war on the whole of classical physics. Seven years later his rejection took more systematic form in his little known theory of matter, expounded in the fourth chapter of *Matter and Memory*. Several important features of modern physics were brilliantly anticipated in this book. James on the other hand at the time of the publication of his *Psychology*, was still dominated by the idea that the qualitative and heterogeneous "stream of thought" covers underlying Newtonian homogeneous time which is an original receptacle of all changes, in-

[14] *Process and Reality*, p. 5.

cluding those which are going on in the brain. Thence also came a rather unconscious tendency to consider physical time with all displacements of the material particles as something more basic and more real than the "apparent" stream of thought. It is true that this tendency crops up only sporadically. Nobody insisted more vigorously than James that in psychology "esse = percipi"; that a feeling is not a phenomenon, but a genuine and irreducible reality. "A material fact may indeed be different from what we feel it to be, but what sense is there in saying that a feeling, which has no other nature than to be felt, is not as it *is* felt?"[15] However, the opposite trend comes up very distinctly, at least in his *Principles of Psychology*, for instance in the attempt to relate the merging of the past into the present to the superpositions of the cerebral processes; while a past dying process is still going on, a present one reaches its maximum intensity.[16] Could that be an adequate explanation of the psychological immanence of the past in the present, if every psychological reality is a genuine and ultimate fact? Bergson pointed out that the successive states of the brain, strictly speaking, cannot mutually overlap in time, being temporally as sharply separated as the successive stages of any material system.[17] The physiological fact that several processes with unequal intensity go on *simultaneously* is something *entirely different* from the psychological merging of the past into the present. On one side we have the physical present which is practically (even if, according to the most recent views, not absolutely) an infinitely thin boundary between past and future; on the other side, the sensible present with a certain temporal breadth.[18]

It would not be just to overemphasize this slightly materialistic, or, better to say, parallelistic, element in James's *Psychology*; but it is no less certain that James, by restricting his analysis of time to the exclusively psychological field and by accepting *en bloc* classical physics of the kinetic type,[19] based on the concept of mathematically continuous time, ran into serious difficulties in formulating his thought. His courageous, but unending struggle with the problem of the interaction of mind and body and the problem of freedom, his tendency to escape the last problem either in a pragmatic way or in espousing a monism of "pure experience" which re-

---

15 *Some Problems of Philosophy*, p. 151; also *Principles*, Vol. I, p. 163.

16 *Principles of Psychology*, Vol. I, p. 635 (about the physiological interpretation of the sense of duration).

17 *Mind-Energy*, p. 70.

18 It is interesting, however, that as early as in 1895 (five years after the publication of *Psychology*) James became aware of the difficulty here. Why the psychological present has a certain minimum span contrasting with the instantaneous character of the physical present *"Why just this amount of time, neither more nor less?"* (*Collected Essays and Reviews*, p. 388).

19 *Principles*, Vol. I, p. 178.

duced the difference of the mental and the physical to simple difference of "context," are the obvious effects of the same gap in his thought. Only in the last stages of his philosophical development and not without the influence of Bergson, he approached the dynamic vision of the physical universe when he noticed that new discoveries in physics are incompatible with the traditional notion of the eternal fixity of elements, and when he began to treat not only mental processes but also "real processes of change" in general "no longer as being continuous, but as taking place by finite not infinitesimal steps."[20] Had James lived longer, there is no doubt that he would have better understood Bergson's *Matter and Memory*, especially its all important Chapter IV. His notion of drop-like, pulsating change shows how close he came to the spirit of modern physics, especially that of quantum theory and indicates how receptive he would have been to the present-day revolution in physics. In this respect James's philosophy remained, as R. B. Perry pointed out with a fully justified regret, "an unfinished task."[21]

A definite point of superiority of the Bergsonian analysis of time in comparison to that of James was its more precise and more intellectual, we almost should say, more deductive character. Horace M. Kallen already pointed out in his comparative study this typological difference between them.[22] Bergson's mind, in spite of all his alleged anti-intellectualism remained definitely Cartesian, while James was always faithful to the empirical English tradition. James restricted himself to the statement of empirical, immediately given psychological data; Bergson, though he claimed to stick to the "données immédiates de la conscience," tried to rationalize the structure of the immediately given and to understand the logical bonds between its individual features. Moreover, he tried to discover the nature of certain intellectual traps which await any attempt at temporal analysis; and he found that the most confusing and distorting effect is exerted by visual and, in particular, geometrical symbols when applied to temporal reality. That helped him to do away with the concepts of homogeneity and mathematical continuity not only of psychological duration, but of duration in general. His strenuous effort to avoid the visualization of time accounts also for one very important, if not essential, difference between his and James's conceptions of the past.

For James as well as for Bergson the present moment is not an infinitely thin instant, but a thick temporal whole tinged with "immediate recency."

---

[20] *Some Problems of Philosophy*, p. 172; p. 149, note James quotes here W. Ostwald and G. Le Bon (*L'Evolution de la matière*.)

[21] *Perry, l.c.*, Vol. II, pp. 666–7.

[22] *William James and Henry Bergson. A Study in Contrasting Theories of Life*, p. 121, note.

According to James," the distinctly intuited present merges into a penumbra of mere dim recency before it turns into a penumbra which is simply reproduced and conceived."[23] In other words, there are two kinds of past for James: one *directly* intuited inside of the "sensible present"; the second outside of the volume of the present, which can be only indirectly *reproduced*, not immediately grasped. A returning recollection is a simple *present* state, symbolizing a more distant past. In this sense Taine called memory "l'illusion vraie," because, though it deceives us in presenting a present mental state as an equivalent of the past one, it gives us nevertheless a certain kind of knowledge, often a very accurate one, *about* the past. James was of course aware that the volume of the present is elastic, that its breadth varies and is different for different categories of sensations; he knew that its "backward edge" is dim and that no sharp cut separates the immediate memory from the absolute past. He, however, judged (or, it is at least distinctly implied in his view) that beyond this dim rim of the "immediate memory" which can condense only twelve seconds of the "public time," the past hopelessly escapes the direct reach of our consciousness. While an impression, or, to speak in a less "atomistic" way, a certain mental quality which occurred eleven seconds ago, still dimly persists on the hazy edge of our specious present, all that is beyond this rim is literally *dead* and perished forever.[24] It is true that even the more distant past survives in the material modifications of the brain; but this type of survival is altogether different from the specific subsistence of the past in the consciousness; it consists in the last analysis of the succession of the instantaneous, or almost instantaneous, cerebral states. Each state like a state of any material configuration, exists exclusively, or almost exclusively, in the present; more accurately speaking, its "specious present" in the physical world (to which the brain belongs) is incomparably shorter than a corresponding mental present and is for all practical purposes equal to a point-like instant. (While a minimum psychological duration is of the order of 10^{-3} sec., that of the material duration is approximately equal to 10^{-24} sec., if we accept the value of the "chronon" of R. Lévi.) The persistence of the material trace is, therefore, only symbolical; it is not equivalent to a true duration, to a real survival of the mental past in the present. A cerebral state is always an almost mathematical present; while a mental past, though merging into a present moment, is, from the point of view of a physicist, a *real* past.[25] (The

[23] *Principles*, Vol. I, p. 636, note.

[24] That is what W. James called a "genuine past."

[25] Thence a persistent tendency of the physiologically oriented psychologists to deny the real persistence of the past and to reduce "the feeling of past" to the present sensations, so called "temporal signs." James (*Principles*, Vol. I, p. 632, note) gives a long survey of the attempts of this kind which all agree in reducing the differences be-

emphasis on a different temporal span in the psychological and material duration is the basis of the Bergsonian theory of matter.)

Because the past beyond the extreme edge of the specious present is dead and gone forever, the stream of consciousness in James's view is a "perpetual perishing" and the only subject of the psychological life is a present "perishing pulse of thought." The successive pulses of thought pervade each other, or, in James's words the latter "appropriates" the preceding one; but this mutual merging welds only two immediately successive pulses together; it cannot overbridge the larger intervals of time. All the past preceding our sensible present exists only symbolically in actual cerebral modification; as far as its true mental ontological status is concerned, it is forever lost in the abyss of non-being.

The difference between this and Bergsonian view of the past is subtle, but essential. It is so subtle that it escaped even Bergson himself; yet, it is the difference between the Heraclitean philosophy of perpetual perishing and the Bergsonian philosophy of true duration. Both philosophies are temporalist; but while James stresses mainly the character of restless change and fluidity, Bergson emphasizes the persistence as much as the flux. Both apparently opposite aspects of temporal reality are only its two complementary features, as we shall see later. Also the choice of the metaphors depicting the reality of time indicates the difference between both views. In James's writings the images of stream and perishing prevail suggesting a past restlessly "flowing away" and disappearing in the abyss of non-being; Bergson, on the other hand, by his metaphors underlines *the preservation* of the past (a melody in which the past tones merge with the present ones; a growing snowball, condensation, etc.). For Bergson not only the past adjacent to the present and condensed by the immediate memory persists, but also that which is more "remote," that outside of "the twelve seconds interval," exists, or better stated, *subsists*, betraying its virtual persistence by un undefinable coloring of the present moment.

tween the present and past feelings to a difference of quality or intensity. Thus a fading trace of a previous sensation which is going on simultaneously with the present sensation and is therefore *present itself*, is interpreted as a past feeling. Why this difference in intensity is "translated" by mind into the corresponding difference of their temporal order is impossible to explain and this impossibility is honestly recognized (by Th. Lipps, for instance).—The most extreme representative of the tendency to reduce all reality to a "knife-edged present" was probably H. Taine, who considered even the shortest sensation as only "apparently simple," being truly composed of an enormous number of successive, almost instantaneous sub-sensations. At this particular point the difference between him and James was only a gradual one, because even for James "the feeling of past time is a present feeling," even if the Jamesian present is much thicker than that of Taine. Bergson's words apply to both of them: "If we make recollection a weakened perception, we misunderstand the essential difference between the past and the present." (*Matter and Memory*, p. 72.)

The irreversibility of time exists only under such a condition. "Memory," says Bergson, "is not a faculty of putting away recollections in a drawer, or of inscribing them in a register. There is no register, no drawer; there is not even, properly speaking, a faculty, for a faculty works intermittently, when it will, or when it can, while the piling up of the past upon past goes on without relaxation. In reality, the past is preserved by itself, automatically. In its entirety, probably, it follows us at every instant; all that we have felt, thought, willed from earliest infancy is there, leaning over the present which is about to join it. . . . Doubtless we think with only a small part of our past, but it is with our entire past, including the original bent of our soul, that we desire, will, and act. . . . From this survival of the past it follows that consciousness cannot go through the same state twice. . . . Our personality, which is being built up each instant with its accumulated experience, changes without ceasing. By changing, it prevents any state, although superficially identical with another, from ever repeating itself in its very depth. That is why our duration is irreversible. We could not live over again a single moment, for we should have to begin by effacing the memory of all that which followed. Even could we erase this memory from our intellect, we could not from our will."[26]

The difference from the Jamesian view is obvious. Again this difference consists in Bergson's more radical attempt to purify our notion of time from all visual and geometrical imagery. James, though he rejected the artificial concept of a point-like instant, did not succeed in getting rid of all subconscious visual representations. His view of a present moment retains definite geometrical characteristics; it is something comparable to a *linear segment*. He did not go as far as Arthur O. Lovejoy who replaced the succession of instants by the succession of contiguous, temporal, sharply separated lengths;[27] nevertheless, in introducing the absolute externality between the more distant moments of time, he also yielded to an obtrusive geometrical analogy. His "perishing pulses of thought" are in this respect analogous to Lovejoy's temporal segments; the only difference is that James admits not only the contact, but also the penetration of the immediately successive moments. But only this kind of the "next-to-next" continuity is recognized by him; the moments separated by more than an interval of twelve seconds are absolutely discontinuous, mutually external like two points in space. More accurately speaking, there is no relation whatever, not even that of discontinuity, between them because the previous terms have completely *vanished*. Even when James stresses the fact that it is

---

[26] *Creative Evolution*, pp. 5–6.

[27] A. O. Lovejoy, "The Problem of Time in Recent French Philosophy," *Philosophical Review*, Vol. XXI, pp. 532–33.

impossible to estimate *when* the past becomes absolutely "dead," he does not change his basic assumption *that* such a death really occurs.

Kant compared the present to a mathematical point floating between two noughts—past and future; Lovejoy extended this point up to a definite length; James insisted that this length has no sharp edges; but all these three views agree in considering the present moment as *the only reality* hovering between two abysses of non-being. Time thus conceived is a "perpetual perishing" in which a spark of the present moment is extinguished in order to be replaced by another spark, equally ephemeral. Properly speaking, it is not so much a reality which gradually grows richer every moment as it is a case of Bergson's cumulative duration. The main difference between Heraclitean and Bergsonian time consists in a difference in *the status of the past.* The difference in structure between two types of time can be illustrated graphically. Any attempt to illustrate the temporal relations graphically, involves the danger of inconsistency; are not all visual symbols of time inadequate? Yes, they are: however, if we keep in mind that the different parts of a diagram represent not *simultaneous*, but successive stages of a temporal process, the danger of misinterpretation is considerably, though not completely, reduced. The following two pictures show approximately the differences in structure between the "stream of thought" and "durée réelle."

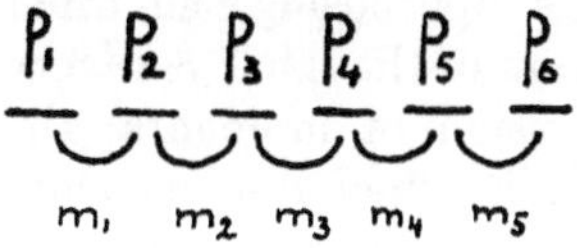

FIG. a. STREAM OF THOUGHT

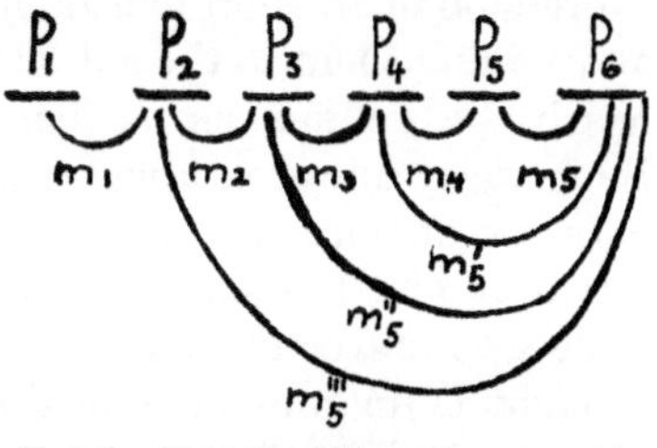

FIG. b. DURÉE RÉELLE

p_1, p_2, p_3, p_4 represent successive specious presents; m_1, m_2, m_3, m_4 represent the links of immediate memory (not recognized by Lovejoy, but admitted by James who insisted upon the "dim backward edge" of the specious present). It is obvious that in the "stream of thought" the successive moments p_1 and p_3 are completely external each to the other, being linked only by the intermediate terms. Bergson's duration shows a much more complex structure. Besides the bounds of "immediate memory" joining together two immediately successive terms, there are a number of relations, or, using Whitehead's term, "prehensions," binding together temporal terms which are not "contiguous." These links are more and more tenuous as the difference in "date" between the successive terms grows larger; but, essentially, there is no difference between this type of memory and immediate "retention." In Bergson's duration the whole past is co-present, though its different parts are present in different degrees; for James only the "immediate past" is real, being perceived at the edge of the "specious present." In Bergson's words: "I should no more seek the explanation of the integral preservation of this entire past than I seek the explanation of the preservation of the three first syllables of "conversation" when I pronounce the last syllable. Well, I believe that our whole psychical existence is something like this single sentence, continued since the first awakening of consciousness, interspersed with commas, but never broken by full stops. And consequently, I believe that our whole past still exists."[28] This passage indicates most clearly that from this point of view the difference between the immediate retention and the reproduction of the past is that of degree, not that of nature.[29]

This is the main difference between the "next-to-next" continuity of Jamesian duration and total continuity of "durée réelle." While the distant phases of the past are entirely separable from the present in the Jamesian view (or, in Bergson's terms, separated by "full stops"), the Bergsonian past is *indivisible* and as such penetrates into a present "occasion." However, this indivisibility does not mean *undifferentiatedness.* On the contrary, the specific characteristics of the past moments find their expression in an enormous number of temporal links superposing each other and joining them to the actual moment. "The stream of thought" can very nearly be defined as a "one-dimensional" succession of moments; the Bergsonian duration appears as an extremely complex "polydimensional" network of temporal bounds of unequal intensity and unequal span; the intensity and the span are "indirectly proportional."

Bergson never put his ideas into such a rigid form as the above, but his thought is unambiguous in this respect, even though it is expressed (and often veiled) in picturesque and poetic language. The difference between him and James came up very distinctly in their early correspondence at the beginning of this century. It concerned the question of subconsciousness which is only a different aspect of the problem of the past. James in denying any ontological status to the distant past except the physiological one, admitted the psychological existence of only *present mental states*, the present being understood in his non-punctual sense as having a certain

---

[28] *Mind-Energy*, p. 70.

[29] A. N. Whitehead who is very near to Bergson in many respects, speaks about the past in almost identical terms: "In memory the past is present. It is not present as overleaping the temporal succession of nature, but it is present as an immediate fact of mind. . . ." (*Concept of Nature*, p. 68). It would require, however, a careful and detailed study to compare Whitehead's idea of the past in *Process and Reality* with that of *Matter and Memory*.

temporal breadth. For him non-present = non-conscious = non-real. Bergson expressed his disagreement in his letter of March 25, 1903: "If we reduce the memories to the categories of things, it is clear that there is no mean for them between presence and absence; either they are unqualifiedly present to our mind, and in this sense conscious; or, if they are unconscious they are absent from our mind and should no longer be considered as present psychological realities. But in the world of psychological realities I do not believe that there is occasion for presenting the alternative "to be or not to be" exclusively. The more I try to grasp myself by consciousness, the more do I perceive myself as the totalization or *Inbegriff* of my own past, this past being contracted with a view to action. 'The unity of Self' of which philosophers speak, appears to me as a unity of an apex of a summit to which I narrow myself by an effort of attention—an effort which is prolonged during the whole of life, and which, as it seems to me, is the very essence of life."[30]

This very characteristic passage shows that Bergson can be regarded as a *personalist* provided that the category of personality is defined in temporalist terms; personality = a condensed past The personality thus defined has not an atom-like simplicity of a self-identical monad, but the dynamic continuity of effort. James, on the other hand, opposed to the artificial notion of a static "Ego" his "perishing Thought" acknowledged as the only "Thinker"; in his Heraclitean "stream of consciousness" where the perpetual perishing was overstressed and the temporal persistence overlooked, there is no place for anything more enduring than a "vanishing pulse of thought." John Dewey's observation of a "vanishing subject" in James's psychology could not have been more accurate.[31]

In the last stage of James's development his position under the admitted influence of Bergson has been considerably modified. His letter on June 13, 1907, to Bergson is revealing in this respect: "The position we are rescuing is 'Tychism' and a really growing world. But whereas I have hitherto found no better way of defending Tychism than by affirming the spontaneous addition of *discrete* elements of being (or their subbtraction), thereby playing the game with intellectualist weapons, you set things straight at a single stroke by your fundamental conception of the continuously creative nature of reality."[32] In a more detailed way James acknowledged the same change of his views in a letter on June 27, 1909: "I think the center of my

[30] R. B. Perry, *Thought and Character of William James*, Vol. II, p. 611. In a more systematic way the problem of unity of personality is treated by Bergson in *Creative Mind*, pp. 202–208.

[31] John Dewey, "The Vanishing Subject in the Psychology of James," *Journal of Philosophy*, 1940. However, the personalist elements in James's thought, especially of his latter period, are ignored by the author.

[32] Perry, *l.c.*, Vol. II, pp. 618–20.

whole *Anschauung* since years ago I read Renouvier, has been the belief that something is doing in the universe and that *novelty* is real. But so long as I was held by the intellectualist logic of identity, the only form I could give to novelty was tychistic, i.e., I thought that the world in which discrete elements are annihilated and others created in their place, was the best descriptive account we could give of things; and if the elements were but minute enough, 'scientific determinism' could be kept, as approximating the appearances sufficiently for practical error to be avoided in our dealings with nature's 'laws'. This sticks in the human crop—none of my students became good tychists! Nor am I any longer, since Bergson's synechism has shown me another way of saving novelty and keeping all the concrete facts of law-in-change."[33]

Obviously, Renouvier's idea of discrete creation and, especially, discrete annihilation, accounts to a large extent for James's "perishing pulses of thought." There is no real perishing and annihilation in the philosophy of true duration, adopted by James under the name of "synechism." Or, in the language of *A Pluralist Universe:* "[Their] changes are not complete annihilations followed by complete creations of something absolutely novel. There is a partial decay and partial growth, and all the while a nucleus of relative constancy from which what decays drops off, and which takes into itself whatever is grafted on, until at length something wholly different has taken its place" (p. 258).

III

It is only just to point out that this final agreement between James and Bergson was virtually contained in Chapter XII of James's *Psychology.* I do not mean Chapter IX, in which the similarity between the "stream of consciousness" and "true duration" is, as we tried to prove, sometimes obvious, sometimes deceptive. In Chapter XII ("Conception"), there is one specific and extremely important point in respect to which James reached the same conclusion as Bergson did years later: the dilemma of

[33] *Ibidem*, Vol. II, p. 656. It is highly interesting how close James under the influence of Renouvier came to the main idea of modern physics, i.e., that of determinism resulting statistically out of the minute discrete novelties. That shows that even in the first period of his thought his acceptance of the classical physics was not without reserve. But a route to a deeper anticipation of the spirit of modern physics was barred to him by his sticking to the *visual* and *corpuscular* ideas of classical physics which also was due to the influence of Renouvier. (*Principles*, Vol. I, p. 178: "But the molecular fact is the only genuine physical fact. . . .") The belief that "outward forces, so far as they are anything, are masses in certain positions or in certain movements, and naught besides" held by James in 1880 (the article, "The Feeling of Effort," reprinted later in *Collected Essays and Reviews*, p. 216), was characteristic for the second half of the nineteenth century.

the continuity and discontinuity of the temporal flux. James proved convincingly in analyzing one concrete case that the antinomy is only apparent arising from an attempt to express the dialectical structure of the temporal flux in isolated and mutually exclusive terms. The temporal fluidity is at the same time continuous and discrete, one and many, enduring and perishing, memory and novelty. All features appear contradictory and mutually exclusive only when they are carved out of concrete temporal experience and expressed in isolated concepts and separate statements, while in a concrete intuition they not only do not exclude each other, but one is a necessary condition of the other. James showed this brilliantly by his analysis of the concept of identity and its psychological conditions. He pointed out that the conception of the sameness can arise, paradoxical as it seems, only in an absolutely irreversible stream of consciousness whose successive sections are never identical. This was overlooked because it was silently assumed that "the conception of the same is the same feeling." But in a mental flux nothing returns; there are no recurrent psychological states. Moreover, if we would have two exactly same mental states referring to the same object, no judgment of identity would arise. "The recurrence of the same idea would utterly defeat the existence of a repeated knowledge of anything. It would be a simple reversion into a preexistent state, with nothing gained in the interval and with complete unconsciousness of the state having existed before."[34] The judgment of sameness implies recognition; the recognition means activity of memory bringing the past idea into the focus of the present moment; but the very effort of "bringing into the present moment" adds the element of novelty which creates the difference between the idea originally perceived and the "same" idea recognized later. The feeling of familiarity which tinges the recognized idea and which was absent in the idea originally perceived, represents the differentiating element of novelty. The reproduced idea, consequently, is no more the same in spite of its recognition; or, better to say, *because* of its recognition. Without the tinge of familiarity (Harald Höfding's "Bekanntsheitsqualität") which has its roots in the survival of the original moment no recognition is possible.

This concrete analysis needed only to be generalized in order to become a true solution of the problem of time. Did not James point out that the retention of the past is inseparable from the novelty of the present and that the novelty of the present and irreversibility of time are possible only when the past is preserved integrally? That the memory is preserving and creating at the same time and that the very act by which novelty emerges from the past is also the act of retaining the immediately preceding moment? Does not the former present moment acquire its character of pastness only

[34] *Principles of Psychology*, Vol. I, p. 481.

by the emergence of the new present? Yet, James failed to grasp the general and far-reaching meaning of his analysis of the recognition. The general treatment of the problem of time in Chapter IX is definitely inferior to the brilliant analysis of the specific problem in Chapter XII. The theory of the stream of thought is definitely influenced by Renouvier's concept of discrete changes involving a radical annihilation and subsequent "ex nihilo creation." James following Renouvier in this point, dissected the continuity of the temporal process into two distinct and successive stages: absolute destruction of the past followed by no less absolute creation of the present. In truth, the vanishing of the past is only relative as is the novelty of the present; both are only two simultaneous aspects of an indivisible temporal flux. The effort to generalize a result obtained by his analysis of recognition, would have saved James from the Heraclitean myth of the "vanishing and perishing moments."

It is true that James criticized Hume's associationism which dispersed the living continuity of the Self into a succession of atomic sensations; it is true that he emphasized "the core of sameness running through the ingredients of the Self."[35] However, in view of his theory of the absolute perishing of the past, this "core of sameness" was hardly more than a figure of speech. Though he recalled John S. Mill who in identifying Ego with memory came very close to the dynamic personalism of Bergson, he finally insisted that "the passing thought is the *only* psychic integer."[36] As for the associationists, for James only a single present sensation is truly real; it is true that he deprived it of its atom-like, rigid character in dimming out its contours. It did not change the fact that the whole past preceding the sensible, non-instantaneous present was considered as absolutely *absent*, that is, non-existing.

Considerably later, in an initial passage of *Creative Evolution*, Bergson also pointed out the connection between *persistence* and *novelty* in psychological duration: "Let us take the most stable of internal states, the visual perception of a motionless external object. The object may remain the same, I may look at it from the same side, at the same angle, in the same light; nevertheless the vision I have now of it differs from that which I have just had, even if only because the one is an instant older than the other. My memory is there, which conveys something of the past into the present. My mental state, as its advances on the road of time, is continuously swelling with the duration it accumulates: it goes on increasing—rolling

[35] *Psychology*, Vol. I, p. 352.

[36] *Ibidem*, Vol. I, p. 338. Significantly it is also called by James an "identifying 'section' of stream."

upon itself, as the snowball on the snow."[37] The difference between the present moment and the immediately previous one just consists in the memory of the immediately past moment which makes the present moment richer in respect to the past one. Obviously, novelty and retention of the past are two aspects of a single dynamic fact—the progress of time. Only when these two aspects become conceptualized do they appear as mutually exclusive and contradictory. In a concrete experience they merge together and one cannot occur without the other; more accurately stated, they are but two names for one single process. More than one decade after James, Bergson also grasped fundamental truth about the paradoxical structure of duration; but, unlike James, he did not fail to understand fully its general meaning, valid for *any* duration, not only for the stream of mental states. James in his *Principles* did not even draw the obvious conclusion from his analysis of the feeling of the sameness: that the irreversibility of time is impossible without a complete preservation of the past.[38] This conclusion is of a rather abstract type; James, being strictly empirical, did avoid it. An integral preservation of the past can hardly be proved empirically, even though the facts of hypermnesia and the amazing completeness of memory in a hysterical state points in this direction. Only in the last period of his

---

[37] *Creative Evolution*, p. 2. But for the first time the retention of the past and the emergence of novelty were linked in Bergson's article *L'Introduction à la métaphysique* (*Revue de la métaphysigue et de morale*, 1903), later reprinted in *La Pensée et le Mouvant* (English translation under the title *Creative Mind*, p. 211): "There is *no consciousness without memory, no continuation of a state without the addition, to the present feeling of the memory of the past moments*. That is what duration consists of." (The italics are mine.) Virtually, though without clear explicitness, this statement is implied already in *Time and Free Will*, p. 200: "The same feeling, by the mere fact of being repeated, is a new feeling." How inconspicuous even now the connection between persistence and passage appears, is inadvertently shown in an interesting article by D. S. Mackay "Succession and Duration" (*University of California Publications in Philosophy, Vol. XVIII*), where succession and duration are considered mutually irreducible.

[38] *Principles*, (Vol. I, pp. 234–5) interpret the irreversibility of the stream of though in purely physiological terms as *an extreme physical improbability*, not as a *logical impossibility* excluded by the nature of time itself: "It is out of the question then that any total brain-state should identically recur."

The importance of this difference between James and Bergson has been fully grasped by A. Ménard, *Analyse et critique des Principes de la Psychologie de W. James*, p. 108: "La description du flot conscient est calquée sur l'image que nous pouvons nous faire des variations moléculaires les plus intimes de la matière cérébrale, elle repose sur les fermes assises de la physiologie; M. Bergson dédaigne des pareilles alliances, qu'il estime être des compromissions . . ." Ménard also noticed that the Bergsonian assumption of the unconscious mental states is theoretically derived from the notion of irreversible duration rather than based on concrete experience (p. 111).

life James influenced by the study of "subliminal or transmarginal consciousness" postulated the immortality of the past in Bergsonian sense.[39]

The relations between the Bergsonian and Jamesian conceptions of duration are obviously manifold and more complex than they appear at first glance. There are the similarities, which when examined closely are only suerficial, and differences which are exaggerated. The most important difference concerns the metaphysical status of the past; but even on this particular point James in his chapter about "Conception" came very close to the conclusion of Bergson. The strictly empirical approach of James to the problem of time made him refrain from anything going beyond a concrete and accurate introspective statement. The problem of time remained for him the purely psychological problem of the perception of time. A slightly stronger effort of analysis and generalization would have brought him already in 1890 to the problem of time in general and to a more critical attitude toward the concept of mathematical time. That was the path of Bergson, who thus arrived, though not immediately, at an entire reformulation of the traditional mind-body problem, involving a new and apparently baffling theory of matter. The development of modern physics indicates that his anticipations were more correct than expected.[40]

Although William James did not see the coming of the wave mechanics and the principle of uncertainty, he lived long enough to witness the collapse of the mechanical models in physics. "The whole notion of the eternal fixity of elements is melting away before the new discoveries," he writes jubilantly in his last posthumous book and states that, with the rejection of the concept of the immutable element, all attempts to reduce the phenomenal change to a simple combination and recombination of the static pre-existing things are no more justified.[41] In other words reality of time is reintroduced into the physical world with all its consequences concerning the novelty and indetermination of the pluralistic universe. Just as Bergson's "true duration," originally purely psychological, became finally a "creative evolution" on the cosmical scale, James's final affirmation of "the everlasting coming of novelty into being" was but an extended vision of his "stream of consciousness." When ten years later A. N. Whitehead, a

---

[39] In the article "A Suggestion about Mysticism" written about six months before James's death. (Reprinted in *Collected Essays and Reviews*, pp. 500–513.) See the footnote on pages 502–3, where the Bergsonian view of subliminal consciousness, carrying "the whole freight of our past," is mentioned and explicitly accepted.

[40] S. Zawirski, "L'Evolution de la notion du temps," *Scientia*, Vol. LV, p. 254; M. Capek, *Bergson et les tendences de la physique contemporaine*, The University of Prague Publications, 1938 (in Czech with the abstract in French).

[41] *Some Problems of Philosophy*, p. 150.

philosopher-physicist, reached the conclusion about "the creative advance of nature," he pursued the path discovered by Bergson and James.[42]

A pesar de sus muchas semejanzas, la idea bergsoniana de la "duración real" y la "corriente de conciencia" de James no eran nociones equivalentes. La verdadera diferencia fundamental entre ellas se refiere a la posición del pasado. Así como James definía la corriente de conciencia como un continuo perecer, en el cual el pasado inmediato solamente se conserva dentro de los estrechos límites del "presente apparente," en la "duración real" bergsoniana persiste la totalidad del pasado, y su íntegra conservación es precisamente la condición de la irreversibilidad del tiempo. Aunque también James sostuvo esta paradójica conexión entre la persistencia del pasado y la novedad del presente (en sus *Principios*, V.I., cap. xii), no llegó a generalizar el resultado de su análisis con el fin de aplicarlo a la estructura del tiempo psíquico y del tiempo en general. Sólo en el útimo periodo de su pensamiento llegó James a una concepción del pasado esencialmente análoga a la bergsoniana, y así coincidieron ambos en atribuir el carácter de verdadera temporalidad a la realidad en general.

[42] *Some Problems of Philosophy*, p. 149; *The Concept of Nature*, pp. 54, 178.

CHAPTER 2

The Elusive Nature *of* the Past

THE MOST CONSPICUOUS feature of any past event is its apparent unreality. A past event does not exist now; it has vanished, passed away, disappeared; it is not present any longer. The alleged unreality of the past is probably the reason why the past itself is rarely an object of systematic philosophical inquiry. Its only feature seems to be a negative one: "to have ceased to exist" or "to have passed out of existence." What else can be said about it? There seems hardly any problem here at all.

Yet, it seems that past events somehow do exist. Although their status is dimmer and less definite than present ones, they seem to have a kind of existence. The Hebrew term "Sheol" and the Greek term "Hades" designate the realm into which the former living beings were thought to "pass away" without, however, being annihilated. They continue to exist, although in an ill-defined, shadowy or ghostly way. Thus, in the early mythologies, the past was pictorially represented as the realm of intangible shadows, of everlasting sleep and silence. The very choice of terms by which the past was usually described suggested that although it lacked the luminous intensity and solidity of the stuff of which the present is made, it still has been regarded as at least semireal. The idea of the primeval chaos of Greek and Middle Eastern cosmogonies as well as the biblical idea of "the darkness above the deep" was another symbol of the vanished past, in this case not of the past of human beings only, but that of the whole universe.

Professor Čapek was born in Bohemia and educated in Czechoslovakia, France, and the United States. He has published two books in Czech, a number of articles in French and American journals, and has finished a third book, "The Philosophical Import of Contemporary Physics."

At first sight, though, it looks as if these two symbols, Hades and the primeval chaos, have altogether different meanings. While the first symbol suggests the view of the past as an abyss into which our present existence sinks, the second symbol apparently conveys the opposite idea: that present existence has gradually emerged from the mists of the past. Does then the present become the past or vice versa? Does time move from the past to the future or from the future to the past? It is clear that the problem of the direction of time was, implicitly at least, present in the prelogical thought, for the nature of the past was symbolized by the two apparently incompatible myths described above. If this problem has not altogether disappeared from the modern thought (philosophers and scientists have become more acutely aware of it in the last hundred years) it is because our approach to it is still tinged by sensory and imaginative elements inherited from the original, mythological approach. We shall see later that the expression "the direction of time" is merely a misleading metaphor, a subtle survival of the prelogical and crudely imaginative approach to problems of time.

When philosophical speculation replaced mythological imagination, it gradually became clear that the belief in the hidden existence of the past was due to the incapacity of the prelogical thought to grasp the abstract idea of nothingness. Because of this incapacity, the unreality of the past was symbolized by visual images suggesting a sort of vaporous, shadowy, diaphanous existence, or by the auditory image of a complete stillness. But it took a considerable time before the abstract concept of nothingness replaced the original imaginative symbols. The decisive step in this direction was taken when Parmenides formulated the correlative concepts of Being and Non-Being, and emphasized the timelessness that Being has: "Never it was or shall be; but the All simultaneously is." Because the past, as something which *was*, but *is* no more, was subsumed under the category of Non-Being, its existence was explicitly denied. On the other hand, the thought of Parmenides was still tinged with sensory associations: "to be," for him, meant "to be in space." Thus neither the concept of Being nor that of Non-Being were sufficiently abstract; the former still had the characteristics of some space-filling material, while the latter still had the geometrical features of empty space. By implication, that meant that the past as a species of Non-Being should have the characteristics of empty space.[1] When the atomists, in opposition to the Eleatic school, affirmed the reality of empty space, the distinction had to be drawn between Non-Being and the spatial void and, by implication, between the past as a species of Non-Being and the void. Only then were the conditions created for grasping the concept of Non-Being in abstract purity, untinged by sensory, especially visual and tactile associations. Thus it is hardly surprising to see Lucretius arguing vigorously against the naive reification of the past events by Stoics and insisting that the past not only has no material reality, but *that it does not even possess the degree of reality which belongs to the void.*[2] The radical identification of the past with nothingness was for the first time proposed.

Until recently, hardly any significant change has occurred in the general approach to the problem of the past. Its reality has been denied as confidently as the reality of the future, with both regarded as equivalent species of Non-Being. St. Augustine wrote in the eleventh chapter of his *Confessions:* "For the past, is not now; and the future, not yet." How little this view has been modified during fifteen hundred years may be seen from the words of the German philosopher and psychologist Herman Lotze, written in the second half of the last century: "The history of the world, is it really reduced to the infinitely thin, forever changing, strip of light, wavering between a darkness of the past, which is done with and no longer anything at all, and a darkness of the future, which is also nothing?"[3] The question mark which Lotze placed behind this simultaneous denial of the past and the future clearly indicated that he was uneasily aware of the inherent difficulties of such a view. We shall see soon how serious these difficulties are.

This is, however, only one side of the picture. For parallel to the increased emphasis on the unreality of the past, through the history of Western thought, we can trace the opposite tendency: the claim that the past as well as the future is as real as the present. As this assertion denies any real difference between past, present, and future, it is equivalent to a denial of time itself. The claim that time is a mere deceptive appearance and that the true reality is timeless may be traced again to Parmenides and Zeno in ancient Greece; from their time on, the worship of changelessness became one of the most persistent features of philosophy. It may be traced through Plato, Plotinus, to the Middle Ages, when the claim that the divine mind embraces the past, present, and future in one timeless insight became the basis of the divine foreknowledge and predestination; we find it again in modern philosophy, in particular in Spinoza, and in a modified form in Kant and modern idealism, especially in Bradley and McTaggart. It decisively influenced the thinking of scientists, in particular physicists, who replaced the medieval omniscient God by the impersonal order of nature, symbolized by the "Laplacean mind." In such an impersonal order of nature, past, present, and future events are as simultaneously contained as in the God of St. Thomas and Calvin. This Eleatic superstition of timeless reality was vigorously attacked in the twentieth century by pragmatism and process philosophy on both logical and empirical grounds. It would be beyond the scope of this paper to restate these arguments in detail, even though we shall not be able to avoid them entirely. Our task is more specific: the status of the past. But I wanted to stress the presence of two contradictory tendencies both in science and philosophy, one of which asserts while the other denies the reality of the past.

Or is it possible that these two tendencies are only *apparently* antagonistic, and that they are based on one common postulate? Indeed, this will be the conclusion to which our subsequent analysis will lead.

§ II

THERE ARE several objections against the alleged unreality of the past. The first one has the form of the following question: If both the past and the future are unreal, is anything real left at all? If our answer is that it is precisely the present which, while hovering between two abysses of nonbeing, still retains the character of reality, we must not overlook the fact that what we call "present" is a stretch of duration which can be ideally subdivided into two nonbeings—a past interval which is no longer and a future interval which is not yet—separated by the mathematical present. This view was clearly stated by E. R. Clay who coined the term "specious present" in order to differentiate it from the genuine durationless present:

> Time, when considered relatively to human apprehension, consists of four parts, *viz.*, the obvious past, the specious present, the real present, and the future. Omitting the specious present, it consists of three . . . non-entities—the past, which does not exist, the future, which does not exist, and their conterminous, the present; the faculty from which it proceeds lies in the *fiction of specious present.* [Italics added.][4]

According to this view, reality is made up of three noughts, the vanished past, the present which endures for a zero time (i.e. does not endure at all), and the future which does not exist either. We seem to be close to the metaphysical nihilism of Gorgias. There are two escape-routes from such nihilism. The first one consists in the claim that what is unreal is merely time, not reality itself which is *beyond* time; the second one accepts the reality of time, but insists that the instantaneous "nonspecious" present is not an ideal fiction, but the only genuinely real part of time. Let us consider each of these solutions separately.

1] We have already mentioned that the tendency to negate the reality of time is one of the most persistent features of the philosophical and, to certain extent, even the scientific tradition. It is not the primary task of this article to deal with the enormous difficulties to which a consistent denial of time leads. But the reality of time cannot be denied consistently. For if we declare time to be "merely illusory," we have to explain the reality of the illusion. Moreover, temporality which is banished from "objective reality" reappears in the consciousness which is a victim of temporal illusion; in other words, a complete *elimination* of time cannot be achieved. If we go farther and claim with Kant that even introspective time is "merely phenomenal," we still do not avoid affirming succession in some form. For if time is regarded as a veil separating the timeless transcendental Ego from the equally timeless noumenal self, we still assume that this veil rises gradually, revealing thus successively what "in itself" is without succession. Temporality is like the Cartesian Cogito which, though denied repeatedly from Leucippus to present day behaviorists and physicalists, always emerges in the very act by which it is denied. This is what Professor Lovejoy called "the paradox of the thinking behaviorist," and I once proposed to call the paradox which we had just analyzed "the paradox of changing Eleatic."[5] If time is nothing but a veil of ignorance, this veil must be lifted gradually in order to uncover the hidden timeless reality. Thus the elimination of time is never successful, because it can never be complete.

It is important to add here that, contrary to Bradley's belief,[6] timelessness does not restore the reality of the past. It is true that the past, present, and future seem to acquire the same degree of reality once they are incorporated into the single timeless whole. But this only means that in absorbing the past into a timeless "Now," we destroy the specific quality of pastness by which it differs from present and future time; that also by making the future timelessly present, we eliminate its very futurity, its potentiality, its "not-yet character." It is a peculiar misunderstanding to believe that we make the past real in transforming it into something different from it. On the contrary, to transform the past into a form of present, means to eliminate it as past.

2] The difficulties of the timeless view are, verbally at least, avoided if the reality of the instantaneous and perpetually changing present is accepted. Then, although both the past and the future are at each particular moment unreal, one being already dead and the other not yet born, there would still be a real change consisting in a perpetual substitution of a new present moment for its immediately preceding ancestor. This perpetual change is a continuous motion of the Now-instant from the past to the future; or, without spatial metaphor, it consists of the continuous succession of instantaneous "Nows." This view was held by the Arabian atomists of Mutakallimun school and by Descartes;[7] and even in this century Alexius Meinong regarded the denial of the mathematical present as a mere prejudice, a mere "horror puncti," inspired by a wrong interpretation of our experience of specious present. For, according to Meinong, the very fact that we experience our specious present as a simultaneous whole, presupposes the reality of an instantaneous act by which the successive content of specious present is brought together into a single unit. Similar views were held by James Ward and C. A. Strong.[8]

But it is highly questionable if, on such a view, the reality of succession and of change is upheld in any other than a verbal way. How can change and succession have its locus in a durationless instant which is in its own nature devoid of change and succession? Both change and succession—both being, contrary to the widespread prejudice, synonymous, for one cannot exist without the other—can take place only within a duration which has a certain temporal thickness. Change implies a differentiation of successive moments; succession is a before-and-after relation, with a specific temporal difference between successive terms; such a relation can never take place within a durationless instant. If we still continue to speak about a "perishing" or "changing" instant, we are simply juxtaposing two mutually exclusive terms; for there is neither perishing nor change of any sort *within* an instant which is by definition static. Thus, in its ultimate consequences, the doctrine of knife-edge instants is as opposed to the reality of an enduring and changing universe as the flat denial of time by Bradley and McTaggart. Both views are based on the assumption that the past and the future are symmetrical and can be treated in the same way. According to the timeless view, the past is *equally as real* as the future; according to the doctrine of *Augenblickexistenz*, the past is *equally as unreal* as the future. But both views ignore the essential asymmetry of time; they ignore the fundamental difference between "not yet" and "never more." This also explains, what seems to be a paradox, why some process philosophers attack classical physics both for equating reality with the succession of perishing instants as well as for claiming that the true reality is beyond time. The latter view, implicitly present in Laplace's determinism, was in particular attacked by Bergson; the first view, clearly formulated by Descartes, before being attacked by Whitehead, was criticized by Bergson who regarded the concept of "instantaneous state" as a distortion produced by "cinematographic mechanism of thought." Both views were implicitly present in classical physics and, as we have seen, were merely two different ways of equating the past and the future. To assert the reality of time means to assert the *unreality* of the future and, as we shall see, also *some kind of reality* of the past.

Besides the futility of attempting to think of change, succession, and duration in terms of elements which are devoid of change, succession and duration, it must not be overlooked that, within our experience, whether psychological or physical, there is nothing that corresponds to a knife-edge dimensionless instant. The unreality of the instantaneous present in psychology has been widely known since William James wrote *The Principles of Psychology*. In his fifteenth chapter, James wrote:

> The unit of composition of our perception of time is a *duration*, with a bow and a stern, as it were—a rearward- and a forward-looking end. It is only as parts of this *duration-block* that the relation of *succession* of one end to the other is perceived. We do not first feel one end and then feel the other after it, and from the perception of the succession infer an interval of time as a whole, with its two ends embedded in it. The experience is from the outset a synthetic datum, not a simple one; and to sensible perception its elements are inseparable, although attention looking back may easily [italics added] *decompose* the experience, and distinguish its beginning from its end.[9]

We have seen that this view was challenged by Meinong and others who claimed that the unity of the specious present requires the existence of an instantaneous unifying act. But Meinong's views are obviously based on two closely related assumptions: *1*] that the notion of *successive unity* is contradictory,—"a monstrosity," according to C. A. Strong's expression; *2*] that the unity can therefore be found only in the unity of instantaneous act of apprehension. The first assumption is plainly incorrect; it is simply not true that succession is nothing but a sheer diversity of external units. This was the view of Hume which Kant as well as neo-Kantians accepted and, which would, were it true, make the simplest perception of succession impossible. In melody, successive tones are perceived in a single durational stretch, not as simultaneous, but as successive, but nevertheless still "at once," provided that we understand the words "at once" not in the sense of durationless instant, but in the sense of enduring present. This, I believe, was convincingly shown by William Stern in his brilliant article "Die psychische Präsenzzeit" written in answer to Meinong and to Strong.[10] If we accept the concept of "successive unity," then the assumption that the only true unity belongs to mathematical instants falls to the ground. The concept of "instantaneous act of apprehension" is obviously heavily tinged by the Kantian distinction between the "organizing activity of reason" and "passive sensory content." According to Kant, the timeless "sensory material" is unfolded by the a priori form of time into a successive pattern of organized perception, while in Meinong's view the opposite takes place: the successive content of specious present is "brought together" by an instantaneous act of apprehension. How such an instantaneous act which, according to C. A. Strong, is placed on the forward edge of the specious present, can reach backwards to gather up the anterior instants of the specious presents which are, by definition, entirely external to it, is a mystery; evidently a sort of action at a distance—to wit, a distance *in time*—is assumed. Why not, instead of such mythological description, assume the temporal coextensiveness of the apprehending act with the durational stretch of the specious present? The idea of psychological experience lasting "zero-time" appeared absurd even to Bertrand Russell in spite of his great sympathy for the concept of mathematical instant and the mathematical continuity, i.e. the infinite divisibility, of time.[11]

But if we admit that the psychological present is not a durationless instant, does not the boundary between the past and the future disappear? For we then may ask: When, *at what particular instant*, does the present become the past? When, *at what particular instant*, does the present impression die and "immediate memory" begin? But such questions are obviously based on *petitio principii*, because the expression "at what particular instant" presupposes the existence of instants. But once the reality of durationless instants is rejected, we have no right to formulate the question in this way. The denial of the mathematical point-like present entails a denial of any sharp distinction between the existing present and the allegedly nonexisting past. Both the present and the past become real, but they are real in a dynamic sense. Their relation is not a static relation of two simultaneous terms, the existing present and nonexisting past, but a dynamic relation of the present to the immediately anterior past. This relation constitutes the very essence of the enduring present which, for the reasons just stated, can be called "immediate memory" as well.

It may be objected that this is true only of *psychological time*, or rather of our *psychological awareness of time*. It may be said that while our awareness of time is *pulsational* in nature, in the sense that there is no mathematical present in our stream of consciousness, the objective physical time consists of continuous succession of durationless instants. Thus, in the physical world at least, a sharp distinction between the past and the present would be preserved. From this view there is only a step to the conclusion that physical time *is the only true time*, while the time of consciousness is merely epiphenomenal. But the situation is different if we turn from classical to modern physics. Shortly before his death Henri Poincaré envisioned the possibility that quantum theory might force physicists to introduce the pulsational view of time into physics. In this respect he was followed by Whitehead who speaks about "quantum of time" in *The Concept of Nature*. In 1925, with the article of J. J. Thomson, this hypothesis appeared in the pages of a scientific periodical, and since then has never entirely disappeared.[12] According to the principle of indeterminacy, it is meaningless to speak of durationless instants in the physical world. Although the question is still far from being settled, the concept of mathematical instant has lost its former prestige even in physics and its adequacy is now seriously in doubt.

But if the distinction between the past and the present cannot be drawn within the time-span of what is called "specious present" or "immediate memory," it still seems legitimate to establish the distinction between the specious (noninstantaneous) present and what E. R. Clay called the "obvious past" or what James called the "genuine past." No matter how large the span of specious present may be (James's value of 12 seconds for its maximum duration was questioned), it seems undeniable that the past beyond the rearward edge of immediate memory is dead and gone forever. The genuine past anterior to our specious present appears to be altogether beyond our reach. Its relation to our specious present is entirely external, if the term "relation" can be applied when one of its terms does not exist. Consequently, the past outside of our immediate memory can be known only representatively, by means of a present image. According to James, "the feeling of the past is a present feeling."[13] The genuine past survives only symbolically, in cerebral traces with which what we call "recollections" are correlated; but as far as its ontological status is concerned, it is forever lost in the abyss of nonbeing. In the act of remembering we are not aware of the past mental events, but only of *their present traces.* As long as the cerebral traces of the past impressions remain dormant, they do not have any mental counterpart; only when they are reactivated, they are accompanied subjectively by mental images which in virtue of their weaker intensity are "rejected from the present" and "projected to the past."

There were no small difficulties in this theory of memory. If, as James claimed, "the feeling of past time is a present feeling," the question arises *why* it is interpreted as *belonging* or *referring* to the past. The answer of associationistic psychology, that an image is judged as belonging to the past because of its weaker intensity, is hardly convincing. For this implies that "image" and "recollection" are synonymous, which is certainly not true. Not every image is interpreted as a recollection; "non-presence" and "pastness" are not synonymous. Indeed, most of our images are *not* dated, even though, like recollections, they have weak intensity. The classical theory of "temporal signs" claimed that the character of pastness arises out of the contrast between two *simultaneous* mental states: one which in virtue of its sensory vivacity has the character of *presentness;* and the second which, because of its faint intensity, is "rejected from the present" and *judged* as belonging to the past. The main difficulty is obviously this: why is the difference between two *simultaneous* states "translated" or "interpreted" by our consciousness as *difference in time?* This difficulty was pointed out by Theodore Lipps even before Bergson more systematically criticized the theory which reduces recollections to "weakened perceptions."[14]

It is even more significant that the classical theory of memory implicitly assumed what it explicitly denied: it assumed the *persistence of the past.* In H. Taine's book *De l'Intelligence* (probably one of the clearest expositions of the classical theory), recollection is regarded as a "veridical illusion." According to Taine, every recollection is illusory, in the sense that in the act of remembering we *interpret* a faint image as belonging to the past, although it is in the present; but it is *veridical* in the sense that it happens to agree with a corresponding past event.[15] In other words, our recollection *resembles* the former impression. But how can there be any relation of "accordance" or "resembling" between an *actual* entity and a completely vanished past? How can there be any dyadic relation when one term is missing? The very assertion of any relation between the present and the past implies that the latter somehow exists. It is a different kind of existence from that of the present. This is the reason why the term "subsistence" was proposed. But it would be utterly wrong to confuse the *absence* of the past with sheer nothingness. The very fact that the term "veridical" may be applied to at least some recollections and that we can differentiate veridical recollections from the illusions of memory indicates that the past is indestructible. This is implied not only in the statements about our personal past, but in those about any kind of past. It is a presupposition, though usually a silent one, of every science dealing with past situations. Whether the past is the past of nation or of whole civilization, whether it is the past of organic life, of our planet and solar system or of the whole universe, the basic assumption remains the same. History, archeology, paleontology, geology, and cosmogony all presuppose the indestructibility of the past events. Without this assumption they would lose their status as sciences and become literary fictions. The very concept of historical truth would lose its meaning.

In spite of the prevailing tradition which regarded the past as not existing, there were still some philosophers who were occasionally aware of the peculiarly indestructible character of the past. Professor Paul Weiss called our attention to one statement by Aristotle to this effect.[16] St. Thomas, who was generally very reluctant to impose any limit on the divine omnipotence, nevertheless conceded that even God is powerless to eliminate past events and refers to the same statement of Aristotle: "Of this one thing God is deprived—namely to make undone things that have been done."[17] For God cannot do anything contradictory and the claim that the past events are destructible is equivalent to an absurd assertion that what has happened has *not* happened: *Praeterita autem non fuisse contradictionem implicat.* Thus the past, which was regarded by mythological imagination as having a sort of vaporous and diaphanous existence, proves to be endowed with a peculiarly stubborn resistance and solidity which even the divine omnipotence is powerless to destroy. It is precisely this solidity and resistance which differentiates the past from the future. This difference between the nonpresence of the past and the nonpresence of the future is essential, and it is not difficult to see its close relation to the basic asymmetry of time. For this reason any *return to the past* is intrinsically impossible and the idea of the returning past cannot even be stated in consistent language. Such a return would require a complete annihilation of the interval which is between the present moment (from which we hypothetically start) and the past moment (to which we are supposedly returning). Or if we symbolize time as returning upon itself in a self-intersecting curve, the point of intersection would show that two moments, successive and simultaneous, are really one. The contradiction is obvious. In this respect the speculation of even such outstanding scientists as Kurt Gödel about the possibility of a "round trip to the past" are on par with the phantasies in H. G. Wells's *Time-Machine.*

§ III

THE "IMMORTALITY OF THE PAST" is increasingly recognized in modern philosophy. Although this term was coined by Bertrand Russell in 1903,[18] Russell's views about time were too conflicting and vacillating to regard him as a supporter for the view he named. Bergson, and, in the last stage of his thought, William James upheld the immortality of the mental past; and C. D. Broad, A. N. Whitehead, and H. Oakley insisted on the indestructibility of the past in general.[19] But we have to be careful not to confuse immortality with static immutability. And this confusion is often found among those who defend the reality of the past. Thus Whitehead's statement that "the past moment is fadeless in the lapse of time" may be misleading if we forget the inseparability of permanence and flux which Whitehead insists on in the statement that "those who would disjoin the two elements can find no interpretation of patent facts."[20] In Paul Weiss's relatively recent essay, the allegedly inert, ineffective, finished, and becomingless nature of the past was strongly emphasized. This view was challenged on various grounds by John E. Smith, Charles Hartshorne, Daniel Leahy, and Nathan Rotenstreich.[21] Although I agree in part with all these critics, in particular with Mr. Rotenstreich, I think the most convincing argument against the immutability of the past has not been stated, at least not with a sufficient explicitness.

The main shortcoming of nearly all theories which recognize the reality of the past is that they speak about it in a *grammatical singular.* But it is fallacious to infer from the grammatical singular to the singleness of the fact which it designates. There is not a single past, but *many pasts.* The term "past" is merely a collective name for the plurality of the past events which differ in their degrees of pastness. The pastness of an event which occurred a second ago is different from the pastness of the events which occurred an hour, a year, or a century ago. But not only are past events differentiated by their degree of pastness, but the pastness of each event is continuously changing. Using a spatial metaphor we say that each past event is continuously "receding" from the present moment. This mode of speech preserves the absolute immutability of the past in Weiss's sense; for while each past event is continuously receding from the present, it nevertheless remains in its content absolutely identical and therefore unchanged. What I intend to show is that such a description is illegitimate and misleading and that, if taken literally, will obscure not only the nature of the past, but the nature of temporal process as well.

The image of past events as moving away from the present is suspect because of its spatial character; and we know, at least we should know since Bergson wrote, how misleading spatial metaphors may be when used for describing temporal process. It would be equally plausible to claim that the past events are stationary while the present moment is moving away from them in the direction of the future. Such ambiguities are inevitable; for as long as we picture the temporal process by the metaphor of spatial movement, a principle of kinematic relativity will hold. When we have two points in space A and B, the increase of their distance may be regarded as due either to the motion of B away from the stationary point A, or due to the opposite motion of A away from the stationary B. The problem of "the direction of time" is a *pseudo-problem* which is due to the spatial symbolism which cannot adequately express the qualitative asymmetry of time. By comparing the qualitative process of the past's fading away to "the increase of distance" between the past event and the present moment, we are applying the same procedure which human intellect usually applies to any kind of qualitative transformation. As Emile Meyerson showed in his classical works on the history and epistemology of science, there is an inherent tendency to reduce all qualitative changes to mere displacements of immutable particles which change only their positions without changing themselves. Are we not precisely applying the same procedure when we claim that a past event, while "moving away" from the present, remains identical and unchanged in all its "temporal position?" The qualitative process of "fading away" or "perishing" of the past is reduced to a mere displacement, even when we call it "displacement in time." But whether we realize it or not, by speaking of the past events as "moving away" from the present or "sinking into the depth of the past," we regard them as corpuscular or quasi-corpuscular entities which in their own nature remain as unchangeable as the particles of classical atomism.

The quasi-corpuscular conception of past events exhibits what Whitehead called "fallacy of simple location." Whitehead differentiated two kinds of this fallacy: that of simple location in space and that of simple location in time.[22] Our case is clearly an instance of fallacious simple location in time. The concept of the immutability of the past evidently implies that the relations by which any past event is linked to the actual present, are *external* relations. Only thus it is possible for any past event to preserve its identity; what is supposedly changing is its relation ("distance") to the present, but not the event itself. Thus we artificially isolate a past event from its relation to the present in order to preserve its allegedly immutable nucleus. But we can do it only if we forget that past and present are correlative terms; not only does any past event acquire its character of pastness in contrast to a new present, but the presentness of any actual event is possible only on the contrasting background of the events which are anterior to it. The relation to the present is thus not an accessory feature of the past, but *constitutes its very nature;* it is inherent in the pastness itself and not external or separable from it. It is then understandable that, as this relation is continually changing (in virtue of the perpetual emergence of novelty), the corresponding past events are changing too. This change is, of course, different from annihilation. Past events remain indestructible, though not immutable; they will forever remain past, but the degree of their pastness is continuously changing; they are continuously perishing, but never completely dead. It is the process of alteration without destruction; it consists of the process in which a temporal link of any past event with the actual present is becoming increasingly more and more tenuous without, however, being completely gone. Complete destruction of the temporal link would destroy the very pastness of a past event.

Our language is ill equipped to express these distinctions; we have no grammatical comparative for the adjective "past" and thus we are bound to express the differences in the degree of pastness by the misleading metaphor of spatial motion. The abstract noun "past" hides from us the wealth and variety which it designates. The fact that the relation of any past event to the present is continuously changing is generally conceded; but much less frequently do we meet the view that this relation is an *internal* one, constitutive of the very nature of what is past. Thus C. D. Broad, who not only recognizes the reality of the past, but also does not overlook the fact that its relation to the present is perpetually changing, still has the tendency to regard this change as *external,* as an addition (almost in the arithmetical sense) of a new entity to the previous sum total of existence. In his own words: "When one event, which was present, becomes the past, it does not change or lose any of the relations which it had before; it simply acquires in addition new relations which it *could* not have before, because the terms to which it now has these relations were then simple non-entities."[23] The superficial plausibility of this passage obscures the fact that by the emergence of a new present a new relation is added not only to an immediately preceding moment, *but to all previous moments no matter how "remote in time" or "deep in the past" they are.* This is expressed by an inadequate and already criticized spatial metaphor that *all* previous moments retreat *en bloc* "pastwards" when a new present is "added" on. Because the relation to the present is an *internal* relation, inseparable from a past event itself, *the whole past* was in some sense retroactively modified by the emergence of a new present; *every* past moment became "paster" after being modified by a new relation to a freshly emerged present.

In his answer to Mr. Rotenstreich's question: "Is it not true that the past becomes different once an event—any event—joins it?" Mr. Weiss uses an apparently convincing argument: "Yes, if by 'different' one means 'different for knowledge, epistemologically different'; no, if by 'different' one means ontologically so. Otherwise Plato's every act, his life . . . could now be changed by what we now do. That life is completed, all that ever can be."[24] Mr. Weiss is unquestionably correct when he insists that there is a permanent recognizable meaning ingredient in Plato's life. Yet, he overlooks one essential ingredient of this meaning: its *pastness*, its *temporal context.* Without this quality of pastness, Plato's life would be present to us now and would not be past at all. For practical purposes it is convenient to isolate Plato's life from its temporal context. This is a legitimate simplifying device, as long as we do not forget that we are dealing with a methodological fiction created for special purposes. As soon as we forget it, we commit a fallacy of simple location by transforming our abstraction into an isolated timeless entity. I am therefore inclined to reverse Weiss's statement: "epistemologically" the past event remains identical, while "ontologically" it is changing in degree because it is becoming more and more pervaded by the quality of pastness. Professor Weiss's concern that any change in the past would destroy the permanency of bygone events is understandable, but not justified. For "permanent" means "enduring," and duration implies the differentiation of successive phases as well as their mnemic continuity. It is precisely the mnemic continuity of duration which makes an enduring pattern *recognizable* in its successive phases. For instance, when a single tone sounds for several seconds, its perception is enriched by immediate memory, and we thus witness its incipient differentation into successive phases, even though throughout, the qualitative content, the tone itself, remains recognizable, i.e. *epistemologically* the same. The fallacy arises by the apparently innocent substitution of the term "same" for that of "enduring," or that of "ontologically identical" for that of "epistomologically the same." Strictly speaking, there can be no absolute identity of successive phases; for their absolute and rigorous identity would be incompatible even with their succession and would make the very act of recognition impossible. In other words, *the feeling of sameness is not the same feeling.*[25] There can be only very close similarity of successive phases in the dynamic continuity of *every* process, including the process of the "passing of the past." The passage of the past is not introduced by our thought, but by nature itself. It is one aspect of the universal becoming, a counterpart of the emergence of new events. The imaginative recovery of the past and the corresponding different interpretations of history is an altogether different problem which is beyond the scope of this paper.

We thus conclude with Miss Oakley that "the past does not find its immortality in a heaven of timelessness."[26] Its immortality is not of static, immutable, and becomingless nature. It is a living, dynamic immortality of past events, perpetually modified by their changing relations to the perpetually emerging novelties of present moments.

NOTES

1. It is interesting that traces of this view may be found in frequently used figures of speech which refer to the past as an abyss or depth into which our present impressions sink or disappear.

2. C. Bailey, *T. Lucretii De Rerum Natura Libri Sex* (Oxford, 1947), II, 676–78.

3. H. Lotze, *Metaphysic*, ed. B. Bosanquet (Oxford, 1884), II, 268.

4. Quoted by W. James, *The Principles of Psychology* (London, 1901), I, 609.

5. M. Čapek, "The Doctrine of Necessity Re-examined," *The Review of Metaphysics*, V (1951–52), 30.

6. F. H. Bradley, *Appearance and Reality* (2nd ed., London, 1899), p. 208.

7. J. Wahl, *Le Rôle de l'Idée de l'Instant dans la Philosophie de Descartes* (2nd ed., Paris, 1953).

8. A. Meinong, "Das zeitliche Extensionprinzip und die sukcessive Analyse," *Zeitschrift fur Psychologie und Physiologie des Sinnesorgane*, VI (1890); C. A. Strong, "Consciousness and Time," *The Psychological Review*, III (1896); J. Ward, "Psychology" in *Encyclopedia Britannica* (9th ed.).

9. W. James, *Principles of Psychology*, I, 609–10.

10. *Zeitschr f. die Psychologie und Physiologie der Sinnesorg,* XIII (1896), 325 ff.

11. B. Russell, "On the Experience of Time," *The Monist,* XXV (1915), 217.

12. H. Poincaré, *Dernières Pensées* (Paris, 1913), p. 188; A. N. Whitehead, *The Concept of Nature* (Cambridge, 1920), p. 162; J. J. Thomson, *Proc. of Royal Soc. of Edinburgh,* XLVI (1925–26), 90; E. T. Whittaker, *From Euclid to Eddington* (Cambridge, 1949), p. 41.

13. W. James, *Principles of Psychology,* I, 627.

14. Quoted, *ibid.,* p. 632.

15. H. Taine, *De l'Intelligence* (16e ed., Paris, 1930), II, 48–49; 231–36.

16. P. Weiss, "The Past; Its Nature and Reality," *The Review of Metaphysics,* V (1951–52), 511.

17. St. Thomas, *Summa Theologica,* Question XXV, art. 4.

18. B. Russell, "Free Man's Worship," republished in *Mysticism and Logic* (London, 1953), p. 59.

19. H. Oakley, "The Status of the Past," *Proc. of the Arist. Soc.,* XXII (1932), 227–50. On the influence of Bergson theory of the past on James, cf. my article "Stream of Consciousness and 'Durée réelle,'" *Philosophy and Phenomenological Research,* X (1950), 351–52.

20. A. N. Whitehead, *Process and Reality* (London, 1929), pp. 513–14.

21. J. E. Smith, "Existence, Past and God," *The Review of Metaphysics,* VI (1952–53), 287–95; D. J. Leahy, "A Pragmatist Theory of Past, Present, and Future," *ibid.,* VI, 369–80; N. Rotenstreich, "The Impact of the Past," *ibid.,* VII (1953–54), 591–603; Charles Hartshorne, "The Immortality of the Past," *ibid.,* VII, 98–112.

22. *Science and the Modern World* (New York, 1926), p. 84.

23. *Scientific Thought* (3rd ed., London, 1949), p. 66.

24. P. Weiss, "The Past: Some Recent Discussions," *The Review of Metaphysics,* VII (1953–54), 303.

25. Cf. W. James, *The Principles of Psychology,* I, 480–81: "The ordinary Psychology of 'ideas' constantly talks as if the vehicle of the same thing-known must be the same recurrent state of mind . . . But this recurrence of the same idea would utterly defeat the existence of a repeated knowledge of anything. It would be *a simple reversion into a pre-existent state, with nothing gained in the interval, and with complete unconsciousness of the state having existed before* [italics supplied]."

26. Oakley, paper cited, *Proc. of Arist. Soc.,* XXII, 235.

CHAPTER 3

The following paper was delivered at the First Conference of the International Society for the Study of Time, Oberwolfach (West Germany), August 31 – September 6, 1969.

The Fiction of Instants

Abstract. The claim that the mathematical durationless instants do exist in a physical sense appeared always as both paradoxical and inevitable. Paradoxical, because it is difficult to find the concept of entity lasting for zero-time intelligible; inevitable, since both experience and logic suggested that it really exists. For all empirically available temporal intervals were, in the classical period at least, divisible into smaller and smaller sub-intervals unless they were durationless instants. Furthermore, the very denial of instants, as proposed by various theories of chronon, surreptitiously postulated their existence. But an attentive analysis of perceptual and, more generally, phenomenal continua shows that they consist neither of instants nor of contiguous atomic segments; their structure clearly transcends the disjunction "instants versus chronons". They exhibit a type of continuity which is different from mathematical continuity and which, as Poincaré's paradox shows, is extremely difficult to be conceptualized. There is a considerable circumstantial evidence that time on the microphysical level has a similar structure. This would make possible to deny the reality of instants without accepting the self-contradictory "atomization" of time.

In his *Essay on the Foundations of Geometry* Bertrand Russell analyzed the so called "antinomy of the point" and concluded that the concept of geometrical dimensionless point is "palpable contradiction, only rendered tolerable by its necessity and familiarity". Now, if we substitute in Russell's analysis the word "instant" for "point", "time" for "space" and "chronometry" for "geometry", we shall obtain "the antinomy of the instant" completely analogous to the antinomy of the point. The passage thus paraphrased will run as follows:

> "We saw, in dealing with measurement, how time must be regarded as infinitely divisible, and yet as mere relativity. But what is divisible and consists of parts, must lead at last, by continued analysis, to a simple and unanalyzable part, as the unit of differentiation. For whatever can be divided, and has parts, possesses some thinghood, and must, therefore, contain two ultimate units, the whole namely, and the smallest element possessing thinghood. But in time this is notoriously not the case. After hypostatizing time, as chronometry is compelled to do, the mind imperatively demands elements, and insists on having them, whether possible or not. Of this demand, all chronometrical applications of the infinitesimal calculus are evidence. But what sort of elements do we thus obtain?

Professor Milič Čapek, Department of Philosophy, Boston University, 232 Bay State Road, Boston, Massachusetts 02215, U.S.A.

Analysis, being unable to find any earlier halting place, finds its elements in instants, that is, zero quanta of time. Such a conception is a palpable contradiction only rendered tolerable by its necessity and familiarity. An instant must be temporal, otherwise it would not fulfill the function of a temporal element; but again it must contain no time, for any finite interval is capable of further analysis. *Instants can never be given in intuition*, which has no concern with the infinitesimal: they are purely conceptual constructions, arising out of the need of terms between which temporal relations can hold. If time be more than relativity, temporal relations must involve temporal relata; but no relata appear, until we have analyzed our temporal relata down to nothing. The contradictory notion of the instant, as a thing in time without temporal magnitude, is the only outcome of our search for temporal relata."[1] (Italics added.)

Russell pointed out clearly the logical correlation between the concept of unlimited temporal divisibility and that of durationless instant: since every finite temporal interval is divisible, the indivisibility can belong only to the zero-intervals, i.e., mathematical instants. He conceeded, at least in this essay, that no empirical data correspond to instants; durationless instants cannot be experienced. Furthermore, they are logically impossible since, according to Russell, they should possess two mutually exclusive attributes – temporality since they are the constitutive elements of every temporal interval, and the negation of temporality since they are supposedly without temporal extension. At the same time, both experience and logic seemingly lead us to postulate their existence. In this way an antinomic situation arises.

Before we shall investigate the alleged empirical and logical reasons which were only in part sketched by Russell, let us say that Russell's attitude to this problem was far from consistent; it kept changing almost from one of his books to another. Only six years later, in his *Principles of Mathematics*, he vigorously upheld the concept of instant and found nothing contradictory in it. This was the main reason why he sided with Zeno against Bergson.[2] Yet, he came remarkably close to Bergson in his article in *The Monist* in 1915 in which he resolutely denied the existence of instants in psychology thus coming back to his original view quoted above (i.e., that nothing in our experience corresponds to instants.[3]) The coming of the quantum physics made him cautious toward the existence of durationless instants even in the physical world.[4] A detailed survey of Russell's successive and changing views about this problem would require a separate essay which is beyond the limits of this paper. In truth, it was not necessary to start our investigation with Russell's

1 Russel, B.: *An Essay on the Foundations of Geometry*. Dover: 1956, pp. 189–190.
2 *The Principles of Mathematics*. New York: W. W. Norton 1964, p. 347.
3 *On the Experience of Time*. The Monist XXV (1915) 217.
4 *The Analysis of Matter*. Dover: 1954, p. 341.

passage quoted above. Equally revealing texts, insisting on the logical correlation between the mathematical continuity of time and the concept of durationless instant can be quoted from Descartes, Galileo, Leibniz, Kant and others.

Two kinds of arguments have been used to justify the belief in the existence of extensionless instants. First the empirical one. Our macroscopic experience suggests that every temporal interval, no matter how short, is divisible into its sub-intervals. It is true that common sense has the tendency to distinguish between temporal intervals which are divisible (the duration of a day, divisible into hours, hours divisible into minutes, minutes into seconds) and *moments* which appear to our spontaneous perception as indivisible: a flash of light, a single sound, a contact of two bodies. This probably led some ancient and medieval thinkers to accept the limit to the divisibility of time; thus Aristoxenus postulated the primordial unit of time, ὁ πρώτος χρόνος which Aristides Quintillianus called "tempus brevissimum" and Isidor of Seville "atomus temporis". Beda Venerabilis computed that each hour consists of 22560 atoms of time. Similar speculations about the atomicity of time can be found in Stoics.[5] In truth, as late as in the seventeenth century we see Gassendi holding the view that the continuity of time is merely a macroscopic illusion hiding to our imperfect perception the reality of the atoms of time, "insecabilia temporis".[6]

Needless to say, such a view proved to be unsatisfactory to the founders of classical physics and the view of Galileo that in every interval of time, no matter how small, there is an infinite number of durationless instants[7] fully prevailed and became the cornerstone of rational mechanics. Furthermore, the more refined experience showed that the distinction between "moments" and "intervals" is only that of degree; what we call moments are merely shorter intervals. Since the time when Wheatstone measured the duration of a lightning which was regarded as a momentary event *par excellence*, the technique of the measuring of small intervals of time increased enormously. It is more than thirty years ago since Magnan showed by the method of ultra-rapid kinematography that the so-called "momentary" event, like a drop of water hitting the ground or a single vibration of insect's wings in a slowed down motion picture appears as a very long history consisting of an enormous succession of shorter sub-events.[8] By a natural extrapolation it was believed that such divisibility belongs to any interval of time, no matter how short; in other words, that only durationless instants are indivisible.

---

5 Lasswitz, K.: *Geschichte der Atomistik vom Mittelalter bis Newton*. Braunschweig: 1890, I, pp. 31–4; Sambursky, S.: *Physics of Stoics*. New York: MacMillan 1959, p. 105.

6 *Syntagma philosophicum* in *Opera omnia*. Florence: 1727, p. 300.

7 *Le opere di Galileo Galilei, prima edizione completa;* XIII. Florence: 1855, p. 158: "in ogni tempo quanto, ancorche picolissimo, sono infiniti instanti."

8 Magnan, A.: *Cinematographie jusqu'à 12000 vues par seconde (Avec application à l'étude du vol des Insectes)*. Paris: Herman 1932.

But far more powerful reasons for the affirmation of the existence of instants were, apparently at least, of logical kind. It was believed on a priori grounds, independently of experience, that the concept of instant is necessarily correlated with the very idea of temporal interval. For no interval of time is thinkable without its limits; and these limits must be without temporal extension, that is, point-like instants. For were they not, they would be just other temporal intervals; and the same reasoning as that above would require the existence of their own instant-like boundaries. The only way to avoid the affirmation of mathematical instants was to insist that they are mere conceptual limits, never attainable in experience, or, in Russell's words, "not given in intuition". This apparently was the view of Kant when in the section on *Anticipations of Perception* he – anticipating Bergson – insisted that points and instants are mere conceptual limits out of which no space and time can be constructed. In other words, what is given are temporal wholes while instants have a mere derivative existence of conceptual constructs. Yet, only a few pages before this passage, in his *Axioms of Intuition*, Kant holds an apparently opposite view according to which temporal intervals are always *aggregates* which their smaller constituent parts – that is, ultimately, instants – precede. This was also the view of Galilei and of Russell in *The Principles of Mathematics.*[9] The discrepancy between two of Kant's views was pointed out by Whitehead in his *Science and the Modern World.*[10] But bearing in mind that the concept of durationless instant and that of infinite divisibility of time are logically correlated, it is fairer to regard the difference between the two passages as that of emphasis rather than that of substance. From the same point of view Russell's vacillations also appear to be less serious; neither he nor Kant doubted the mathematical continuity of time and the existence of the mathematical extensionless present. From the time of Galilei this was one of the basic features of classical physics and one of the most fundamental ideas of rational mechanics.

The fact that the mathematical instant is never experienced did not shake the belief in its existence in any way. This explains why the psychological present which is the only present, the only "moment" which we can experience and which is always of non-zero duration, was called "specious present". The choice of this term is significant: it was coined to suggest that the only *true* present, the only true instant is without duration, a mere temporal point; in contrast to this true mathematical present our psychological present, which we experience as "minimum sensibile", is necessarily only "specious". Again experience seemingly substantiated this claim: are there not countless physical events whose duration was measured to be incomparably shorter than the duration of our psychological present? Does not this show that the unper-

9 Russell, B.: *The Principles of Mathematics*, p. 144. On Galileo, cf. Note 2.

10 Whitehead, A. N.: *Science and the Modern World.* New York: MacMillan 1926, pp. 183–84; Kant, I.: *Critique of Pure Reason*, transl. by Norman Kemp Smith. London: MacMillan 1953, pp. 198, 204.

ceivability of the true mathematical present cannot be used as a decisive argument against its existence?

The final and seemingly decisive argument for the existence of mathematical instants was the *apparent impossibility of their denial.* More specifically, their very denial could not be phrased without surreptitiously asserting their existence. This is what made "the chronon theory" so suspect in the past and it remains its main shortcoming even today. By "chronon theory" I mean the thesis according to which there are the minimum intervals of time which are not further divisible; they are the physical analogues of our "specious present" only on a vastly reduced temporal scale. This is the common feature of all atomistic theories of time from the Stoics to some contemporary physicists and it is of secondary importance whether the atoms of time are called ὁ πρῶτος χρόνος, *tempus brevissimum*, *tempusculum*, or *chronon*. Now the theory of the minimum temporal segments affirms that time consists *not* of the succession of durationless instants, but of the succession of finite temporal intervals. But since each interval has its own boundaries and since these boundaries are necessarily durationless, are we not surreptitiously re-introducing the very concept of instant which we purported to eliminate?[11] And since within each chronon no successive instants can be distinguished, do we not implicitly assert that within each chronon time is "standing still", in other words, that the minima of time are themselves "holes in time"? Thus the chronon theory seems to be burdened by the same shortcoming as the doctrine of instants: it tries to build time out of non-temporal elements and, furthermore, its very formulation implies the assumption of mathematical, durationless instants.

So much in favor of the doctrine of instants. Now let us survey the facts and reasons which make this doctrine extremely implausible, especially in the light of contemporary physics; and then let us explore whether it is possible to phrase the chronon theory in a language free of contradiction, i.e., that which would not involve a tacit assumption of what it denies.

In the first place, no matter how strongly the concept of instant imposes itself on our mind, its intrinsic difficulties remain. It is the same basic difficulty as that concerning the geometrical points and the extensionless atoms of Boscovich: how can we build temporal intervals out of durationless instants, spatial segments out of extensionless points, material volumes out of the atoms devoid of volume? It is true that from the standpoint of formal mathematics this difficulty is not insurmountable: elementary calculus taught us that the product of zero by infinity can be equal to any finite value. Thus in assuming an actually infinite number of instants any finite interval of time can be built. But we have to be on guard against the controversial concept of actual infinity

11 On the intrinsic difficulties of the chronon theory cf. Čapek, *The Philosophical Impact of Contemporary Physics.* Nav Nostrand, 1964, pp. 40–41; 231f. Cf. also G. J. Whitrow; *The Natural Philosophy of Time.* London, Edingburgh: Thomas Nelson 1961, Ch. "Temporal Atomicity", pp. 153–57.

by which various magical results can be obtained which are plainly absurd from the point of view of a physicist. This may be illustrated by the astonishing conclusions reached by Tarski and Banach from the discovery made by Hausdorff:

> "It is possible to divide a large sphere (say, of the size of the sun) into a finite number of mutually disjoint parts which together exhaust the volume of the large sphere, and to move each one of these parts (without changing its size or shape) into a small sphere (say, of the size of a pea) in such a way that the moved parts remain mutually disjoint and exhaust the volume of the small sphere. This statement means that, if a man could only break up the large sphere in the proper clever way, he could put the whole of it into his pocket".[12]

Karl Menger, who mentions this curiosity, is certainly right when he says that a physical application of this mathematical discovery is impossible and he points out explicitly the reason of this impossibility: the required construction involves an infinite number of operations and as such it is inapplicable to any physical object. Thus man will never be able to put the sun into his pocket, – not even the moon unless we mean it as a figure of speech. Since the concept of instant requires the actual infinity to be useful for physical science, its physical applicability is as suspect as that of the actual infinity itself.

The general trends of contemporary physics only strenghten the doubts stated above. In the whole area of physics the actual infinities, whether the infinitely large or infinitely small, are on retreat. Since the time of Riemann we know that the infinitely large space of Euclid and Newton is not the only logical possibility; beginning with Einstein a number of cosmologists believe that the cosmic space is *finite*, though without limits. We know today that Boscovich, despite his other remarkable anticipations was wrong in his conception of the extensionless atoms; all elementary particles apparently have a finite radius of the order of 10^{-13} cm. The assumption of the electron with the zero radius was always loaded with difficulties since it led to the infinite density of electric charge compressed into a rigorously mathematical point; it was this difficulty that led to the postulation of the minimum length or hodon. Similarly, infinitely large velocities which implied that a certain physical action can be literally *ubiquitous*, that is, present at one and the same instant in *all* the positions of its instantaneous trajectory, began their departure from physics in 1675 when Olaf Römer discovered the finite velocity of light. Gravitation seemed to be the only exception, but the recent discovery of the gravitational waves, if it is confirmed, would strengthen one of the central ideas of the relativity theory about the finite velocity of *all* physical actions.

---

12 Menger, K.: *Theory of Relativity and Geometry*. In: Albert Einstein: *Philosopher and Scientist* (ed. by Paul Schilp). Evanston, Illinois: 1949, pp. 469–70.

It is only logical that the physical applicability of the concept of instant is now equally questioned. The postulate of the minimum time-interval or chronon was frequently made jointly with the assumption of hodon; this was only natural since the discovery of quantum phenomena made doubtful the physical applicability of the concept of spatio-temporal continuity. More specifically, it was the same doubts about the possibility of the infinite density of energy which led Lévi in 1927 to postulate that the proper times of the electron consist of the succession of chronons whose duration is of the order of 10^{-24} sec.

Epistemological as well as psychological considerations make the concept of zero-duration even more suspect. For there is no problem how this concept psychologically originated. We already pointed out that the concept itself is correlated with that of spatio-temporal continuity, that is, infinite divisibility of both space and time. There can be no doubt that this type of continuity is, as Schrödinger observed and as even the mathematicians Hilbert and Bernays conceded, "an enormous, exorbitant extrapolation" of what is empirically given in our perception and experience.[13] Furthermore, there is no doubt that the concept of instant was created in analogy to the concept of point. As Bergson observed, "as soon as we make a line correspond to duration, 'segment of duration' must correspond to segments of line, and an 'extremity of duration' must correspond to an extremity of line; such will be the instant...".[14] One does not have to be a Bergsonian to see how much Bergson's observation about the spatialization of time is correct; let us only listen to the following words of Kant:

> "We represent the time-sequence by a line progressing to infinity, in which the manifold constitutes a series of one dimension only; and we reason from the properties of this line to *all* the properties of time, with this one exception, that while the parts of the line are simultaneous, the parts of time are successive. (Italics added.)
> Even time itself we cannot represent, save in so far as we attend, in the drawing of a straight line (which has to serve as the outer figurative representation of time), merely to the act of synthesis of the manifold whereby we successively determine inner sense..."[15].

From this to the affirmation of infinite divisibility of time and of the mathematical continuity of all changes in *The Anticipations of Perception* there is only a small step. The reality of the mathematical instant was accepted in both editions of *Critique of Pure Reason* not only on the physical, but even on the psychological level. That was only natural; if time is symbolized by a geometrical line and since every line contains points, time must consist of instants.

---

13 Hilbert, D., Bernays, P.: *Grundlagen der Mathematik.* Jena: 1931, pp. 15–17. Schrödinger, E.: *Science and Humanism.* Cambridge: University Press 1952, pp. 30–31.
14 Bergson, H.: *Durée et simultanéité.* 3me ed. Paris: 1926, pp. 68–69.
15 Kant, *op. cit.*, pp. 77, 167.

Bergson's diagnosis of the origin of the concept of instant in our tendency to geometrize is clearly illustrated by the above quotation from Kant. This is certainly one illustration among many. Even if we, for a moment, do not question the legitimacy of the concept of the geometrical point, the tacit assumption that within any temporal sequence there must be instants as the counterparts to the points is nothing but an effect of the misleading spatial associations. Bergson's lasting merit is to point out the distortions which spatializing habits cause in our representation of time.

The inadequacy of the concept of instant for physical considerations was pointed out after Bergson by Whitehead. In the very first pages of his *An Enquiry Concerning the Principles of Natural Science* he showed that "a state of change at a durationless instant is a very difficult conception".[16] Some essential physical quantities such as velocity, acceleration, momentum and kinetic energy are meaningless without some reference to the past and the future. (It is interesting to note that Russell, despite his vigorous defense of the durationless instant in his *Principles of Mathematics* in which he sided with Zeno against Bergson by saying that "we live in an unchanging world, and that the arrow, at every moment of its flight is strictly at rest", in another passage of the same book virtually agreed with Whitehead in rejecting the concept of *the instantaneous state of motion.*[17]) Whitehead clearly saw that the concept of instantaneous state applies even less to organism: "In biology the concept of organism cannot be expressed in terms of a material distribution at an instant. The essence of organism is that it is one thing which functions and is spread through space. Now functioning takes time." "There is no such thing as life 'at an instant'; life is too obstinately concrete to be located in an extensive element of instantaneous space." But in this respect the difference between the organic and inorganic nature is merely that of degree: "This is no special peculiarity of life. It is equally true of molecules of iron or of a musical phrase".[18] Here we have the root idea of Whitehead's organic conception of nature.

Whitehead's reference to musical phrase has a far deeper significance than it appears at the first glance. For the perception of melody, or of any succession of sounds, possesses a certain structure showing that there are certain continua which in spite of their conspicuous temporal character are radically different from what we call 'mathematical continuum'. The awareness of melody represents a phenomenal field whose temporal character is not only very pronounced, but essential; for the individuality of its constituting sounds exists

16 *An Enquiry Concerning the Principles of Natural Knowledge.* Cambridge: University Press, 1955, p. 2.

17 *The Principles of Mathematics*, p. 473. Contrast it with the passage on p. 347 referred to in Note 2.

18 *An Enquiry* ..., p. 3, 196.

only *because* of their succession. At the same time, melody is literally nothing at an instant as such diverse thinkers as Whitehead and Norbert Wiener observed; even a single tone has a certain duration.[19] *The auditory phenomenal field clearly is devoid of instants;* this is true, though less conspicuously, of other phenomenal fields. In this way, certain paradoxical features of modern physics can be made more intelligible. We know that the relativistic rejection of absolute simultaneity means that, to speak with Eddington, there are no "world wide instants"; in other words, that the physical world is nothing at an instant. This appears paradoxical only as long as we believe that *every* temporal continuum must be mathematically continuous. But this is clearly not true as the case of auditory continuum shows. In the light of what I would call the "melodic structure" of time the second form of Heisenberg's principle, according to which to pin down a microphysical event at a mathematical instant would make the corresponding energy completely undetermined, also becomes more intelligible. The only reason why the formula $\Delta E \cdot \Delta t \geqq h$ appears to us arbitrary is that the three centuries of calculus and classical mechanics firmly established the belief that it is meaningful to speak of 'energy at a certain instant'. Now, as Whitehead observed, "nature is nothing at an instant" in a similar sense that "melody is nothing at an instant". "Instantaneous cross-sections" are as illegitimate on the microphysical as on the macrophysical level.

The structure of the auditory phenomenal field is revealing also in another sense; it shows clearly that the dichotomy "instants versus chronons" is *not* logically exhaustive; it is a false dichotomy. As pointed out above, one of the most decisive arguments for the theory of instants was that the only alternative view was the chronon theory which surreptitiously assumed the very concept of instant which it overtly denied. But if there is anything certain, it is the fact that the structure of any phenomenal field – not only that of auditory continuum – is *not* atomic. This is – or at least *should* be – known since the times of William James and of Gestalt psychology when it was convincingly shown that the basic fallacy of the associationist psychology was an artificial atomization of the introspective data. The phenomenal field, which James called "stream of thought" and Bergson "durée réelle", contains no instants and no edges, nor does it consist of the succession of well defined atomic entities; in other words, it is neither mathematically continuous nor atomic in the sense of the chronon theory. It is neither divisible *in infinitum*, nor divisible into finite segments; though its successive phases are qualitatively differentiated, their diversification must not be confused with the mutual externality of the contiguous segments as the geometrical symbolism of the chronon theory wrongly suggests. In the same way as, to use Wittgenstein's words, "the visual field

19 Whitehead, A. N.: *Science and the Modern World*, p. 54; Wiener, N.: *I am a Mathematician*. M. I. T. Press, 1964, pp. 105–107.

has no limits", the concretely intuited present has no boundaries.[20] Or, in the words of Bertrand Russell, unwittingly agreeing on this point with Bergson, "the present has no sharp boundaries, and no constituent of it can be picked out as the earliest".[21]

It was not only William James and the Gestalt psychologists, but also some outstanding mathematicians and physicists, who became aware of the distinction between mathematical continuum and qualitative (phenomenal) continuum. The doubts of Schrödinger and Hilbert about the universal applicability of the concept of spatio-temporal continuity had been already referred to. Herrman Weyl stressed the difference between what he called "intuitive" ("anschauliches") continuum and mathematical continuum; he even gave credit to Bergson for establishing this distinction. Poincaré used a less appropriate term – "le continu physique" – for what Weyl called "intuitive continuum", but he was equally aware of its difference from mathematical continuum. Both Weyl and Poincaré pointed out that the term "mathematical continuity" is misleading since in the continuum of real numbers the elements are as external with respect to each other as natural numbers. This again was in agreement with Bergson who correctly recognized that the so called "mathematical continuity" is nothing but "discontinuity infinitely repeated", i. e. infinite divisibility.[22] There is no doubt that much confusion could be avoided if the same term – "continuity" – were not used in two radically different senses.

But while both Weyl and Poincaré stressed the difference between mathematical and qualitative continuum, they were – unlike Gestalt psychologists – less explicit in stressing the non-atomic character of the intuited continua. In other words, they failed to stress that qualitative temporal continuum does not consist of the mutually external, contiguous segments. In truth, Poincaré in one of his last writings suggested the possibility of the atomic structure of time[23] which would be nothing but a contiguum of finite intervals in the sense defined above. Nor did he suggest the possibility that on the microscopic – or rather *microchronic* – scale the temporal continuum may have a similar structure as intuitive, qualitative continuum.

---

20 Wittgenstein, L.: *Tractatus Logico-Philosophicus*, 6.43311. Cf. also Ephron, R.: *The Duration of the Present.* Annals of the New York Academy of Science, vol. 138 (February 6, 1967), p. 714: "The onset of a perception cannot be perceived for it is not an object of perception ... Analogously, we do not *perceive* the 'edge' of our visual field or the 'borders' of our blind spot."

21 Russell, B.: *On the Experience of Time*, p. 223.

22 Weyl, H.: *Das Kontinuum und andere Monographien.* New York: Chelsea Publishing Co., n. d., pp. 65–71; Poincaré, H.: *Science and Hypothesis.* In: *The Foundations of Science*, transl. by G. B. Halsted. Lancaster: The Science Press 1913, p. 43; Bergson, H.: *Creative Evolution*, transl. by Arthur Mitchell. New York: Random House 1944, p. 170: "the intellectual representation of continuity is negative, being, at bottom, only the refusal of our mind, before any actually given system of of decomposition, to regard it as the only possible one."

23 Poincaré, H.: *Dernières pensées.* Paris, 1913, p. 188.

I suspect that there were two reasons why he did not suggest it. First, such suggestion would lead him to concede the existence of some rudimentary qualities even on the physical or microphysical level – and he was too deeply immersed in the classical habits of thought to do it. Second, Poincaré was the first who explicitly analyzed the paradoxical and in a sense "logically scandalous" structure of qualitative continua; the idea that something analogous to it could occur on the objective physical level simply did not occur to him. He showed that the most paradoxical feature of such continua is that *the transitivity of the relation of equality* apparently does not hold in them: while two contiguous terms are indistinguishable from each other, the non-contiguous are:

$$A = B,\ B = C,\ A \neq C.$$

He illustrated it by the perception of weight increase; while we do not perceive the difference between 10 grams and 11 grams, or between 11 and 12 grams, we do perceive the difference between 10 and 12 grams. This has been known since the time of Gustav Theodore Fechner. Similar examples can be found in different sensory or introspective continua, such as different shades of the same color or different intensities of the same pitch, etc. Bertrand Russell pointed out in 1915 that the perceptual temporal continuum exhibits the same structure: the relation of *psychological simultaneity* is not transitive:

> "Suppose that I see a given object A continuously while I am hearing two successive sounds B and C. Then B is simultaneous with A and A with C, but B is not simultaneous with C".[24]

There seems to be an obvious way out of this difficulty: to assume that qualitative continuum, whether its terms are simultaneous or successive, is merely "apparent", "phenomenal" or "illusory" and that the alleged logical difficulty stems from its intrinsic "haziness". In other words, the difficulty disappears when we consider the underlying physico-mathematical continuum as "the only real". Illustrated by a concrete example: when we gradually increase weight from ten to twelve grams, the only thing we have to consider is the continuous range of magnitudes through which the physical stimuli pass; within this continuum each term is sharply distinguished from each other and the logical absurdity of the non-transitivity of equality can never arise. The whole difficulty is removed if the above scheme $a = b$, $b = c$, $a \neq c$ is replaced by the following: "a is *indistinguishable* from b, b is *indistinguishable* from c, but c is *distinguishable* from a". In other words, the paradox arises merely out of the limited capacity of consciousness to discern the minutely different stimuli. Contradiction is thus confined to our experience; it is not inherent in reality. The same is true of the intuited temporal continuum: the underlying mathe-

24 Poincaré, H.: *The Foundations of Science*, p. 46; Rusell, B.: *loc. cit.* p. 228.

matical continuum of successive physical stimuli is *the only real;* in it the transitivity of simultaneity is fully preserved and the apparent non-transitivity of "temporal togetherness' is due entirely to the haziness of our experience, more specifically, to the fact that our psychological present is merely "specious", without sharp boundaries. The only true present is the mathematical, "knife-edge" instant of the physical world.

This explanation is in line with the centuries old philosophical tradition which from Parmenides to Bradley opposed the logically flawless "real world" to the "haziness", "confusions" and "contradictions" of our immediate experience. Yet, a closer scrutiny shows that this explanation faces two serious objections.

First, nobody claims that the non-transitivity of equality exists on the level of physical stimuli. Nobody denied that two stimuli whose difference is imperceptible are *physically* different, though indistinguishable psychologically. But it is clearly meaningless to call two sensations "different, though indistinguishable". The sensations resulting from two minutely different physical stimuli are qualitatively the same *qua* sensations; to postulate their difference despite their unperceivability, does not make sense. An unperceived difference, that is, that which is neither sensed nor felt, simply does not *psychologically* exist; if we continue to say that two sensations are "really different" in spite of their "apparent" identity, we are not speaking of sensations *qua* sensations, but of their external stimuli. In other words, we are unconsciously slipping from the language of perceptual data to the language of physical stimuli. The paradox of intuited continuum is not dismissed when we insist that it is absent on the physical level; it continues to exist on the psychological level whose paradoxical structure it reveals.

This leads us to the second objection. Are we really sure that the physical reality down to its deepest microphysical level is adequately described in the terms of mathematical continuum? We mentioned this before when we pointed out that the applicability of the concept of spatiotemporal continuity on the quantum level is more than questionable. In truth, there is a considerable circumstantial evidence that the physical world, at least in its deepest microphysical strata, does not possess such sharp edges and clear cut contours as the last century physics hopefully expected. This is perfectly compatible with the fact that nature on the macrophysical level of physiological stimuli *is for all practical purposes* continuous in a mathematical sense. The main reason why it is so difficult to give up the applicability of mathematical continuity to the microphysical level is that it would imply the admission that there is an irreducible qualitative element in nature which resists a complete mathematization or formalization. We are still unconsciously committed to the dogma of bifurcation of nature which relegates all qualities into the subjective realm and eliminates them from the allegedly homogeneous, "purely quantitative" realm of matter. If we give up this dogma, as Whitehead did in his organic philosophy of nature, our reluctance will disappear together with other difficulties which Cartesian dualism created.

Let us sum up. Mathematical continuity is *very approximately* applicable on the macroscopic – and macrochronic – level; it ceases to be applicable on the microphysical, i. e. quantum level. In other words, time on this level is not infinitely divisible; consequently, durationless instants have no physical existence. Nor do they possess psychological existence since the psychological present has a certain duration. This absence of instants on both the psychological and physical level suggests that perhaps the microphysical time may have the same paradoxical structure as qualitative temporal continua. From this point of view, the failure of the "chronon theory" would become as understandable as the failure of the associatinistic, atomistic psychology. Whether it will ever be possible to construct a consistent formal calculus which would adequately express the paradoxical features of such continua, is an open question. Karl Menger's "topology without points" as well as A. L. Zadeh's "fuzzy set theory" are serious attempts in this direction. But it is also possible that we are here reaching the ultimate limits of formal analysis.[25]

25 Zadeh, L. A.: *Fuzzy Sets.* In: Information and Control, 8 (1965) 338–353; Menger, K.: *Topology without Points.* Rice Institut Pamphlets, vol. 27 (1940), No. 1, p. 107; Bergmann, G.: *Duration and Specious Present.* Phil. of Science, vol. 27 (1960) 47.

CHAPTER 4

TWO TYPES OF CONTINUITY

In this paper I am going to deal with two very different kinds of continuity. One is of *mathematical* kind and it is familiar to every student of calculus; the other was named by Poincaré – not very appropriately, as we shall see – *physical* continuity (*le continu physique*). While the obvious contrast between these two different types of continuity is fairly well known, its deeper philosophical significance is rarely analyzed. This lack of interest in it is not accidental; it is due to the persistent influence of the intellectual tradition generated by the three centuries of classical science (1600–1900). We shall see that a more subtle epistemological approach together with the emergence of some new and quite unexpected problems in contemporary physics requires another fresh look at the contrast between both types of continuity and the way it was interpreted both by classical science and classical philosophy.

As far as mathematical continuity is concerned, one illustration from elementary calculus will make its meaning clear. The function $f(x)$ is defined to be continuous at the point x_0 of its interval when for any arbitrarily small number ε another sufficiently small number ϑ can be found such that the following inequalities hold:

$$|f(x) - f(x_0)| < \varepsilon \quad \text{when} \quad |x - x_0| < \vartheta$$

It is clear that this definition of continuity assumes the continuity of the argument, i.e. of an independent variable which is represented by the horizontal x-axis: to every point of this line corresponds a certain real number, representing the value of the argument, and each such value is correlated with a particular value of the function which is visually represented by the segment of the vertical line erected at the corresponding value-point of the argument. The definition thus presupposes the *continuum of real numbers*, each of which is correlated with one particular point of the x-line. Since it is in the nature of continuum that between any of its terms another intermediate term can be inserted, the continuum of real numbers as well as the continuum of the points on a straight line is 'everywhere dense or compact' ('überall dicht' in Cantor's words). Consequently, there is no smallest number since between zero and an arbitrarily small ε there is always some real number. For the same reason there is no minimum interval since every linear segment and, more generally, every geometrical interval, no matter how small, contains a subinterval and so on, *ad infinitum*. This shows clearly that continuity in the mathematical sense of the word is synonymous with *infinite divisibility*: real numbers as well as geometrical intervals are divisible without any limit for the reasons stated in the previous sentence.

Such 'infinite divisibility' is clearly a conceptual construct and its artificial character will become even more conspicuous when we compare it with what Poincaré called 'physical continuity' and whose more appropriate term is 'perceptual' or 'intuitive' continuum. Let us take the example above of a continuous function represented by a smooth continuous curve. The visual perception of its continuity contrasts by its simplicity with the complexity of the mathematical definition in which there occur two inequalities and which refers to such ideal entities as dimensionless points. Perceptual continuity is an indivisible *Gestalt* whose components can be only artificially isolated by the effort of analytical attention; as Berkeley and, after him, Hume recognized, in our perception of space there are no points, no infinite divisibility, but concrete *minima sensibilia*, indivisible quanta of extension which can be subdivided only in our imagination without being actually divisible. Points, instants and actual infinity are mere figments of mathematical imagination.

Confronted with such objections, the mathematicians have not remained silent, especially when theoretical physicists hurried to their rescue. They concede to their sensualist opponents that there are no points and no instants in our sensory experience. But, they said, this is only because our experience is hazy and confused; the incapacity of our perception and even of our imagination to attain the dimensionless points and durationless instants cannot rule out their existence. In this respect the whole weight of the philosophical tradition or at least of its major part was on the side of the mathematician. No theme in the history of philosophy was more recurrent than the unreliability and haziness of our sensory perception; in truth, man did not have to become a philosopher to be aware of various sensory illusions. An enormous extension of our visual field by telescope and microscope in the seventeenth century still further discredited the authority of our spontaneous perception and showed its original limits. Furthermore, in the same century, Newton's identification of Euclid's infinite space with the objective physical space was used as a powerful argument by the mathematician in his polemic against sensualistic philosophers. For the Euclid-Newtonian space was not only infinite, but continuous in the mathematical sense, that is, divisible without limits. One of the first theorems which Euclid established in his *Elements* is that *every* straight segment can be bisected; and by 'every' is meant *any* distance in infinite space, whether huge or minute, whether it is the distance of the earth from the sun or the radius of the minutest atom. In truth, as Sir Thomas Heath pointed out[1], the continuity of Euclidean space is postulated in the third postulate of Euclid which removes any restriction on the size of the circle; such restriction is clearly incompatible with the continuity of space. This idea *of relativity of magnitude*, according to which any geometrical figure can be constructed on any scale, underlay one of the basic assumptions of classical physics about the basic *similarity* (in a geometrical sense) of the microcosmos and macrocosmos. Hence the repeated attempts to construct intuitive and mechanical models of the atom and ether in the last century; in truth, even the original twentieth century model of the atom as a planetary system on a very minute scale was guided by the same – false, as we know today – assumption.

Such then was essentially the answer of the mathematician, supported by the classical physicist, to the sensualistic objections of Berkeley and Hume. To sum it up in one sentence: infinite divisibility of space as well as the existence of dimensionless points is not *disproved* by the fact that they both are inaccessible to our sensory experience which is notoriously hazy and inadequate. As indicated above, the classical physicist followed the line of the centuries old philosophical tradition from Parmenides to Bradley which posited the logically flawless realm behind the chaotic and 'imperfect' flux of our hazy and confused sensory experience.

It was not only the classical belief in the reality of continuous physical space which encouraged the mathematician in his confidence that his concepts are not mere artificial constructs, but that they have their objective ontological counterparts. Time, which Isaac Newton regarded as objectively existing as space, was endowed with the same property of homogeneity, that is, with infinity and continuity. Galileo affirmed that in any interval of time, no matter how small, there is an infinite number of instants, and Bertrand Russell reaffirmed the same claim at the beginning of this century. This assumption was inherent within the very structure of the infinitesimal calculus and without it no rational dynamics would have existed. For the classical concept of motion presupposed the classical concepts of space and time and it inevitably shared their continuity. Without spatio-temporal continuity, it would be impossible to speak of the identity of the material particle in different points of its trajectory, that is, in different points of space and different instants of time. When the quantum theory indicated the possibility that this may not be so, it came as a real shock and it indeed led to the crisis of the classical concept of particle. For, if the trajectory of a certain corpuscle is not continuous, that is, if the corpuscle does not exist in all the points intermediate between its successive positions, how can we be sure that it is the *same* particle? Thus the persistence of the material particle through time, its very identity – what Kurt Lewin called *Genidentität* – is assured by the possibility of tracing its *continuous* displacement through space and time, i.e. to its existence in *every* point and *every* instant of the corresponding spatio-temporal interval.

But let us not anticipate the discussion which belongs to the second part of this paper. For our immediate purpose let us add that the absence of the instants on the sensory and, more generally, psychological level did not imply their non-existence on the physical level. The fact that the psychological present has a certain duration has been known for a considerable time; but the very name by which it was designated – 'specious present' – indicated that it was not regarded as the true, mathematical present which is strictly durationless and does exist on the physical level. Again it was assumed that only the haziness of our perception prevents us from perceiving the true instants of the physical time in their precise, sharp-edged unambiguity. In this sense the psychological or mental present is indeed only apparent, 'specious', being merely a fuzzy, imperfect representation of what in the physical world is an enormous number of successive, rigorously point-like presents. In truth, this number is infinite since every temporal stretch, no matter how short, must be regarded as an infinite aggregate of durationless instants. Thus the function – or rather malfunction – of our temporal perception is to condense spuriously the infinity of the knife-edged, infinitely thin instants into the deceptive phenomenal unity of the 'specious present'.

This is clearly a *dualistic* view in which objective reality possessing a clear-cut logico-mathematical structure, is opposed to the qualitative realm of appearance, what Kant called a 'rhapsody of sensations', seemingly devoid of any order. It was precisely the haziness of our sensory and introspective experience which led philosophers to regard it as a mere 'appearance'. But on this point we have to be on our guard against the haziness of our own terminology; for *appearance* does not mean *unreality*. In this respect those who uphold the reality of the logico-mathematical structure of nature, independent of our perception and awareness, are guilty of certain haziness of language which then inevitably affects the clarity of their own thought. For, no matter how uncertain and questionable appearance may be, it still *does* exist; it cannot be dismissed as unreal. Otherwise the whole distinction between 'appearance' and 'reality' would be impossible. This was the old problem of Parmenides: if reality is One, where does the illusion of diversity and change come from? If it comes from 'the finite perspective of individual minds', do we not concede *ipso facto* the existence of 'individual minds' or 'finite perspectives', that is, something *alongside* or *outside* the Eleatic One? And by conceding the reality of something else besides One, are we not departing from our original rigorous monism? In a similar way, we can call the quality of yellow color of the sodium spectral line a mere 'appearance' of what in the objective physical world is the electromagnetic wave of the length of 5890 ångströms; but this does not change the fact that the quality of yellow is *actually* and *indubitably* experienced and cannot be dismissed as simply unreal. What is wrong and must be rejected is the *hypostasizing* of the quality of yellow, that is, its projection into the physical realm; there is no question that 'yellow' *qua* quality does not have any physical status. In this respect neorealists and John Dewey were grievously wrong in claiming that "things are what they are experienced as" and it is difficult to understand how some persons with their knowledge of elementary college physics could make such a claim. But it is equally wrong to conclude from the non-existence of qualities on the objective, physical level their non-existence in general; yet, this is the common fallacy of behaviorism and similar epistemologically innocent trends. For the self-evidence of introspective quality cannot be questioned; for if *esse est percipi*, and, while I may be wrong in *interpreting* this quality, I cannot err in *having* it. This should be known since the time of Descartes, in truth since the timeof St. Augustine. To use Professor Feigl's words, who certainly cannot be accused of any sympathy for Descartes or Berkeley, the mental qualia are not *evidenced*; they are *evident.*[2] For, let us repeat it, without the mental qualia, without sensory qualities being *in some sense real,* the whole distinction between 'reality' and 'appearance' loses its meaning.

As mentioned above, it was the so called 'haziness' of the sensory and introspective qualities which arouses the distrust of not only the behaviorists, but also of nearly every quantitatively minded scientist. But again we have to be on guard against our linguistic habits. 'Hazy' is a relative term as much as 'disorder'; as Bergson, incidentally, one of the most misunderstood and unjustly abused thinkers, convincingly pointed out, we speak of 'disorder' only when we do not find the type of order which we expect.[3] We complacently speak of the vagueness of music or poetry and claim that they both are 'less definite' than the 'normal language'. This is a mistake or at least a misunderstanding; there are not a few philosophers and musicians who claim that music is *more definite* than the words since the spoken language cannot express the passing individual nuances of our inner life which possess their specific qualities despite the fact that our object-oriented language does not have the ready-made symbols for designating them[4].

These preliminary remarks lead us to our main topic – that of qualitative or intuitive continuity. The fact that this second type of continuity is radically different from the mathematical continuity does not mean, that it does not have its own structure and its own specificity which can be analyzed. All depends on the tools by which we analyze it.

Let us then analyze as closely as possible what is that second type of continuity. We touched upon it before when we recalled that in our perception of extension there are no points and in our perception of time there are no instants. This was the meaning of Berkeley's and Hume's *minima sensibilia*; on the sensory and introspective level there is no mathematical continuity, no infinite divisibility. Let us focuss our attention on the problem of temporal continuity since temporality is a more pervasive feature of our experience than extension; extensionality or spatiality is characteristic of the sensations of touch and sight and only indirectly, by association with optical and tactile sensations, of other categories of sensations; it is altogether absent in imageless thought and emotive experience whose temporal aspect is still very pronounced. Kant expressed this fact in his own jargon by saying that "space is the form of outer sense while time is the form of both outer and inner sense." There is thus no reason to fear that the results of our analysis will lack in generality if we confine our attention to the analysis of psychological time.

Our analysis will start with our first negative conclusion: that perceptual and introspective continuum does not consist of mathematical durationless instants. On this point the agreement is nearly general; even Bertrand Russell, whom nobody can suspect of any antipathy against mathematical continuity, explicitly conceded it. It is true that C. A. Strong and Alexius Meinong tried by a very artificial way to establish the existence of the point-like psychological instants; I dealt with their views in my other papers and do not intend to repeat my criticism here.[5] A great majority of philosophers and psychologists side with William James and George H. Mead about the non-existence of knife-edge present. Now if we agree on this point, it is very easy and tempting to accept what seems to be the only alternative; that qualitative continuum, being not divisible *in infinitum*, must be atomic. This was the step which Hume and after him all associationistic philosophers took. For associationism is nothing but psychological atomism. Thus the perceptual and psychological time would be a succession of *minima sensibilia* – which were named differently by different thinkers – *impressions, ideas, sensations, Vorstellungen* and finally *elements* by Mach. (He used the term *Empfindungen*, i.e. sensations, too.) Even William James, in spite of his vigorous criticism of psychological atomism, comes very close to it when he speaks of the specious present as "elementary sensation of duration" or as "the unit of composition of our perception of time."[6] Thus the atomism of duration seems to be present on the sensory and psychological level; the second type of continuity would then be no continuum at all, but a *discretum*, the succession of contiguous temporal segments.

But such representation of qualitative continua is only slightly less inadequate than the theory of instants. Everybody acquainted with the development of psychology since about 1890 knows how the inadequacy of psychological atomism were exposed by Gestalt psychologists and holistically oriented philosophers. It has been pointed out that associationism was based on inadequate psychological observation which mistook artificial conceptual constructs for authentic 'elements' of introspective experience. What we call 'sensations', 'impressions', 'images' etc. are nothing but artificial entities carved out from the continuity of 'stream of thought'. William James in the passages of unsurpassed subtlety in his *Principles of Psychology* showed how psychological atomism focussed its attention on the sensory nuclei of the psychological self and ignored almost entirely more elusive imageless fringes and relations, although it is precisely in these dynamic non-sensory links that the true continuity of consciousness is located.[7] Today we would say that our object-oriented language is ill suited to express adequately the structure of the processes to which the category of 'thinghood' clearly does not apply. Bergson's and Dilthey's criticism of atomistic psychology were pointing in the same direction.

But it was not only psychological experience which militated against the atomistic theory of the intuitive continuum. Equally serious was the intrinsic, logical difficulty which was inherent in the very concept of *contiguum theory*, as we may call it. This difficulty will become immediately obvious when we approach this theory on a more abstract level. For the main reason for postulating the atomistic structure of psychological time was the rejection of mathematical continuity, that is, the rejection of instants. But a very cursory glance at the doctrine of temporal atomicity will show that it *covertly reintroduces the very concept which it purportedly rejects*. For in assuming that time consists of contiguous, successive segments, it implicitly postulates the boundaries by which these segments are separated; and such boundaries must be instant-like, if we want to avoid an infinite regress. For were they not, were they constituted by shorter segments interposed between the successive atoms, the same question concerning their boundaries would re-emerge. It is rather interesting to observe that even such an outstanding and serious thinker as Lovejoy did not perceive this difficulty.[8]

This difficulty is far more serious since the only alternative to the atomicity of time seems to be its mathematical continuity. Both alternatives are unacceptable since the intuitive, qualitative continuum consists neither of durationless instants nor of sharply delimited, contiguous segments. Are we not facing a contradiction here? Or is the dilemma 'atomism versus continuity' falsely stated in the sense that it is *not* logically exhaustive?

Before considering this possibility, we must take a look at two serious attempts which have been made to conceptualize the structure of the intuitive continuum. The first one was made by Henry Poincaré in his book *Science et hypothèse* in 1902. It was he who coined the rather misleading and narrow term 'physical continuum'. (What he was really dealing with was *the perceptual continuum* which is one of the cases of qualitative continua.) He showed in a precise way the paradoxical and in a sense 'logically scandalous' structure of such continuum when we try to express it in a rigorous way. Its most paradoxical feature is that *the relation of equality does not seem to be transitive in it*; while its two contiguous terms are indistinguishable from each other, the non-contiguous are:

$$A = B, \quad B = C, \quad A \neq C$$

Poincaré illustrated it by the observation known since the time of Gustav Theodore Fechner: while we do not perceive the difference of weight between 10 grams and 11 grams or between 11 grams and 12 grams, we do perceive it between 10 and 12 grams. Similar examples can be found in different sensory or introspective continua, such as the scale of different shades of color, of different intensities of the same pitch etc.[9]

Poincaré obviously dealt with that type of sensory continuum whose terms are given simultaneously; it is therefore even more striking to see that Bertrand Russell arrived at the same results when in 1915 he tried to conceptualize the temporal qualitative continuum whose terms are successive. (This is the second attempt in this direction which I have in mind.) Russell showed that the relation of psychological simultaneity or 'temporal togetherness' is *not* transitive. Let us hear his own *ipsissima verba*:

> Suppose, for example, the sounds *A, B, C, D, E* occur in succession, and three of them can be experienced together. The *C* will belong to a total experience containing *A, B, C*, to one containing *B, C, D*, and to one containing *C, D, E*. ... In the above instance, *C* is at the end of the specious present *A, B, C*, in the middle of that *B, C, D*, and at the beginning of that *C, D, E*.

Evidently, when we designate the relation of 'being in the same specious present' as 'psychological simultaneity', then this relation is not transitive. This is what Russell explicitly states when he considers a slightly different concrete example:

> Suppose that I see a given object *A* continuously while I am hearing two successive sounds *B* and *C*. The *B* is simultaneous with *A* and *A* with *C*, but *B* is not simultaneous with *C*.[10]

There seems to be an obvious way out of both Poincaré's and Russell's paradox and this way was already indicated before. All we have to assume is that qualitative continuum, whether its terms are simultaneous or successive, is merely 'apparent' or 'illusory', and that the alleged logical difficulty stems from its 'haziness'. In other words, this difficulty disappears when we focuss our attention on the underlying physico-mathematical continuum which is allegedly 'the only real'. Illustrated by a concrete example: when we gradually increase weight, pressing on our hand, from 10 to 12 grams, the only thing we must consider is the continuous range of magnitudes through which the physical stimulus passes; within this continuum each term is sharply distinguished from any other and the logical absurdity of the non-transitive equality can then never arise. The whole difficulty is removed, if the scheme above $a=b$, $b=c$, $a\neq c$, is replaced by the following one: '*a* is *indistinguishable* from *b*, *b* is *indistinguishable* from *c*, but *c* is *distinguishable* from *a*'. In other words, the paradox arises merely out of the limited capacity of consciousness to differentiate the minutely different stimuli. The same is true of the intuited temporal continuum: the underlying *mathematical* continuum is *the only real* and in it the transitivity of simultaneity is fully preserved. The apparent intransitivity of 'temporal togetherness' or 'psychological simultaneity' is due to the haziness of our temporal experience, more specifically, to the fact that our psychological present is merely 'specious', without definite boundaries. The only true present is the mathematical, instantaneous, 'knife-edge' present in the physico-mathematical continuum of events.

I hope that I was not unfair to the convential explanation of both paradoxes which both Poincaré and Russell suggested. It remains to be shown that this explanation simply will not do.

In the first place, nobody claims that the intransitivity of equality exists on the level of physical stimuli. Neither does anybody deny that two stimuli whose difference is imperceptible are physically different, though indistinguishable psychologically. But it is clearly meaningless to call two sensations *qua* sensations 'different, though indistinguishable'. For the sensations resulting from two minutely different stimuli are *qualitatively the same*; to affirm their difference despite their imperceivability, does not make sense. A difference which is neither sensed or felt, simply does not exist *psychologically*; if we continue to say that two indistinguishable sensations are 'really' different, we are not speaking of sensations *qua* sensations, but of their *external stimuli.* In other words, we are unconsciously slipping from the language of perceptual data to that of physical stimuli. The paradox of the intuited continuum cannot be dismissed when we insist that it does not exist on the physical level; it continues to exist on the psychological level whose paradoxical structure it reveals. And this level can be called 'appearance' only with respect to the physical stimuli, but not with respect to itself; for as William James stated unanswerably: "A material fact may indeed be different from what we feel it to be, but what sense is there in saying that a feeling, which has not other nature than to be felt, is not as it *is* felt?"[11]

Are we then back to the dualism of two continua, one mathematico-physical, other qualitative-psychological – to what Whitehead called "bifurcation of nature"? This seemed to be true still at the beginning of this century, but it is rather doubtful today. Today we cannot assert dogmatically that physical nature down to its deepest microphysical level is describable adequately in the terms of mathematical continuum. In truth, there is a considerable circumstantial evidence that the applicability of the concept of spatio-temporal continuity to the quantum level is very questionable. Not only a number of physicists, but even some outstanding mathematicians admit that the belief in infinite divisibility of space and time is nothing but 'an enormous extrapolation' of our experience to which nature has no obligation to conform.[12] Various theories postulating the minimum intervals of time (chronons) and motion (hodons) are symptoms of the growing distrust of physicists to the traditional concept of continuity. It is therefore quite possible, indeed probable, that the physical world, at least in its deepest microphysical – and microchronical – strata does not possess such sharp edges and clear cut contours as the last century physics hopefully expected; in other words, that there is, to use Professor Margenau terms, an "elementary diffusion"[13] in nature. This is perfectly compatible with the fact that nature on the macroscopic level of the physiological stimuli is *for all practical purposes* continuous in the mathematical sense. The main reason why we find it so difficult to give up the applicability of this concept on the microphysical level is that this would be tantamount of the admission of the irreducible *qualitative* element in nature. We are still very strongly committed to the dogma of bifurcation of nature which divides reality into two completely different realms – the homogeneous and quantitative realm of matter and the mental or phenomenal realm of qualities. If we give up this dogma, as Whitehead did in his organic philosophy of nature, our reluctance will disappear together with other difficulties which Cartesian dualism created.

What does this mean with respect to our problem? It means nothing more, but also nothing less than the possibility that the physical and intuitive continuum are intrinsically not so different as it was believed; more specifically, that the microchronic structure of the physical continuum is not *essentially* different from that which we call with Weyl "intuitive continuum."[14] As stated above, it is neither the dense continuum of instants nor the contiguum of the atomic segments. This is not as irrational as it appears. The main reason why the dilemma 'atomism versus continuity' appears to us as logically exhaustive, is that we are habitually inclined to think in the visual terms; as long as we symbolize time by a geometrical line, there are clearly only two possibilities: either this line is divisible *in infinitum*, and then the only indivisible elements are zero-intervals, that is, extensionless points-instants; or it is not – and then it must consist of finite segments. Needless to say that our natural tendency will be to prefer the first alternative. But if we consider some other continuum, for instance the succession of sounds in melody, we see clearly that it transcends the opposition between atomism and continuity. This was already touched upon when I briefly discussed the inapplicability of both the concept of instant and of atomization in psychology. A more detailed discussion would require a more extensive digression into Gestalt psychology and phenomenology which would be beyond the scope of this paper.

It must be further added that the above mentioned anti-thesis 'atomism versus continuity' is *spurious*, as long as we mean by the latter term the continuity in a mathematical sense. Again, there is a considerable agreement among philosophers and mathematicians that the so-called 'mathematical continuity' is a *disguised discontinuity*, that is, discontinuity, so to speak, infinitely repeated. In Bergson's words, the "intellectual representation of continuity is negative, being, at bottom, only the refusal of our mind, before any actually given system of decomposition, to regard it as the only possible one."[15] In the words of Bertrand Russell quoting approvingly Henry Poincaré, "the continuum thus conceived is nothing but a collection of individuals arranged in a certain order, infinite in number, it is true, but *external to each other*... Of the famous formula 'the continuum is unity in multiplicity', the multiplicity alone subsists, the unity has disappeared."[16] In other words, the points and instants in the mathematico-physical continuum are *as external* to each other as are discontinuous finite segments.

But such discontinuity is clearly absent in intuitive continua. For instance, the temporal continuum is not the succession of the well defined atomic segments. Wittgenstein observed correctly that "visual field has no limits"; similarly, as Dr. Efron noted, we do not perceive the onset of perception, we are not aware of its incipient temporal edge; therefore such mathematical instantaneous edge does not *psychologically* exist.[17] Several thinkers then hastily concluded – Lovejoy, for instance, – that by eliminating the boundaries between successive specious presents we eliminate also their differences. Such claim is based on the assumptions that every qualitative difference implies the *externality* of the qualities in the question – the assumption whose falsity is shown by a simple perception of melody or by any temporal *Gestalt*. In truth, the simplest perception of difference exhibits in its very structure the co-presence of two different, though not mutually separated qualities. But we have to be on guard against our own habitual language: this 'co-presence' is not 'co-instantaneity' nor should 'two' be understood in the usual arithmetical sense of two mutually external units. Qualitative continua are heterogeneous, qualitatively differentiated, and not consisting of additively distinct elements. For this reason the paradox of intransitivity of equality is merely apparent, a mere pseudo-paradox since it arises only when we try to translate the structure of intuitive continua into the language of distinct, atomic elements. This atomistic language is conspicuous in both Poincaré and Russell; they both treat the sensory continuum as a *contiguum* of distinct, atomic elements, each of which can be designated by a distinct and self-identical symbol: *A*, *B*, *C*,..., etc. The paradox which crops up merely shows the inapplicability of such atomistic language to the experienced continua.[18]

Let us say, in conclusion, that mathematical continuity is *very approximately* applicable on the macroscopic level since on that level time and space are *practically* continuous in the mathematical sense. Very probably it is inapplicable on the microscopic, i.e. quantum level and it is manifestly inapplicable on the sensory and introspective level. In other words, on either of these levels neither time nor space are infinitely divisible; extensionless points and durationless instants have neither physical nor psychological existence. This suggests that the structure of the microphysical time-space is perhaps not *essentially* different from that of the directly experienced qualitative continua.

NOTES

[1] Thomas L. Heath, *The Thirteen Books of Euclid's Elements*, Dover, New York, 1956, pp. 199–200.

[2] Herbert Feigl, 'The 'Mental' and the 'Physical'', *Minnesota Studies in Philosophy of Science* II (1958), p. 466.

[3] H. Bergson, *Creative Evolution*, Random House, New York, 1944, pp. 240–44.

[4] William James, *The Principles of Psychology*, Dover, New York, 1950, pp. 251–2: "Which is to say that our psychological language is wholly inadequate to name the differences that exist .... *But namelessness is compatible with existence*...", p. 254: "It is, in short, the re-instatement of the vague to its proper place in our mental life which I am so anxious to press on your attention." (Italics added.)

[5] Cf. my articles 'The Elusive Nature of the Past', in *Experience, Existence and the Good*, Essays in Honor of Paul Weiss (ed. by Irwin C. Lieb), Southern Illinois Univ., Carbondale, 1961, pp. 130–33; 'Memini ergo Fui?, in *Memorias del XIII Congresso Internacional de Filosofia*, Mexico 1963, esp.pp. 422–26.

[6] W. James, *op. cit.*, pp. 609–10.

[7] W. James, *op. cit.*, pp. 243ff.

[8] A. O. Lovejoy, 'The Problem of Time in Recent French Philosophy', *Phil. Review* **21**, 532–33.

[9] *Science and Hyphothesis* in *The Foundations of Science* (tr. by G. B. Halsted), The Science Press, Lancaster, 1946, p. 46. Cf. also *The Value of Science*, *ibid.*, pp. 240f.

[10] B. Russell, 'On the Experience of Time', *The Monist* **25** (1915), 218–228.

[11] W. James, *Some Problems of Philosophy*, Longmans & Green, London, 1940, p .151; *The Principles of Psychology* I, p. 63.

[12] D. Hilbert and P. Bernays, *Grundlagen der Mathematik*, Jena 1931, pp. 15–17; E. Schrödinger, *Science and Humanism*, Cambridge Univ. Press 1955, pp. 30–31.

[13] Cf. H. Margenau, 'Methodology of Modern Physics', *Phil. of Science* **2** (1935), 164–187, esp. 174–75.

[14] H. Weyl, *Das Kontinuum und andere Monographien*, Chelsea Publishing Company, New York, n.d., pp. 65–74.

[15] H. Bergson, *Creative Evolution*, Random House, New York, 1944, p. 170.

[16] B. Russell, *The Principles of Mathematics*, The Norton Co., New York, 1964, pp. 346–347; H. Poincaré, *The Foundations of Science*, p. 46.

[17] L. Wittgenstein, *Tractatus Logico-Philosophicus*, 6.4311; R. Efron, 'The Duration of the Present', *Annals of the New York Academy of Science* **138** (February 1967), 714: "The onset of a perception cannot be perceived for it is not an object of perception.... Analogously, we do not *perceive* the edge of our visual field or the 'borders' of our blind spot."

[18] F. B. Fitch in his penetrating article 'Physical Continuity', *Philosophy of Science* **3** (Oct. 1936), 486–493, rejects the alleged contradiction between atomism and continuity: "The reputed contradiction arises only if we try to employ a Democritean theory of solid, unbreakable, definitely determined atoms of material substance." (p. 490). When he says that "indistinctness may well be a fundamental clue to the nature of physical continuity" (p. 487), he is very close to David Bohm who in his *Quantum Theory* (Prentice Hall, 1951) pointed out the striking similarity between the mental processes and quantum phenomena, especially their indivisibility: "In any thought process, the component ideas are not separate but flow steadily and indivisibly, An attempt to analyze them into separate parts destroys or changes their meaning." (p. 170.)

CHAPTER 5

Process and Personality in Bergson's Thought

If we want to understand the present significance of Bergson's thought one century after his birth and nearly two decades after his death, we have not only to restate the central ideas of his philosophy, but also to sketch briefly the main phases of his philosophical development. His first book appeared in 1889, and both its original French title — *Essai sur les données immédiates de la conscience* — and the title of its English translation, *Time and Free-Will,* clearly indicate the problems which he faced at that time. Since then the three problems — that of theoretical self-knowledge, that of time, and that of freedom — remained closely related in his thought; in truth they were merely three aspects of one and the same problem. In the three successive chapters of his first book Bergson applied to the problem of *self-perception* or *introspection* the method which he himself characterized as "inverted Kantianism". While Kant, and the classical epistemology in general, tried to answer the question to what extent are the subjective elements present in our perception of the external world, Bergson raised a similar question concerning our introspection: Is not our self-perception colored by imaginative elements borrowed from the *sensory* perception in a similar way as our sensory perception is colored by "psychic additions"? In other words, is not our introspection tinged and even distorted by unconscious interpretations which our imagination, shaped by the perpetual contact with the external world, *adds* to the immediate data of consciousness?[1] In order to recover the immediacy and purity of the introspective data, we must free them from all imaginative admixtures borrowed directly or indirectly from the public world of senses. Bergson's widely misunderstood term "intuition" designates the effort to recapture the introspective data in their undistorted immediacy. As their most serious distortion is, according to Bergson, due to their imaginative *spatialization,* the main task of intuition is to de-spatialize them by being constantly on guard against the spontaneous ten-

[1] *Time and Free-Will,* p. 223.

dencies of our visualizing imagination. Bergson repeatedly emphasizes that such a task cannot be achieved effortlessly; nothing is more unfair than to assimilate the Bergsonian intuition to passive contemplation or to a mere confused emotional or semi-emotional state.[2] For this reason, I used the term "effort of intuition"; it is an effort to re-educate our introspection by eliminating from it all irrelevant imagery, especially the imagery of visual type. For this reason, and for other reasons, some of which will be stated in this paper, it is unjust, contrary to the prevailing *clichés,* to label Bergsonism as "irrationalism".

What then is, according to Bergson, the content of such re-educated introspection? It follows from what had been said that this content must be expressed in *negative* statements; there is no other way to grasp its positive significance than to state explicitly *what it is not.* The most obvious feature of our introspective contents is unquestionably their *dynamic, successive* character: our mental states *follow* one another in time. But the very wording of the last sentence is profoundly misleading. It is misleading because the expression "in time" wrongly suggests the Newtonian notion of time as homogeneous and static container which is independent of its dynamic content; as we shall see, Bergson rejects this notion. Second, for the refined introspection, purged of parasitic associations, there is no such thing as a "mental state". The concept of "mental state" was introduced by associationistic psychology which was a form of psychological atomism; in the same way as physics and chemistry explain the variety of physical nature by the combination and re-combination of physical atoms, associationism aimed at reducing the wealth of introspective qualities to the aggregates of the basic "psychic elements" which, whether they were called "impressions" by Hume, "sensations" by Condillac or *Vorstellungen* or *Empfindungen* by German psychologists, possessed the same quality of permanency and identifiability as the atoms of classical physics. Bergson rejects associationism because, in his view, "psychological elements" are mere conceptual entities, at best mere methodological fictions; they are products of artificial analysis, which arbitrarily dissects the dynamic continuity of our stream of thought into the series of contiguous, mutually external blocs, each of which remains

[2] *The Creative Mind,* pp. 103, 217.

the same during the interval of its duration. Bergson rejects psychological atomism for the same reasons for which it was rejected by his contemporary, William James, and by Gestalt psychology which was born at the same time. One year after the publication of *Essai*, *The Principles of Psychology* of William James as well as the famous article of Ch. Ehrenfels, *Uber Gestalt-qualitaten*, appeared; in both these publications the unitary character of psychic wholes against the artificial atomization of associationistic psychology was stressed. As James observed:

> A permanently existing "idea" or "Vorstellung" which makes its appearance before the footlights of consciousness at periodical intervals is as mythological an entity as the Jack of Spades.[3]

Bergson rejects the concept of "separate mental state" for similar reasons; not only because such a "state" is arbitrarily carved out of the dynamic continuity of the stream of thought, but also because of its alleged immutability:

> The truth is that we change without ceasing, and that the state itself is nothing but change. This amounts to saying that there is no essential difference between passing from one state to another and persisting in the same state. If the state which "remains the same" is more varied than we think, on the other hand the passing from one state to another resembles, more than we imagine, a single state being prolonged: the transition is continuous.[4]

From this point of view, the traditional distinction between "succession" and "duration" disappears; in the concretely experienced process of change the differentiation of its successive phases is as essential as the mnemonic persistence of the anterior phase in its successor. But it is important to realize that the adjective "continuous", which both Bergson and James join to the nouns like "change", "transition", "stream of thought", etc., should not be understood in the sense of *mathematical* continuity. For the latter concept is synonymous with that of "infinite divisibility" and as

---

[3] *The Principles of Psychology*, I, p. 236. Ehrenfels' article appeared in *Vierteljahrschr. für wissenschaftl. Philosophie*, v. XIV (1890), pp. 249-292.

[4] *Creative Evolution*, p. 4.

such it is closely correlated with the concept of durationless instant. Both James and Bergson insist vigorously that the concept of "knife-edge present" is a mere conceptual artifact to which nothing in our concrete experience corresponds.[5] Our belief in infinite divisibility of time and in the existence of durationless instants arises inevitably from our habit of symbolizing temporal process by a geometrical line; as the latter is divisible *ad infinitum* and as its divisibility cannot stop short of dimensionless points, it is believed that what is true of the symbol, must be true of its referent; consequently, in the same way as a geometrical line is made of points, time must consist of durationless instants. This is merely another instance of fallacious spatialization. The living continuity of the "stream of consciousness" cannot be reconstituted by the discontinous solid "elements" whether they are assumed to be of finite duration or devoid of any duration.

But, like his contemporary and friend, William James, Bergson was really waging battle on two fronts simultaneously; if he was opposed to the artificial atomization and fragmentation of our inner experience, he with equal vigor protested against its artificial unification by means of some fictitious substantial "Ego". He pointed out that both fallacious procedures — that of artificial atomization as well as that of artificial unification — are closely correlated:

> But as our attention has distinguished and separated them [i.e., "mental states"] artificially, it is obliged next to reunite them by an artificial bond. It imagines, therefore, a formless *ego,* indifferent and unchangeable, on which it threads the psychic states which it has set up as independent entities. Instead of a flux of fleeting shades merging into each other, it perceives distinct and, so to speak, *solid* colors, set side by side like the beads of a necklace; it must perforce then suppose a thread also itself solid, to hold the beads together. But if this colorless substratum is perpetually colored by that which covers it, it is for us, in its indeterminateness, as if it did not exist, since we only perceive what is colored, or, in other words, psychic states. As a

[5] W. James, *op. cit.*, p. 609; *The Creative Mind,* p. 178; *Durée et Simultaneité,* pp. 68-69.

matter of fact, this substratum has no reality; it is merely a symbol intended to recall unceasingly to our consciousness the artificial character of the process by which the attention places clean-cut states side by side, where actually there is a continuity which unfolds. If our existence were composed of separate states with an impassive ego to unite them, for us there would be no duration. For an ego which does not change does not *endure,* and a psychic state which remains the same so long as it is not replaced by the following state does not *endure* either. Vain, therefore, is the attempt to range such states beside each other on the ego supposed to sustain them: never can these solids strung upon a solid make up that duration which flows. What we actually obtain in this way is an artificial imitation of the internal life, a static equivalent which will lend itself better to the requirements of logic and language, just because we have eliminated from it the element of real time. But, as regards the psychical life unfolding beneath the symbols which conceal it, we readily perceive that time is just the stuff it is made of. *There is, moreover, no stuff more resistant nor more substantial.* (The last italics mine.) [6]

In this passage we find in a condensed form several features of what may be termed the *dynamic personalism* of Bergson, which is clearly opposed not only to the psychological atomism of Hume and his followers, but to the substantialist view of Descartes and Kant as well. The only distinction between sensualistic associationism and the rationalism of Descartes and Kant is that, while the first did not go beyond the statement of the complete discontinuity of successive mental states, the second tried to re-establish the unity of experience by introducing the artificial unifying link, whether it was called *res cogitans* by Descartes or *Transcendental Ego* by Kant. Thus the artificially disjoined phases of mental life were brought together by an equally artificial substantial link, fabricated expressly for this purpose. For Kant experience is a sheer diversity of "sensual material"

[6] *Creative Evolution,* pp. 5-6.

as much as for Hume; all its unity comes from its extra-empirical subject and its categorical equipment. Both substantialistic rationalism and atomistic sensualism ignore the living continuity of our inner experience as well as its personal character. Descartes implicitly destroyed the dynamic and personal character of our self by equating it with a *thing* (*res*) which only by its attribute of thought (*res cogitans*) is differentiated from a material object (*res extensa*). The elimination of the dynamic character of the self followed immediately from its definition as "thinking substance"; as substance is by its own nature unchanging, its static, timeless character is thus attributed to the inner self. The impersonal character of the Cartesian substance is less obvious, but only as long as we do not focus our attention on its synonym which Descartes used — "thing" (*res*) — whose impersonal connotation is so obvious that it is usually regarded, especially by personalists, as the very opposite of personality.[7] The impersonal character of the Cartesian *res cogitans* became much more obvious in his successor — Spinoza — who, after clearing the system of Descartes of inconsistencies, replaced the Cartesian *Cogito* by the impersonal and passive *Deus cogitat* or simply *Cogitatur;* this translated into modern languages became the *"es denkt"* of Ernest Mach or the "it thinks" of John Dewey.[8] But this depersonalization of the human self was inevitable as long as the concept of *thing* or *object,* formed by our imagination under the impact of sensory experience, in particular of our experience of solid bodies, was unconsciously applied to our introspective data. In the light of Bergson's epistemology, the very concept of "thinghood" is inapplicable to our introspection and should be replaced by that of process. Thus the Bergsonian *Cogito* is significantly different from the *Cogito* of Descartes; instead of *Cogito ergo sum* we should say, according to Bergson, *Cogito ergo devenio* — "I think therefore I become."

From this point of view there is little difference between the Cartesian substance and Hume's sensations; the latter are nothing but the Cartesian Ego cut into tiny pieces, which are no less impersonal and no less static than the substance from which they

[7] The antithesis of "person" and "thing" appears in the very title of the German personalist, W. Stern, *Person und Sache* (Leipzig, 1906).

[8] E. Mach, *Contributions to the Analysis of Sensations* (tr. by C. M. Williams, Chicago, 1897), p. 22; John Dewey, *Experience and Nature,* p. 232.

are carved out; they are, so to speak, Cartesian substances on a minute scale. Hume himself was clearly aware of the substantial character of his "impressions".[9] Thus substantialist rationalism and atomistic sensualism both replaced the living personal continuity of "I" by the artificial and impersonal "It" or "its" (in plural). Kant was equally aware of the inadequacies of sensualism as well as of the limited applicability of the concept of substance; he was aware that sensualism does not do justice to the unitary character of knowledge and he also opposed the application of the category of substance to the introspective flux.[10] But his attempt at saving the unity of the self was eventually a failure as the subsequent development of post-Kantian philosophy clearly showed. Believing with Hume that our experience, both sensory and introspective, is nothing but a sheer diversity of disjoined elements, he was really forced to look for the unity of consciousness *beyond* the limits of introspective experience; hence his Transcendental Ego, residing in the extra-empirical realm of noumena beyond space and time. In locating his Transcendental Ego in the timeless realm of noumena, Kant unwittingly substantialized it in spite of his verbal claims that the category of substance applies only to sensory experience. In doing this, Kant deprived his Ego of both dynamic and personal character and transformed it into a diaphanous and ghostly entity whose relation to the concrete "empirical Ego" remains completely unintelligible. William James observed, only one year after Bergson's *Essai*, that "Transcendentalism is only substantialism grown shamefaced and the Ego is a 'cheap and nasty' edition of Soul . . . Ego is simply *nothing*; as ineffectual and windy abortion as Philosophy can show."[11] Its impersonal character became much more obvious in Kant's successors. As Bergson observed in his *Creative Evolution:*

True, when he [i.e., Kant] speaks of human intellect,

[9] D. Hume, *A Treatise on Human Nature,* bk. I, pt. IV, sec. V: "My conclusion . . . is that since all our perceptions are different from each other, and from everything else in the universe, they are also distinct and separable, and may be considered as separately existent, and have no need of anything else to support their existence. *They are therefore substances,* as far as this definition explains a substance." (*Italics mine.)* Note that Hume's definition of substance is the same as that of Descartes.

[10] Cf. *Critique of Pure Reason,* Transcendental Dialectic, ch. "Paralogisms of Rational Psychology".

[11] W. James, *op. cit.,* I, p. 365.

> he means neither yours nor mine: the unity of nature comes indeed from the human understanding that unifies, but the unifying function that operates here is impersonal. It imparts itself to our individual consciousnesses, but it transcends them. It is much less than a substantial God; it is, however, a little more than the isolated work of humanity. It does not exactly lie within man; rather, man lies within it, as in an atmosphere of intellectuality which his consciousness breathes. It is, if we will, a *formal* God, something that in Kant is not yet divine, but which tends to become so. *It became so, indeed, with Fichte.* (The last italics mine.)[12]

Thus the development of Cartesianism from Descartes to Spinoza is paralleled by the development of Kantianism from Kant to Fichte; in either case, the abstract and impersonal "I" transformed itself naturally into the cosmic or divine "It". This transformation was implicit in the abstract and impersonal nature of both *res cogitans* and the Kantian Ego. In the Absolute Ego of Fichte, and, even more so, in the impersonal Hegelian or neo-Hegelian Absolute, the depersonalization of human mind was as complete as in the naturalistic psychology of Mach and Dewey. The impersonal "it" or "it thinks" replaced the concrete, personal, and living "I"; both in naturalism as in neo-Hegelianism the personal self was dissolved into an impersonal stuff, whether its name was the Bradleyan Absolute or the neural machinery of behaviorists.

It is only a matter of historical justice to admit that Bergson's simultaneous fight against psychological atomism and rationalistic substantialism had been anticipated by William James who in this respect almost certainly influenced his younger French colleague. The view of human personality as transcending both the abstract homogeneous unity and the Humean atomistic plurality was defended by James in his early article in *Mind* "On some omissions of introspective psychology" which, as its title indicates, was thematically similar to the first book of Bergson. Large portions of the article were printed with little modifications in James's *Principles* and only in this way they became more

[12] *Creative Evolution*, p. 388.

widely known, very probably to Bergson himself, who at the time of the publication of his *Essai* (one year before the publication of James's *Principles*) was acquainted only with James's essay "The Feeling of Effort".[13] But the problem which James faced reappears on the pages of Bergson's essay *An Introduction to Metaphysics* and from the following quotation it is clear that his solution was the same as that of James.

> To these detached psychic states, to these shadows of the ego, the sum of which was for the empiricists the equivalent of the self, rationalism, in order to reconstitute personality, adds something still more unreal, the void in which these shadows move — a place of shadows, one might say. How could this "form", which is in truth formless, serve to characterize a living, active, concrete personality, or to distinguish Peter from Paul? Is it astonishing that the philosophers who have isolated this "form" of personality, should, then, find it insufficient to characterize a definite person, and that they should be gradually led to make their empty ego a kind of bottomless receptacle, which belongs no more to Peter than to Paul, and in which there is room, according to our preference, for entire humanity, for God, or for existence in general? . . . The distance, then, between a so-called "empiricism" like that of Taine and the most transcendental speculation of certain German pantheists is very much less than is generally supposed.[14]

It would be, however, a mistake to conclude that there was no orginality in Bergson's approach toward the problem of personality. On the contrary, Bergson, while accepting James's view that neither the concept of abstract unity nor that of arithmetical multiplicity are applicable to psychological self, offers a positive clue to this apparent paradox in insisting on *the dynamic nature of self*. It is true that even this aspect was stressed by James who coined the famous term "stream of thought". But in spite of superficial similarities between Bergson's "true duration"

[13] Bergson's letter to James on January 6, 1903; reprinted in *Ecrits et Paroles*, I, p. 192-193.
[14] *An Introduction to Metaphysics*, tr. by T. E. Hulme (New York, 1912), p. 34, reprinted in *The Creative Mind*, pp. 205-206.

and James's "stream of consciousness" there were also some important differences. As far as the problem of freedom was concerned James's attitude around 1890 was still almost entirely pragmatistic; it is true that he rejected determinism, but mainly on ethical, not logical grounds.[15] Bergson, on the other hand, had already claimed that the temporal character of psychological self entails its indetermination. This is Bergson's famous thesis about the incompatibility of time and the rigorous determinism of the Laplacean type.[16] If time is genuinely real, that is by its own nature *incomplete,* then the indetermination of the future follows truistically from its futurity; only when the future is regarded as a *hidden present* which exists *already now* independently of whether we know it or not, is rigorous determinism possible. In this respect it was Bergson who influenced James; for James indeterminism remained a mere "willed belief" and only in the last period it became a rational conviction which found its expression in *Pluralistic Universe* and *Some Problems of Philosophy* where his intellectual debt to Bergson is fully acknowledged.

For those who read Bergson's first book attentively, it is clear that although it allegedly deals with the problem of *psychological* duration, by its implications it transcends the limits of psychology as it tries to determine the nature of reality in general. For already Bergson had claimed that homogeneous time is merely a verbal time because every authentic temporal process is by its own nature *heterogeneous.* "Eliminate the qualitative heterogeneity of successive phases", says Bergson, "and you will eliminate the fact of succession itself". What will remain, will be a homogeneous geometrical line which is labeled "time". Similarly, if genuine temporality implies indetermination of the future, then wherever time flows, there must be an element of contingency. Thus Bergson unwittingly passed over from the problem of psychological self to the cosmological problem of reality in general: if the whole of reality is dynamic in nature, then determinism must be false even within the physical realm. The problem of human freedom is thus subordinated to the

15 Cf. James's essay, "The Dilemma of Determinism" in: *The Will to Believe and Other Essays in Popular Philosophy.* Cf. also my article "The Reappearance of the Self in the Last Philosophy of W. James", *Phil. Review,* V. LXII (1953), pp. 526-544.

16 *Creative Evolution,* pp. 43-45.

more general problem of indetermination of the future. At first Bergson hesitated to take such a radical step for obvious reasons. For around 1890 determinism in physics seemed to be established beyond any reasonable doubt and practically nobody paid any attention to a few isolated prophets of indeterminism in physics like Renouvier, Boutroux and C. S. Peirce. In accepting strict determinism in the physical world, Bergson was driven to rather an uncomfortable and queer conclusion that the physical world is timeless:

> Thus, within our ego, there is succession without mutual externality; outside the ego, in pure space, mutual externality without succession. . .[17]

But although this denial of time in the physical world was in agreement with the Laplacean physics and with the perennial static preferences of human intellect, it entailed an intolerable dualism of the temporal world of consciousness and the timeless world of space and matter. Bergson was aware of the queerness of such a view, and this explains why his attitude to the problem of reality of physical time was then rather confused and floating:

> What duration is there existing outside us? The present only, or, if we prefer the expression, simultaneity. *No doubt external things change, but their moments do not succeed one another,* if we retain the ordinary meaning of the word, except for a consciousness which keeps them in mind. We observe outside us at a given moment a whole system of simultaneous positions; on the simultaneities which have preceded them nothing remains. To put duration in space is really to contradict oneself and place succession within simultaneity. Hence we must not say that external things endure, but rather that there is in them some *inexpressible reason* in virtue of which we cannot examine them at successive moments without observing that they have changed. (The first and the last italics mine.) [18]

But how can change exist without succession? What is that "inexpressible reason", to which Bergson refers, and which accounts

[17] *Time and Free Will,* p. 108.
[18] *Ibid.,* p. 227.

for the strange parallelism between the succession of our mental states and the changes in the external world? Moreover, Bergson overlooks that if it is contradictory to place duration within space (more accurately, as Whitehead would say, into instantaneous space), there is nothing contradictory in placing space within duration. This is what Bergson did in his later books, *Matter and Memory* and *Creative Evolution* in which the reality of duration on the cosmic level is explicitly asserted.

But there was another difficulty, closely related to the first one just mentioned, which led Bergson to modify his original sharp dualism of the mental and the physical. It was basically the same difficulty which Descartes faced; if the mental events, constituting our psychological dynamic self, are undetermined, how can they be related to completely determined physical events in the brain? It is clear that no psycho-physical interaction is possible because no contingency and no novelty in the physical world is possible as long as the strict determination of physical events is retained. It is clear that the value of freedom asserted by Bergson in his first book, is, pragmatically speaking, very limited because it is confined to the realm of private data; although we are free to *think* and to *feel,* we are not free to *act* and not even to *talk* as all overt reactions, including our vocal reactions, are determined by our neural machinery which is a strictly determined part of the strictly determined physical whole. In his last book *Creative Mind,* Bergson conceded that he was painfully aware of this difficulty and that it was this difficulty which inspired him to modify his original sharp dualism:

> When I set myself the problem of the reciprocal action of mind and body upon one another, it was solely because I had met it in my study of "the immediate data of consciousness" (*Essai sur les donneés immédiates de la conscience*). At that time freedom appeared to me to be a fact; but on the other hand the affirmation of universal determinism, established by savants as a rule of method, was generally accepted by philosophers as a scientific dogma. Was human freedom compatible with the determinism of nature? As freedom had become for me an undoubted fact, I had dealt with it to

> the exclusion of almost everything else in my first book: determinism could come to terms with it the best it could; it would have to do so, as no theory can resist a fact for long. But the problem I had avoided throughout my first work now presented itself as inescapable. Faithful to my method, I tried to get the problem stated in less general terms and even, if possible, to give it a concrete form, to shape it to certain facts upon which direct observation could be based. It is not necessary to relate here how the traditional problem of "the relation of mind and body" contracted before me to the point where it became no more than that of the cerebral location of the memory.[19]

It is beyond the scope of this paper to deal with *Matter and Memory* and *Creative Evolution.* But at least let us state their leading ideas in order to show in what sense and to what extent the original sharp dualism of Bergson has been modified:

1. Even the physical world *endures;* in other words, its temporal character is as real as the temporal character of self.
2. Although matter endures, its duration is of *different kind* from the psychological duration in the sense that the physical events have an incomparably shorter temporal span than the "specious present" in psychology.
3. Even the physical duration in virtue of its dynamic character possesses an element of contingency or indetermination.

It would be interesting to explore more extensively various aspects of Bergson's theory of matter as it was exposed in *Matter and Memory,* in particular in its fourth chapter; it would be equally interesting to compare it with the contemporary trends in physics and Whitehead's organic theory of nature. But this is beyond the scope of this paper. For our present purpose it is sufficient to state that beginning with *Matter and Memory* the original sharp distinction between enduring mind and timeless

[19] *The Creative Mind,* pp. 85-86.

matter is dimmed, being replaced by the difference of temporal span between the psychological and the physical duration. It would be beyond the scope of this paper to analyze Bergson's attempt to show how the extensive (spatial) character of the physical world, as well as its approximate determinism, are correlated with the particular rhythm of the physical duration. This aspect of Bergson's philosophy was generally neglected and only recently it became the object of more detailed analysis.[20] There is no question that without understanding Bergson's theory of different rhythms of duration the whole of *Matter and Memory* as well as the main part of *Creative Evolution* remains completely unintelligible.

For the dualism of mind and matter in the sense defined above was extended by Bergson in 1907 to the cosmic dualism of life and matter. Again, time is lacking to confront Bergson's theory of creative evolution with the mechanistic doctrines of evolution, in particular with the present fashionable form of mechanism — Neo-Darwinism. Let us only say that as soon as we reject mechanism on philosophical grounds, we must reject also its particular applications in various sciences, including biology. For Bergson the creative character of organic evolution was a mere consequence of the rejection of mechanism in general. But it is less frequently stressed that Bergson rejected equally vigorously finalism, at least in its radical form, according to which organic evolution is regarded as a realization of the preconceived program which is being fulfilled in every detail. Such radical finalism, whose most famous representative was Leibniz, not only ignores or makes unintelligible all tragic and evil features of nature, but is basically as deterministic as radical mechanism: "It substitutes the attraction of the future for the impulsion of the past."[21] For this reason radical finalism is merely an "inverted mechanism"; genuine creation and novelty are denied by Leibniz and Driesch as much as by Spinoza and Laplace. Life, however, "transcends finality as it transcends the

20 Cf. my articles "La Theorie bergsonienne de la Matiere et la Physique contemporaine". *Revue Philosophique,* CXLIII (1953), pp. 41-46; "Le Genese ideale de la Matiere chez Bergson". *Revue Metaphysique et de Morale,* pp. 325-348; "Bergson et l'esprit de la Physique contemporaine", *Actes du Xe Congres des Societes de Philosophie de langue francaise* (Paris, 1959), pp. 53-56.

21 *Creative Evolution,* p. 45.

other categories." For this reason it is possible to speak about the "term" or the "goal" of evolution only in a very special and provisional sense.[22]

By "other categories" which do not apply to life Bergson means especially the category of multiplicity and unity. The inapplicability of these categories to the psychological reality has been pointed out in the first book of Bergson and, later, in his *Introduction to Metaphysics;* it is hardly surprising that Bergson emphasizes again their inadequacy when he deals with life which to him is not reducible to a mere physiological mechanism, being a reality of the *psychological* order. But "life" is for him not a mere collective term; it is more than a mere arithmetical sum of individual consciousnesses associated with individual organic bodies. Nor is life an abstract, homogeneous and undifferentiated unity. Life is *Gestalt* in a similar sense as the personal Self is a Gestalt; although it contains the individual selves, it transcends them at the same time, being more than their sum total. In this respect it is analogous to the individual self which also is neither an aggregate of atomistic entities nor is it a bare and empty Ego; it is a dynamic synthesis of unity and multiplicity which only duration can realize. This idea of supra-personal or supra-individual life became later the basis of Bergson's philosophy of religion.

In this last stage of Bergson's philosophy the influence of William James again played an important role. Thus the contacts between both thinkers may be divided into three distinct periods. In the first stage James's early psychological studies provided young Bergson with stimulating topics and problems, sometimes with solutions, or at least with the hints of solutions. In the second stage, it can hardly be doubted that the transformation of James-pragmatist and James-psychologist into James-metaphysician of process would not have been possible without the impact of Bergson's thought. Between the publication of *Varieties of Religious Experience* (1903) and that of *A Pluralistic Universe* there is the publication of *Creative Evolution* (1907), and the decisive influence of the latter on James is acknowledged by James himself. *Varieties of Religious Experience* is still dominated by psychological curiosity, at best by

22 *Ibid.*, p. 289.

pragmatist interest in religious experience, while the interest in the cognitive claims of that experience is in background and hardly appears before the conclusion of the book. In *A Pluralistic Universe* there is a definite shift from the psychological to the metaphysical aspect of the problem. The main question for James at that time was: What is the true meaning of religious experience? Is such experience merely a private affair without any objective referent or should it be interpreted as an intensification of the hidden links between the individual selves and the supra-individual Self? The second alternative is obviously meaningless within the framework of mechanistic naturalism which interprets *all* kinds of religious experience as delusions and emotional disturbances which are traceable to the physiological factors *within* individual organisms. Obviously, if no supra-individual Self exists, there is no objective referent for religious experience, or at least for the religious experience of Western kind. James was clearly encouraged in his rejection of the latter alternative by Bergson's previous rejection of mechanistic naturalism in *Matter and Memory* and *Creative Evolution*. On the other hand, *Creative Evolution* created a mere *possibility* of a theistic interpretation of evolution without explicitly proposing it. It has been frequently said that Bergson's "elan vital" is more akin to Schopenhauer's blind Will than to the personal God of Western religious tradition. I believe that the comparison of Bergson's "elan vital" with Schopenhauer's Will is unfair; there are important differences which Bergson himself stressed even before the First World War. It is, however, true that even after the religious interpretation of "elan vital" given in *Two Sources*, the difference between Bergson's God and the traditional God still remained. Following in this respect both William James and the implications of his philosophy, Bergson did not regard his God as omnipotent and omniscient. For, as the history of Western religious thought clearly showed, omniscience implies predestination and therefore determinism; omnipotence implies, whether we want it or not, the negation of evil and pantheistic monism. Like naturalistic determinism, the theological determinism transforms the cosmic history into a pre-arranged farce completely devoid of any ethical meaning. Such a view is clearly incompatible with the "temporalist view of God" of James and

Bergson and, later, of E. S. Brightman and Whitehead. To sacrifice the power-attributes of God is not too heavy a price for the preservation of his moral qualities.

It is significant that the term "temporalist view of God" was used prior to the publication of *Two Sources* by J. A. Leighton and E. S. Brightman.[23] This shows clearly the affinity of personalism and Bergsonism. Some more explicit reasons were given already in a brief essay of Mary Calkins, "Henry Bergson: Personalist", in *The Philosophical Review* in 1924. It is important, however, to stress again that the reality of the self which Bergson defends against the purely physiological view of man as well as against the sensualistic associationism is not the substantial Ego of Descartes or the timeless entity of Kant, Herbart and MacTaggart. Similarly, what can be called with some qualifications "Cosmic Self" of Bergson (Bergson himself used the term "supra-consciousness" in *Creative Evolution*)[24] is not the timeless Being of St. Thomas, Spinoza and Bradley, but the enduring and living process which in virtue of its temporal character is akin to the finite human selves, although the term "personality" must be used only with a great deal of caution. In other words, Bergson's God is certainly not anthropomorphic and even less caesaromorphic; the whole second chapter about "static religion" in *Two Sources* shows how reserved was Bergson's attitude to the religious traditionalism and institutionalism. This fact is too lightly overlooked by those who tried to annex Bergson to Catholicism. But neither is Bergson's God an impersonalistic, all-pervading substance of pantheism. He is not static for obvious reasons; he is not all-pervading because Bergsonism always retained — more than James, more even than Whitehead — the dualistic character so clearly expressed in *Matter and Memory* and *Creative Evolution*. In spite of its common panpsychic and dynamic ground, reality remains essentially *bi-polar* and the material and mental processes, though not so sharply separated as in Cartesianism, remain recognizably different to justify the dramatic and almost Manichean world-view which Bergsonism advocates. If the dynamic and incomplete universe is the only

[23] J. A. Leighton, "Temporalism and the Christian Idea of God", *Chronicles*, V. XVIII (1918), pp. 283-288; 339-344; E. S. Brightman, "A Temporalist View of God", *Journal of Religion* (1932), pp. 544-555.

[24] P. 284.

place in which human action can be meaningfully called "action", then certain polarity within the universe is necessary to save ethical judgments from becoming inconsequential emotive reactions.

CHAPTER 6

RUSSELL'S HIDDEN BERGSONISM

The title above certainly sounds strange and even facetious; for Russell's attitude to Bergson was not only that of philosophical disagreement, but of positive, almost personal dislike. This dislike accounts for Russell's frequent misunderstandings and misrepresentations of Bergson's thought – the misrepresentations which often border on caricature. It is true that this caricaturing was due more to Russell's inattentive reading than to a conscious desire to ridicule. Russell's own reading of Bergson was accurately characterized by Russell himself when he wrote that "to read an author in order to refute him is not the way to understand him." (OKEW, 47)*). Sometimes, however, the desire to ridicule is clearly discernible; for instance when, ignoring all the distinctions which the author of *Creative Evolution* draws between instinct and intuition, he confuses them, adding with humor that intuition is strongest "in ants, bees and Bergson." (PB, 3.) In any case, inattentive reading is as much a sign of intellectual indifference or hostility as a distorting caricature. Whether the touch of personal animosity in Russell's attitude was due, as it was submitted, to his suspicion that Bergson "lured" Whitehead away from him, is not certain[1], but it would not be too surprising; philosophers are human beings too, Russell more than any other.

Although we pointed out a number of times the deep differences separating Bergson's thought from that of Russell, let us briefly recall those which are the most basic. In this way we shall have a contrasting backdrop against which the unintentional agreements between them will appear even more striking. One of Russell's sentences in *Our Knowledge of the External World* (1914) summarizes the contrast between his and Bergson's philosophy in the most concise way: "Both in thought and in feeling, to realize the unimportance of time is the gate of wisdom." (OKEW, 167). In *Mysticism and Logic* he repeated the same sentence, but somehow more cautiously: "Both in thought and in feeling, *even though time be real*, to realize the unimportance of time is the gate of wisdom." (ML 21–22; italics mine). But this note of caution disappears altogether in the sentences which immediately follow:

That this is the case may be seen at once by asking ourselves why our feelings toward the past are so different from our feelings toward the future. The reason for this difference is wholly practical: our wishes can affect the future, not the past, the future is to some extent subject to our power, while the past is unalterably fixed. But every future will some day be past: if we see the past truly now, it must, when it was still future, have been just what we now see it to be, and what is now future must be just what we shall see it to be when it has become past. The felt difference of quality between past and future, therefore, is not an intrinsic difference, but only a difference in relation to us; to impartial contemplation, it ceases to exist. And impartiality of contemplation is, in the intellectual sphere, that very same virtue of disinterestedness which, in the sphere of action, appears as justice and unselfishness. Whoever wishes to see the world truly, to rise in thought above the tyranny of practical desires, must learn to overcome the difference of attitude towards past and future and to survey the whole stream of time in one comprehensive vision.

It certainly would be difficult to find in the philosophical literature a passage which would be more anti-Bergsonian in spirit as well as in letter. It is a perfect illustration of the view that "all is given" (*tout est donné*) – the view which eliminates becoming, transforms the future into a concealed present and wipes out the qualitative differences between the successive phases of time. It is the view of all strict determinists from Democritus to Laplace, and Russell is merely consistent when he says that "it is a mere accident that we have no memory of the future". (OKEW, 234). It eliminates the concept of causation in its original and dynamical sense by substituting for it the relation of logical co-implication in which the future is deducible from the past and *vice versa*; thus there is not such a thing as "direction of time" or "asymmetry of becoming". In Russell's words: "We shall do better to allow the effect to be before the cause or simultaneous with it, *because nothing of any scientific importance depends upon its being after the cause*." (OKEW, 226.)

This fundamental difference between Russell's and Bergson's views shows itself clearly in their attitude toward Plato and Zeno. While for Bergson, "the intelligible world" of ideas resembles the world of solids in its essential character except that its constitutive elements are "lighter, more diaphanous, easier for the intellect to deal with than the image of concrete things", for Russell in 1912 Plato's doctrine of ideas is one of the most successful attempts to solve the problem of the universals which he accepted with some terminological modifications. He was aware that this view leads to a very sharp kind of dualism:

Thus thoughts and feelings, minds and physical objects *exist*. But universals do not exist in this sense; we shall say that they *subsist* or *have being*, where 'being' is opposed to 'existence' as being timeless. The world of universals, therefore, may be also described as the world of being.

It is true that he somehow softens his commitment to Platonism by the following rather sober and remarkably impartial passage:

The world of being is unchangeable, rigid, exact, delightful to the mathematician, the logician, the builder of metaphysical systems, and all who love perfection more than life. The world of existence is fleeting, vague, without sharp boundaries, without clear plan or arrangement, but it contains all thoughts and feelings, all the data of sense, and all physical objects, everything that can do either good or harm, everything that makes any difference to the value of life and the world. *According to our temperaments*, we shall prefer the contemplation of the one or of the other. The one we do not prefer will probably seem to us a pale shadow of the one we prefer, and hardly worthy to be regarded as in any sense real. But the truth is that both have the same claim on our impartial attention, *both are real*, and both are important to the metaphysician. (P. 100; italics added.)

The passages just quoted show clearly the complexity of Russell's mind as well as the resulting instability of his views. While explicitly admitting his preference for the realm of being, he still conceded then the reality of becoming only to deny – only after two years – the reality of time altogether while at the same time ridiculing Kant for degrading time to a mere appearance. (OKEW, 116–117). But in the passage just quoted another note creeps in – an uneasy awareness that the metaphysical preferences for either being or becoming are perhaps mere personal idiosyncracies, due to individual differences in temperament. A tendency to prefer the metaphysics of Being, together with an underlying note of radical scepticism are two characteristic features of Russell's thought.

Comparison of Bergson's and Russell's attitude toward Zeno's paradoxes will show again the basic contrast in their philosophical views, but at the same time will bring out in the most unexpected way certain hidden affinities. Bergson's view of Zeno's paradoxes was consistently held through all his books: the paradoxes arise from the fallacious assumption that motion and time are divisible *in infinitum*, that is, that the only parts of them which are indivisible are geometrical points and durationless instants. This assumption is based on the confusion of the movement itself with its motionless trace in space; it is this motionless trace, not the act of moving (*la mobilité, le mouvant*), which is infinitely divisible. "At bottom, the illusion arises from this, that the movement, *once effected*, has laid along its course a motionless trajectory on which we can count as many immobilities as we will. From this we conclude that the movement, *while being effected*, lays at each instant beneath it a position with which it coincides. We do not see that the trajectory is created in one stroke, although a certain time is required for it; and that though we can divide at will the trajectory once created, we cannot divide its creation, which is an act in progress and not a thing."[2]) Russell's comment on Zeno in his *Principles of Mathematics* (1903) was significantly different:

After two thousand years of continual refutation, these sophisms were reinstated, and made the foundation of a mathematical renaissance, by a German professor, who probably never dreamed of any connection between himself and Zeno. Weierstrass, by strictly banishing infinitesimals, has at last shown that *we live in an unchanging world*, and that the arrow, at every moment of its flight, is truly at rest. The only point where Zeno *probably* erred was in inferring (if he did infer) that, because there is no change, therefore the world must be in the same state at one time as at another. This consequence by no means follows. (PM, 347; italics mine.)

In other words, Russell agrees with Zeno that we are living in "an unchanging world"; but against Zeno he claims that the world is not in the same state at every moment. How an unchanging world can be different at different successive moments, he does not explain. The only plausible explanation of what appears to be a glaring contradiction is that by 'change' Russell meant the dynamic passage, the transition, the overflow of one moment into the subsequent one; he rejected 'change' understood in this sense, since it is incompatible with the mutual externality of instants in the mathematically continuous time, and since in mathematical continuum there is no "next" element with respect to the "preceding" one. In any case, it is certain that in 1903 Russell regarded the mathematically continuous space and time as the *truly real*, as "the world in which we are living", as he says. What we call 'change' was for him nothing but "diversity in time", time being unconsciously regarded by him in a mathematical fashion as the axis of independent variables on which "successive" instants with corresponding different states of the world exist or rather *coexist*. This reconstruction of Russell's thought is the only possible way in which his strange view that "the world is unchanging without being identical in its successive moments" could be made at least *psychologically* understandable – without becoming any more convincing. Needless to stress that the adjective 'successive' loses its meaning in Russell's scheme which is a perfect illustration of what Bergson called the "fallacy of spatialization". This came up again eleven years later in a strangely ambiguous passage of *Our Knowledge of the External World* where Russell wrote:

> The contention that time is unreal and that the world of sense is illusory must, I think, be regarded as based on fallacious reasoning.

Was Bergson then right? Not quite; here is the immediately following sentence:

> Nevertheless, there is some sense – *easier to feel than to state* – in which time is an unimportant and superficial characteristic of reality. *Past and future must be acknowledged to be as real as the present*, and a certain emancipation from the slavery to time is essential to philosophic thought. The importance of time is rather practical than theoretical, rather in relation to our desires than in relation to truth. ... But unimportance is not unreality... (OKEW, 166–167; italics added.)

One must agree with Russell: it is "easier to feel than to state" how time can be real – though unimportant – when past and future are as real as the present. It is ironical to see the thinker who so severely accused Bergson of vagueness, make an appeal to such a diffused feeling and to get entangled in transparent contradictions.

But in 1914, when the book just mentioned above was published, Russell's views on Zeno were to some extent modified. While in 1903 he agreed with Zeno that the arrow at every instant of its flight is "truly at rest", in 1914 he disclaims it: "we cannot say it is at rest at the instant, since the instant does not last for a finite time... Rest consists in being in the same position at all the instants through a certain period, however short..." (OKEW, 136). Furthermore, there is another modification of Russell's view: he does not insist any longer that we live in the unchanging world of Zeno. He admits explicitly that "the theory of mathematical continuity is an abstract logical theory, not dependent for its validity upon any properties of actual space and time". But he is clearly aware that the applicability of such continuous series to the world of experience is another matter. He concedes that "interpenetration", that is, the "transition which is not a matter of discrete units" is a datum of immediate experience, but he attempts to escape this uncomfortable fact by differentiating fictitiously "appearance" from "reality". Such distinction is clearly meaningless on the level of immediate experience where – as

Russell himself conceded at another place of the same book (72, 85–86) appearance and reality coincide. But while insisting that the world of senses *may* be continuous, he concedes that there is no sufficient reason for it either. In other words, while the theory of mathematically continuous series is *compatible* with experience, it is not *necessitated* by it.

From what has just been said it follows that the nature of sense data cannot be validly used to prove that they are not composed of mutually external units. It may be admitted, on the other hand, that nothing in their empirical character specially necessitates the view that they are composed of mutually external units. This view, if it is held, must be held on logical, not on empirical grounds. I believe that the logical grounds are adequate to the conclusion. They rest, at bottom, upon the impossibility of explaining complexity without assuming constituents. (OKEW, 145).

In other words: Russell's *logical atomism* still makes him lean toward accepting the infinite divisibility of space and time and the actual existence of points and instants, even though he is aware that they are unverifiable empirically. While he was then closer to Bergson in admitting the conceptual constructive nature of "instants", his insistence that every multiplicity must be of the *atomistic* type, built of the mutually external units, ignores completely the qualitative multiplicity which constitutes immediate experience.

It was one year later, in two articles which appeared in *The Monist*, that Russell – without realizing it – came closest to Bergson's thought. In the article 'On the Experience of Time' he explicitly gave up the reality of the durationless mathematical present in psychology:

Suppose, to fix our ideas, that I look steadily at a motionless object while I hear a succession of sounds. The sounds A and B, though successive, may be experienced together, and therefore my seeing of the object while I hear these sounds need not be supposed to constitute two direct experiences. But the same applies to what I see while I hear the sounds B and C. Thus the experience of seeing the given object will be the same at the time of the sound A and at the time of the sound C, although these two times may well not be parts of one specious present. Thus our definition will show that the hearing of A and the hearing of C form parts of one experience, which is plainly contrary to what we mean by one experience. Suppose, to escape this conclusion, we say that my seeing the object is a different experience while I am hearing A from what it is while I am hearing B. Then we shall be forced to deny that the hearing of A and the hearing of B form parts of one experience. In that case, the perception of change will become inexplicable, and we shall be driven to greater and greater subdivision, owing to the fact that changes are constantly occurring. We shall thus be forced to conclude that one experience cannot last for more than one mathematical instant, *which is absurd.* (ET, 216–217); italics mine.)

This is the very opposite of the view held by Russell only a year before when he claimed that "it is perfectly reasonable to suppose that the sense data of a given type ... really form a compact series". The passage above could well have been written by Bergson himself; in truth, it is not difficult to find similar passages in Bergson's works. It is not only the concept of durationless instant which Russell here rejects; he concedes also, however reluctantly, the inapplicability of the concept of arithmetic multiplicity to sensory and introspective experience. It is true that he tries valiantly to define "one momentary total experience", as well as the "specious present"; but as it is clear from the text above, he arrives at the conclusion that the relation "belonging to the present" is not transitive and "that two presents may overlap without coinciding". But what else is this than Bergson's "*pénétration mutuelle*", i.e. "qualitative multiplicity"? (ET, 223, 214).

No less Bergsonian, but equally consistent was Russell's acceptance of the direct perception of succession and change, in other words, of "immediate memory". Thus he says that "succession may be directly experienced between parts of one sense datum, for example, in the case of a swift movement" which is the object of one sensation. (227) "It is indubitable that we have knowledge of the past, and it would seem, though this is not logically demonstrable, that such knowledge arises from acquaintance with past objects in a way enabling us to know that they are past." (222) He admits that we know the past by acquaintance, that is, directly; like Bergson, he claims that "immediate memory is intrinsically distinguishable from sensation" (226) and therefore should not be confused with the present trace of the past sensation:

There is first what may be called "physiological" memory, which is simply the persistance of a sensation for a short time after the stimulus is removed... This fact is irrelevant to us, since it has nothing to do with anything discoverable by introspection alone. Throughout the period of "physiological memory", the sense datum is actually *present*... We will give the name of "immediate memory" to the relation which we have to an object which has recently been a sense datum, but is now felt as past, though *still given in acquaintance*. (224–225; last italics added).

He also fully realizes that the direct knowledge of the past presupposes the Bergsonian *immanence of the past within the present*:

At first sight, we should naturally say that what is past cannot also be present; but this would be to assume that no particular can exist at two different times, or endure throughout a finite period of time. It would be a mistake to make such an assumption, and therefore we shall not say that what is past cannot also be present... *The present has no sharp boundaries*, and no constituent of it can be picked out as certainly the earliest. (222–223; italics mine).

All these passages have their nearly exact counterparts in Bergson's writings and they have been quoted in this book. It is true that Russell's language is still static, atomistic and spatializing in its characteristics; the terms like 'object', 'constituent', 'part of an object' show it clearly. Thus instead of speaking of "duration", he speaks of "one particular existing at two times". But the substance of his view is the same. The fact that Bergson is hardly quoted in this article should not deceive us; Russell refers to William James's analysis of the perception of time (225) and we do not need to repeat how close James's "stream of thought" was to the Bergsonian *durée réelle*.

In the second essay, 'The Ultimate Constituents of Matter' (reprinted in *M.L.*), Russell gave up the applicability of the concept of durationless instant even to physics. He attacks the assumption that the ultimate constituents of matter must be permanent and indestructible; the allegedly permanent "thing" is a logical construct based on the perceptual illusion of fusing together the successive series of momentary states:

Each of these [i.e. of tables, chairs, the sun, moon, stars] is to be regarded, not as one single persistent entity, but as a series of entities succeeding each other in time, each lasting for a very brief period, though probably not for a mere mathematical instant.... A true theory of matter requires a division of things into time-corpuscles as well as into space-corpuscles. (M. L., 129).

Russell thus constructs the world out of momentary particulars which, he insists, "are to be conceived, not on the analogy of bricks in a building, but rather on the analogy of notes in a symphony. The ultimate constituents of a symphony (apart from relations) are the notes, each of which lasts only for a very short time. We may collect together all the notes played by one instrument: these may be regarded as the analogues of the successive particulars which common sense regard as successive states of one 'thing'" (*M.L.* 129–130). It is the same auditory model as that used by Bergson and Whitehead. This attack on the basic assumptions of classical atomism and the tendency to reduce physical existence to the succession of events is obviously similar to Bergson's and Whitehead's "vibratory theory of matter" whose consonance with the philosophical implications of the quantum theory we stressed before. Russell took it over again in his *Analysis of Matter* where he spoke about the atomicity of space-time and "the quantized geodesic routes between two events". (A.Ma., 304, 341).

A close analysis of Russell's view will show the limits of its affinity with the views of both Bergson and Whitehead. This can be stated in one single sentence: Russell gave up *physical*, not *logical* atomism. Russell's idea of corpuscular time may appropriately be characterized as an atomistic translation of the pulsational time of James and Bergson; Russell's "corpuscles of time" are apparently externally related. It is true that this view was not consistent with Russell's view upheld nearly at the same time that the present does not have sharp boundaries. But he apparently overlooked it. Otherwise he would not have characterized his view of the universe as "cinematographic". He recalls that his first visit to a cinema was motivated by his desire to verify Bergson's statement that the mathematician conceived the world after the analogy of a cinematograph, and he found it "completely true" (M.L. 128). There was, however, a significant difference between both philosophers. For Bergson the successive projection of the static picture on the screens symbolizes "the cinematographic mechanism of thought" which tries to reconstruct change out of changeless entities; for Russell, it is a correct and adequate analysis of the physical processes. For Bergson the experienced continuity of change, succession and motion is real, the successive static "moments" are illusory; for Russell the very opposite is true, "the cinema is a better metaphysician than common sense, physics or philosophy". (M.L. 129) Bergson would agree with Russell that the persistence of the physical objects is an illusion since Bergson's matter is also constituted by the momentary (though not instantaneous) events.

Another basic difference is that Russell's "atoms of time" apparently have the same, no matter how minute, duration; the idea of variable temporal span, which is the cornerstone of Bergson's metaphysics, is altogether absent from his thought. This explains his view that he – again unlike Bergson – does not recognize the genuine continuity of psychological duration:

> The real man too, I believe, however the police may swear to his identity, is really a series of momentary men, each different one from the other, and bound together, not by a numerical identity, but by continuity and certain intrinsic causal laws. (M.L. 129)

This is nearly altogether a Humean view; I say 'nearly', since in the Humean framework, there is no place for 'intrinsic causal laws' at all.

In the subsequent development of Russell's thought, his 'logical atomism' came to the fore far more clearly. In *The Analysis of Mind* he upheld the associationism of Hume with hardly any improvement. Memory is reduced by him to the occurrence of a *present* image accompanied by the belief in past existence: "this existed". He insists on the absolute externality of the present remembering and the event remembered; for this reason he condemns "Bergson's interpenetration of the present by the past, Hegelian continuity and identity-in-diversity, and a host of other notions which are thought to be profound because they are obscure and confused" (A.Mi. 180), apparently forgetting all the subtle analyses of his essay 'On Our Experience of Time'. His amnesia on this point continues in *The Analysis of Matter* (1927): "For my part, I do not think logical interpenetration can be defined without obvious self-contradiction; Bergson, who advocates it, does not define it." (A.Ma. 387) In his *Human Knowledge* he does not stop short of the most extreme form of "the fallacy of simple location in time", as Whitehead would call it, when he claimed that not only is a man private from other people, but he is also private from his own past. It is not "here" alone that is private, but also "now". (HK. 90) Obviously, if there is such complete externality of the past with respect to the present, the supposition that the world might have come into existence a few moments ago with all things as they are now, including my fallacious recollections, is not absurd and is, strictly speaking, irrefutable. (HK. 212) Russell, however, concedes that nobody takes such supposition seriously. But if it is so, if nobody takes seriously what cannot 'logically' be refuted, then there must be something radically wrong with a logic of this kind.

It is precisely, such atomistic logic, – an inadequate conceptual tool for dealing with the dynamic and elusive structure of time – which prevented Russell from agreeing with Bergson. For this reason his "incipient Bergsonism" of the year 1915 remained so well hidden – to him as well as to his commentators.

NOTES

* The references and the abbreviations refer to the following of Russell's books:
The Principles of Mathematics (PM), W. W. Norton & Co., New York, 1964.
The Problems of Philosophy (PP), Galaxy Book, Oxford Univ. Press, New York, 1959.

The Philosophy of Bergson (PB), Macmillan, London, 1914; originally in *The Monist* **22** (1912).
Our Knowledge of the External World (OKEW), Allen & Unwin, London, 1914.
'On the Experience of Time' (ET), *The Monist*, 1915.
Mysticism and Logic (ML), W. W. Norton, New York, 1929.
The Analysis of Mind (AMi), Allen & Unwin, London, 1921.
The Analysis of Matter (AMa), Dover, New York, 1954.
Human Knowledge (HK), Simon & Schuster, New York, 1962.

[1] Cf. H. C. McElroy, *Modern Philosophers: Western Thought since Kant*, Russell F. Moore Co., New York, 1950, p. 141. Bergson referred to Russell's attack in his conversation with Jacques Chevalier on May 30, 1933 when he expressed a different theory about the origin of Russell's animosity toward him: "Bertrand Russell has never forgiven me the refutation which I made once of his too material interpretation of the Platonic Ideas. He revenged himself by saying that the evolution culminated on one side in the intellect which found its complete development in mathematicians, and on the other side, in instinct which is at best in ants, bees and Bergson." (J. Chevalier, *Entretiens avec Bergson*, Paris, 1959, p. 197.) Bergson's mildly ironic quotation of Russell clearly referred to the discussion which took place in the Aristotelian Society, 1911, and to which Russell refers in *My Philosophical Development*, Simon & Schuster, New York, 1959, p. 161. The unfairness of Russell's attack was recognized by Alan Wood in his book *Bertrand Russell, the Passionate Sceptic*, Simon & Schuster, New York, 1958, pp. 197–198, where he pointed out that while Russell criticized Bergson for the confusion of subject and object, in the next chapter he praised William James for denying the same distinction!

[2] *C.E.*, p. 336.

PART II

MATTER, CAUSATION, AND TIME

CHAPTER 7

THE DEVELOPMENT OF REICHENBACH'S EPISTEMOLOGY

1. *From Kant to the Biological Theory of Knowledge.*

As late as in 1917 Reichenbach was still nearly an orthodox Kantian;[1] three years later in his book *Relativitätstheorie und Erkenntnis Apriori* he resolutely rejected, or at least profoundly modified Kantian apriorism. What happened in the interval between 1917 and 1920 is not difficult to guess: the spectacular confirmation of the general theory of relativity in May 1919, when Einstein's prediction of the curvature of the light rays in the gravitational field of the sun was fully confirmed, attracted the attention not only of physicists and philosophers but of the general public as well. The theory of relativity was, using Gaston Bachelard's expression, "born of an epistemological shock"[2] experienced when various experiments showed the impossibility of detecting any influence of the motion of the earth on the propagation of light. But "an epistemological shock" caused by experimental confirmations of the same theory was no less severe, especially among Kantian and neo-Kantian philosophers who were so profoundly imbued with the ideas of the Newtonian physics. It is then hardly surprising that Reichenbach's philosophical orientation was radically changed by the impact of the relativity theory.

It is generally agreed that Kant's first *Critique* was merely a codification of the Newtonian physics. Kant not only had no doubt about the principles of classical mechanics, but he even tried

[1] In his doctoral thesis "Der Begriff der Wahrscheinlichkeit für die mathematische Darstellung der Wirklichkeit," *Zeitschrift für Philosophie und philosophische Kritik*, Bd. 161, 162, 163 (1916-1917). Hereafter referred to as BW.

[2] Gaston Bachelard, "The Philosophic Dialectic of the Concepts of Relativity," in *Albert Einstein: Philosopher and Scientist*, ed. P. Schilpp, p. 566.

to prove that no other principles of physics are possible. According to the principles of his epistemology, no matter how much the "material" of experience may increase, its *form* will remain forever the same, since it is determined by the fixed and static character of the perceiving subject. More concretely: it is inconceivable that in some distant future facts will be discovered which would contradict the Euclidian character of space, the principle of the constancy of material substance, the principle of spatio-temporal continuity of all changes, or the principle of strict causality. Why this alleged impossibility? Kant's answer was simple: because the principles just mentioned are *not* empirical generalizations; they do not stem from experience at all. On the contrary, they *precede* experience because they make every experience possible; or, in Kant's jargon, they are *transcendental conditions of experience*. They belong to the innate structure of the epistemological subject, and it is due to their action that the structured and organized perception arises. It is true—and Reichenbach did not fail to notice it—that Kant hypothetically considered the possibility of beings endowed with a different cognitive apparatus. This, according to Reichenbach, would imply the possibility of some cognitive forms intermediate (in the evolutionary sense) between such hypothetical beings and human beings.[3] But this evolutionist interpretation of Kant is very questionable; the fact that this was done by Helmholtz should not blind us to the fact that it is completely foreign to the whole spirit of the Kantian philosophy. Evolution and change in general were regarded by Kant as legitimate concepts only within our sensory and introspective experience; from the transcendent realm of "things-in-themselves" (to which the transcendental Ego belongs) they were completely excluded. This view was a natural and inevitable consequence of the Kantian claim that time is merely a form of our *perception*, whether external or internal; hence the timelessness of the transcendental Ego and the utter impossibility of any kind of evolution in our cognitive faculties.

Kant's implicit prophecy about the definitive and final character of the classical picture of the world was not fulfilled.

---

[3] *Relativitätstheorie und Erkenntnis Apriori*, pp. 53 f. Hereafter referred to as REA.

The relativity theory and, even more recently, the theory of quanta and wave-mechanics were severe shocks to all those who shared with him the belief in the immutability of human intellect. Not only are other geometries than that of Euclid logically possible, but one of them probably represents a more economical description of the real physical space than the classical geometry. Mass is not constant any longer and, although the law of conservation of mass is formally saved by attributing mass to every form of energy, it is evident that the classical—and allegedly a priori—concept of material substance is transformed now beyond recognition. Even when mass is replaced in its function of substance by energy, nothing is ultimately gained, as, according to Heisenberg's principle, it is meaningless to speak about the exact values of energy on the microphysical level. Even the principle of mathematical continuity of all changes and the classical principle of causality are now questioned in the light of new discoveries. Although in 1920 the full extent of the revolutionary changes in physics was not yet apparent, Reichenbach grasped the full significance of the changes already known and drew one important epistemological conclusion from this knowledge: there *is no a priori structure of human intelligence which is eternally changeless and forever independent of experience.* Kant's claim that there are a priori principles of knowledge is, according to Reichenbach, only a sanctification of human common sense ("Heiligsprechung des gesunden Menschenverstandes" [4])—which is an extremely unreliable guide in both science and epistemology.

Yet, the classical Newton-Kantian principles still retain their validity in our daily experience. Space is very nearly Euclidian not only on the terrestrial scale, but even within the limits of the solar system; classical mechanics is valid for velocities small in comparison with the velocity of light; the quantic and microphysical indeterminacy practically disappears on the macroscopic scale. In other words, the classical concepts have a limited sphere of validity within the world of middle dimensions; but beyond these limits their applicability does not exist. This does not mean that there is a sharp discontinuity between the world of our daily

[4] REA, p. 70.

experience and the physical realm beyond these limits; on the contrary classical mechanics, which holds for the world of the middle dimensions, is a limit-case of both the relativistic mechanics and the wave-mechanics; similarly, the geometry of Euclid is a limit-case of the Riemannian geometry when the spatial regions are small enough in comparison to the cosmic space. From the world of our daily experience we pass over to the world of quanta or to the Riemannian universe by "the principle of continuous extension." [5] All this was clearly stated by Reichenbach only one decade after his criticism of Kant. At that time it was already clear that not only our classical picture of the large-scale universe had to be modified in the sense indicated by Einstein, but also that our model of the microcosmos should be radically revised under the impact of the quantum theory and wave-mechanics:

> In the large-scale world, where we investigated the doctrine of time and space, we found this idea realized in the passage from Euclidian geometry to the much more complicated conceptions of non-Euclidian geometry; and in the small-scale world we saw it exemplified in that renunciation of the old concepts of substance and rigorous law which the comprehension of the elementary structure of matter requires. The idea that the general concepts of space, time, substance and causality, which have been borrowed from the middle-sized world, should have unlimited validity for the extensive fields of large-scale and small-scale phenomena must be regarded as the last consequence of the metaphysical picture of the world, which has been overcome for the first time by physics of our generation. [6]

It is not entirely clear what Reichenbach means here by "the metaphysical picture of the world." If he means by it the classical mechanistic picture of the universe which was an extrapolation of our naive sensory perception of ordinary solid bodies, then he is undoubtedly correct. But then he should have specified that what was discredited by modern physics was a simple-minded, materialistic or mechanistic metaphysics rather than metaphysics in general. But this anti-metaphysical effusion apart, the quoted passage is certainly a correct characterization of the present epistemological situation in physics. A large amount of philo-

---

[5] *Atom und Cosmos*, p. 38, 237. Hereafter referred to as AC.
[6] AC, p. 288.

sophical confusion stems, in my opinion, from the fact that Reichenbach's epistemological diagnosis is still not sufficiently appreciated. The opposition to this diagnosis comes from two different sources: first, from the dogmatic, materialistic metaphysics to which we referred above; then also from classical rationalism of the Kantian or neo-Kantian type. Paradoxical as it may seem, they both agree on the *essential homogeneity of nature*, of physical reality, though on different epistemological grounds. Nature is one, according to Democritus, Lucretius, Lamettrie, and Büchner. In their view, the universe is made of solid bodies of different sizes and as the difference between atoms, billiard balls, and celestial bodies is that of dimensions only, the impenetrability and inertia of matter as well as the mechanical laws of impact are the same, no matter what the scale of magnitude. The idea that physical reality is made of heterogeneous strata is altogether foreign to classical science as well as to the metaphysics which grew out of it. But Kant and rationalism in general stressed the unity of nature with an equal vigour, though on a different ground: nature is *one* in the Newtonian and Euclidian sense in virtue of the unity of the human cognitive apparatus which organizes the sensory material always into the same spatio-temporal pattern. Reichenbach's polemic is mainly directed against "the rigid apriorism" ("starre apriorism" [7]) of Kant, but he does not hesitate to attack, and to a certain extent even to ridicule, an almost religious and emotional attachment with which some mechanistically minded scientists adhere to the classical principle of causality.[8] Reichenbach remained convinced that the indeterminacy of the microphysical events is not due to mere technical limitations of our observational procedures but is one of the basic and constitutive features of reality.

With this clear formulation of the belief that the classical forms of thought apply only to the world of middle dimensions, Reichenbach's epistemology acquired the final shape to which its author adhered until the end of his life. His last books do not

---

[7] REA, p. 68.

[8] *Ziele und Wege der heutigen Naturwissenschaft* (1931), pp. 30-31. Hereafter referred to as ZW. Reichenbach expressed his first doubts about the universal validity of the principle of causality in REA, p. 77.

show any significant change in this respect. In his *Philosophical Foundations of Quantum Mechanics*, published in 1948, the same epistemological principle is adhered to, though it is exhibited from a different angle, and some of its consequences, previously not explicitly mentioned, are stated. The basic principle remains the same: the applicability of the classical modes of thought is restricted to the world of middle dimensions, and any attempt at applying them to the microphysical "inter-phenomena" leads to "causal anomalies." To such causal anomalies lead both the corpuscular language and the wave-language; both are based on concepts borrowed from the world of our daily experience. The resulting "causal anomalies" [9] are thus nothing but the result of the inapplicability of these concepts to the microphysical zone situated below the lower boundaries of the zone of middle dimensions. It is evident that the basic idea remains the same as that expounded in *Atom und Kosmos* nearly two decades before: the existence of the heterogeneous strata of reality; the impossibility of extending automatically our inherited modes of thought outside of the zone to which they were originally adjusted; the necessity by "the principle of continuous extension" to broaden our classical forms of thought in order to make them more *flexible* instruments for dealing adequately with the enormously widened experience of contemporary physics.

The reason why the usefulness of the classical modes of thought is restricted to the zone of middle dimensions is nowhere stated by Reichenbach explicitly; but from brief and sketchy remarks occurring in his *Ziele und Wege der heutigen Naturphilosophie* (1931), as well as *Rise of Scientific Philosophy* (1951), it is clear that his views in this regard were fairly close to the *biological theory of knowledge* which, under the impact of the evolution theory, gained ground among some philosophers of the recent past. For several reasons Reichenbach remained unaware of this affinity. First, his interest in the history of ideas was very limited. Of this he was not only aware but even proud; once he wrote that originality of thought is almost incompatible with

---

[9] *Philosophic Foundations of Quantum Mechanics*, pp. 26, 34, 160. Hereafter referred to as PFQM.

historical scholarship.[10] This lack of interest, bordering on contempt, extended not only to the past systems of thought but also to contemporary or almost contemporary philosophers—with the single exception of the Vienna circle. This distrust of philosophy was only strengthened by his great disappointment in Kant's apriorism; the disappointment was all the more severe because of his early allegiance to Kant's philosophy. Finally, the only recent, biologically oriented philosophy with which he was acquainted, was the philosophy of John Dewey. From his critical analysis of Dewey's thought in 1939 it is clear how much he disagreed with him. The biological approach to the problems of knowledge did not prevent Dewey from defending his own version of naive realism under the name of "immediate empiricism." Nothing was more unacceptable to Reichenbach than "the postulate of immediate empiricism," according to which "things are what they are experienced as." It would be beyond the scope of this article to analyze in detail this controversy between Reichenbach, who refused to accept sensory perception at its face value, and Dewey, who asserted in a neorealistic fashion the objectivity of secondary and even tertiary qualities. It was difficult, not only for Reichenbach but for everybody who took modern physics seriously, to agree with Dewey that "things are what they are experienced as" [11] in the case of sensory illusions or to believe that the emotional quality of fearsomeness of a sound is really in the objective situation and not in the perceiving subject.[12]

---

[10] ZW, p. 13.

[11] John Dewey, "The Postulate of Immediate Empiricism," *Journal of Philosophy*, II (1905), No. 15. This principle was upheld by Dewey throughout his life. It was reasserted in his *Experience and Nature* (1925) and defended against Reichenbach in 1939 (in *The Philosophy of John Dewey*, ed. P. Schilpp), pp. 157-192; 534-543.

[12] In the quoted article Dewey explicitly insists on the objectivity of the emotional qualities: "Empirically, that noise *is* fearsome; it *really* is, not merely phenomenally or subjectively so" (p. 393). To Reichenbach's question whether a stick immersed in water really *is* broken, Dewey answered that at least the light rays are broken; but in saying this, he really departed from his postulate of immediate empiricism, according to which "things are what they are experienced as." For in the perception of an immersed stick, the deflection of light rays is *inferred*, never immediately disclosed! What in the immediacy of perception is presented as

But John Dewey was not the only thinker with a biological orientation. There are several others whose epistemology is in several respects close to the implicit meaning of Reichenbach's theory of knowledge. Before investigating these affinities, it is first necessary to characterize briefly the leading idea of Reichenbach's epistemology. It may be stated as follows: *the classical—that is, Newton-Euclidian—form of the human intellect is a system of intellectual habits acquired in a process of evolutionary adjustment to a limited segment of reality.* This limited segment of reality, situated between the macrocosmos and microcosmos, is called by Reichenbach "the world of the middle dimension;" its core is the world of our daily experience. Thus it is evolutionary adaptation which accounts for the triumphs of classical science in the zone of the middle dimensions, and it is the *lack* of this adaptation which explains its failure outside of this zone. It is thus understandable that Reichenbach could not accept the Kantian belief in the immutability of human reason. What Kant regarded as a timeless, a priori structure is, according to Reichenbach, merely a certain evolutionary form which, under the pressure of new experience, has to be replaced by a new one. The classical structure of human intellect is a priori only in a very restricted and non-Kantian sense, defined in the second half of the last century by Herbert Spencer: though it is a priori for an individual, it is a posteriori for the whole species.[13] The feeling of familiarity with which the propositions of Euclidian geometry as well as of classical mechanics impose themselves on our mind results from the ancestral experience of countless generations by which certain unbreakable associations were formed. Thus the experience of the whole species is, so to speak, condensed in the present structure of the human mind, or, using a more naturalistic language, of the human brain. Such was the view of Spencer, Hippolyte Taine, and, more recently, of Henri Poincaré.[14]

---

broken is the stick and not the rays of light. (*The Philosophy of John Dewey*, p. 171, 574).

[13] *The Principles of Psychology*, II, Ch. XIV, p. 195.

[14] Ibid., II, pp. 428-430. Also *First Principles*, where the *inconceivability of the opposite* is regarded as a reliable criterion for the axiomatic character of the laws of conservation of matter and energy (Chapters IV to

But here the similarity between the classical biological theory of knowledge and that of Reichenbach ends. The essential difference between them is that, according to the classical theory, the evolutionary process of adjustment of the cognitive forms to objective reality is completed; in other words, the present structure of human intellect reflects adequately and without essential distortion the general structure of reality. According to Reichenbach, this adjustment is only *partial*; the classical modes of thought are adjusted only to the "mesocosmos," that is, the realm of the middle dimensions. In order to grasp the significance of this contrast in all its concrete implications, a brief and sketchy outline of the classical theory is necessary. The clearest exposition of this theory may be found in Herbert Spencer's *Principles of Psychology*, since Spencer possessed the rare combination of a thorough knowledge of the science of his time and a capacity for clear and orderly exposition.[15]

Spencer anticipated pragmatism and instrumentalism in showing that the original function of knowledge was purely utilitarian and practical: it was a weapon in the struggle for life. Better knowledge meant a better chance for survival. In the majority of animals "knowledge" is synonymous with sensory perception. It may be observed that in the ascending hierarchy of organic beings the field of perception was gradually widened as the sense organs gradually grew in their complexity and responsiveness to an increasing number of external conditions. The environment to which the organism became adjusted was gradually made wider and wider; at the same time, with the appearance of memory and imagination—that is with the possibility of preservation and rearrangement of the traces of the past impressions (if we accept the language of Spencer's physiologically oriented psychology), animals ceased to live in a narrow present moment and were able to retain the influences of past events and even to anticipate some future situations. In order words, the

VI of Part II). See also H. Taine, *De l'intelligence*, II, p. 456; and H. Poincaré, *Science et Méthode*, p. 120.

[15] Spencer expounded his epistemology in his *Principles of Psychology*, I, Pt. III ("General Synthesis").

area to which the organism became adapted was not only wider in space—but in time as well.

This adaptational evolution culminates in man. The field of human perception is, in general, incomparably wider than that of animals, since it is artificially extended by ingenious devices such as the telescope, microscope, spectroscope, seismograph, etc., by which the natural limitations of the human senses are overcome. Thus the spatial limits of the human sensory field are pushed far beyond our earthly and even planetary surroundings. But even more significant is the extent to which man, by means of his imagination and reasoning, transcends the narrow confines of his present moment. By his ability to reconstruct past situations, not only human and earthly history but even the cosmical past is opened to his imaginative insight. By means of his inductive generalizations he is able to anticipate important aspects of the future. It is hardly surprising, then, that Spencer and his contemporaries, dazed as they were by the spectacular successes of astronomy and cosmogony, naturally assumed that the wide segment of the spatio-temporal reality to which human intellect is adjusted is practically co-extensive with *the whole objective reality*; this means that the process of adjustment came to a natural end in the species of *homo sapiens*, or, more specifically, *in the mind of a positivist philosopher of the last century*.

This silent assumption of Spencer's is common to all positivists of the last century and to many of their contemporary descendants. It is true that Spencer's theory of knowledge, like all other positivistic epistemologies of his time, was outwardly very modest and verbally agnostic. He repeatedly stressed his exclusive interest in *phenomena* and *relations between phenomena*, while insisting again and again on the impossibility of knowing "Ultimate Reality" or the "Unknowable." Metaphysics was forbidden by him as strictly as by his twentieth century descendants, at least in words. But underlying this purely verbal agnosticism was an almost unlimited confidence in the adequacy of the classical modes of thought, the confidence which was so characteristic of the nineteenth century science. For Spencer, the alleged fact that human intelligence is a *final* product of the evolutionary process was a *sure guarantee* of the objective validity of human knowledge.

This meant concretely the belief in the applicability of the classical, i.e., Euclid-Newtonian, modes of thought to the universe in its whole spatio-temporal extent. The characteristic belief of the last century in "the unity of nature in space and time" was merely another expression of the same basic conviction. Nature seemed then perfectly transparent. Everywhere there is the same matter and the same space, no matter in what direction or how far from our earthly habitat we look; the telescope and spectroscope were continuously confirming Giordano Bruno's magnificent vision of the infinite and homogeneous universe. The same homogeneity existed on the microscopic scale; the paradoxical world of electrons and photons still lay hidden beyond the reach of experimental detection. Finally, the same homogeneity was to be found in time; the past reconstructed from the present features of the universe as well as the past directly seen in the form of the distant galaxies, seemed to indicate that the universe is an eternally functioning mechanism, cyclically moving "from nebula to nebula" and exhibiting unchanging laws in a rich variety of successive patterns.

This claim, that nineteenth century science possessed a final and definitive picture of the universe, was made by the majority of the last-century scientists and especially loudly and explicitly by such materialists as Büchner and Haeckel. But this claim, though strangely contrasting with the verbal agnosticism and the phenomenalistic language of the positivists, was nevertheless implicitly present also in their thought. The case of Herbert Spencer's mechanistic monism, thinly disguised by the agnostic terms and pantheistic effusions about the "Unknowable," is fairly representative of the last century positivism. Thus by a curious detour evolutionary positivism arrives at the same conclusion as the rationalism of Kant. For Kant any future departure from the Euclid-Newtonian picture of the universe was logically excluded by his assumption about the rigid and timeless character of the human cognitive apparatus. For Spencer and evolutionary positivists no significant change in the cognitive functions can ever occur for the simple reason that the present structure of human intellect represents a *final* adaptation to an allegedly "unknowable reality." Any important modification of the basic categories of human thought would represent, according to Spencer, a step backwards, an epistemological retrogression. Thus the idea of the *absence of evolution* in Kant leads to the same dogmatic conclusion as the idea of the *already completed evolution* in evolutionary empiricism.

Such dogmatism is now untenable. The world of physical experience is much less transparent than Kant and Spencer believed. Using the Kantian terminology, we may say that the "empirical material" enormously increased since the end of the last century. But, contrary to previous expectations, it does not fit the a priori framework of the Kantian forms and categories nor the alleged "first principles" of Spencer; the allegedly timeless framework fits only the realm of the middle dimensions, while outside this realm its usefulness is very restricted. Thus it is impossible today to uphold the basic assumption of Spencer's positivism according to which the intellectual forms embodied in nineteenth century science are adjusted to the whole of reality. But the biological theory of knowledge does not stand or fall with one of its particular historical forms. It may be divorced from the specific form given to it by Spencer and, contrary to a wide-spread prejudice, it is not necessarily tied up with the mechanistic interpretation of evolution. Its essential idea is that the cognitive functions of mind are not static entities, but that they are subject to evolution; it is by no means necessary that this evolutionary process be ruled by the laws of the last century physics. William James, Henri Bergson, and pragmatists in general retained this theory without becoming mechanistic monists. It would be beyond the scope of this article to discuss the main differences between the philosophy of "creative evolution" of Bergson and James and the "false evolutionism" of Spencer.[16] For our present purpose it is sufficient to stress one difference only: while the adaptative process of human intelligence is completed, according to Spencer, Taine, Mach, Avenarius and other positivists, it is not so according to the process philosophy. As early as 1878 James wrote a penetrating analysis of Spencer's epistemology in which he pointed out that even the humblest

[16] The term applied by Bergson to Spencer's evolutionism in *Creative Evolution*, Ch. IV.

sensation is not as passive as Spencer seemed to suggest. Instead of being a passive photographic replica of the physical reality, it has a *selective* character by virtue of which certain external stimuli are registered, others ignored.[17] Our eye does not perceive all wave-lengths of the electromagnetic vibrations; our ear does not register the ultrasonic waves; scents which we perceive are only a very small portion of those perceived by other animals. On the whole the zone of the physical reality *not* perceived by senses is at the same time a zone *not* possessing biological importance for the organism in question. In this precisely consists the significance of *the threshold of sensation:* the external stimuli are perceived only when they reach certain intensity. While they remain *below* that intensity they do not affect significantly even the special sensitive parts of the organism, i.e., the "sense organs," and the corresponding sensation does not arise. What does not affect the organism is biologically irrelevant and is not perceived; in this sense, to perceive means to eliminate the features practically unimportant. For this reason the world of microphysics lay, until the beginning of this century, below the threshold of our sensory perception.

Today only an artificial extension of our sensory field by means of ingenious technical devices brings human beings into contact with the subatomic events. This contact is very indirect and radically different from ordinary perception. We never perceive electrons or other microphysical entities directly; the only thing that we do perceive is their *macroscopic effect*, from which we *infer* their existence. Thus we perceive directly only tiny sparks on the fluorescent screen of the spinthariscope, and we *interpret* them as impacts of the particles emitted by some radioactive substance; what we perceive directly in Wilson's chamber are luminous streaks formed by the illuminated drops of condensed vapour, but we *interpret* these streaks as the traces of the alpha-particles which have passed through. But although we still use the corpuscular language when we speak about these

---

[17] W. James, "Remarks on Spencer's Definition of Mind as Correspondence," *Journal of Speculative Philosophy*, IX, pp. 1-18. Reprinted in *Collected Essays and Reviews* (1920).

inferred microphysical entities, we know today that such language is thoroughly inadequate for describing *all* their properties. Whence this inadequacy? It seems to me that only the biological theory of knowledge, reformulated by Bergson and pragmatism and implicitly present in Reichenbach's writings, supplies a satisfactory answer: our imaginative and linguistic habits, which have been shaped for geological periods by the pressure of our macroscopic environment, simply fail outside the area to which they were originally adjusted. It would be a miracle as well as an absurd lack of biological economy if it were otherwise, i.e., if our mental structure were adjusted to the events which never entered our sensory perception. All incoherencies, contradictions, and "causal anomalies" which constitute the present crisis in physics are merely symptoms of the inadequacy of our inherited modes of thought, when they are illegitimately extrapolated beyond the domain of their original application. Thus the biological theory of knowledge accounts both for the triumphs of classical physics within the realm of the middle dimensions and for its failure outside of that realm.

Kant's phrase "reason prescribes its laws to nature" may be accepted only with the following qualifications: first, we have to realize that if reason finds its structure in nature it is only because nature imposed its own structure on the human mind. In this respect Spencer was right; but he was wrong in his dogmatic (though in the last century very natural) assumption, according to which the *whole* order of nature produced its accurate replica in the Newton-Euclidian form of human reason. Today it is evident that only the biologically important zone of the middle dimensions is with a great degree of approximation mirrored in the classical Newton-Euclidian form of human intellect. This form of intellect, which Kant confused with rationality in general, prescribes its laws only to a limited portion of reality. This is epistemologically the most significant implication of modern physics, formulated correctly by Reichenbach and anticipated by the pragmatically reformulated biological theory of knowledge.

2. *Reichenbach's residual Kantianism.*

Although the development of Reichenbach's epistemology may be characterized as a steady movement away from Kant, the traces of the Kantian influence never completely disappeared from his thought. We have seen that as early as 1920 Reichenbach definitely rejected the basic Kantian assumption about the static nature of human cognitive functions and thus began to move in the direction of the biological theory of knowledge. Yet, in the concluding part of his book *Relativitätstheorie und Erkenntnis Apriori* Reichenbach still speaks about "the unsurpassed greatness" of Kant's philosophy, and even thirty years later he still thinks that Kant is the last philosopher worth studying.[18] This was something more than a persistence of a youthful respect for his early master; it was a genuine adherence to certain aspects of Kantianism. For instance, in the book mentioned above he distinguishes two meanings of the term a priori in Kant's philosophy: *a*) universally valid, and *b*) constituting the concept of object ("Gegenstandbegriff konstituierend"). While he rejects the first meaning for the reasons already mentioned, he retains the second.[19] However, for Kant the meanings were inseparable, or, more specifically, the first meaning depended on the second. Our knowledge is universally valid precisely because the object of knowledge, instead of being imposed from the outside, is constructed conceptually in the mind. This was the meaning of Kant's answer to Hume: the causal relation, instead of being *abstracted* from experience, is *imposed* on it by the never changing category of the epistemological subject. Thus in Kant the terms "universally valid" and "Gegenstandkonstituierend" are bound together. Reichenbach, in giving up the immutability of the epistemological subject, had to give up the universal validity of classical modes of thought and thus he freed himself from the inflexibility of the Kantian epistemology and its virtual hostility toward the trends of recent physics.

But Reichenbach still retained the second meaning of a priori ("Gegenstandkonstituierend"), and this had a decisive influence

---

[18] *The Rise of Scientific Philosophy*, pp. 121-122. Hereafter to be referred to as RSP. See also REA, p. 107.

[19] REA, pp. 46-47.

on his epistemology. In the concluding part of his book about relativity and apriorism he summarized his attitude in the following words: "A priori means: before knowledge, but not: valid independent of experience." But if it is so, why keep the term "a priori" at all? Reichenbach claims that even if we reject the Kantian analysis of knowledge, it cannot be denied that experience contains rational elements. But as he has already stated that "the fact that a priori elements exist does not mean that they are static and independent of experience," one wonders why Reichenbach still uses the term with such misleading associations. For, according to usual definition, "a priori" means "independent of experience" while everything which depends on experience is called "a posteriori." Nobody can forbid Reichenbach to define the term a priori in a new way and to claim that the so-called rational and a priori elements in our experience are in truth empirical and a posteriori; but such use of the words is certainly misleading, and their puzzling effect is only increased when Reichenbach continues to set up his redefined term *in opposition* to the empirical elements. Had he been acquainted with Spencer's thought, he could have used his clarifying statement that what is a priori for an individual is still a posteriori for the whole human species. This was undoubtedly the implicit meaning of Reichenbach's own view, but his language does not have the clarity and explicitness of the great English positivistic thinker. But as there is no doubt that Spencer's epistemology is ultimately empiricist—because in his view all knowledge is ultimately a posteriori—so there is none regarding Reichenbach's own theory of knowledge. By accepting the biological theory of knowledge, Reichenbach had to accept its inevitable consequence: what we call "reason" is merely a disguised form of experience, i.e., "a priori" exists only in the ontogenetic, not in the phylogenetic sense.

Yet, Reichenbach's use of the word "a priori" was based on something more than semantic inertia. Though rejecting the Kantian idea of the static character of the epistemological subject and its cognitive functions, he retained another Kantian presupposition in its entirety: the distinction between the *form* and the *content* (Stoff) of cognition. Perception, according to Reichenbach, does not furnish the object but merely the material (*Stoff*) of which the object is constructed.

> Die Anschauung ist die Form, in der die Wahrnehmung den Stoff darbietet, also gleichfalls ein synthetisches Moment. Aber erst das begriffliche Schema, die Kategorie, schafft das Objekt; der Gegenstand der Wissenschaft ist also nicht ein "Ding an sich", sondern ein durch Kategorien konstituiertes, auf Anschauung basiertes Begriffsgebilde.[20]

Here we face Reichenbach's assumption that there is such a thing as "bare experience" completely devoid of any structure, consisting of isolated and relationless elements on which our reason imposes, by means of its forms and categories, the rational network of relations. Knowledge consists of a correlation of the innate rational structure with the perceptual content, which, in its originally experienced ("erlebt") immediacy, is completely surd and devoid of any structure. The organised and structured perception is a combined result of the blind sensory content and the rational structure of the mind. As mentioned above, this rational structure is of a mathematical nature, and Reichenbach even seems, in this respect, to adhere to Kant's maxim that knowledge is scientific only to the degree to which it is mathematized. Reality (*Wirklichkeit*) thus understood, i.e., reality in the sense of the perceptual sensory content *prior* to its organization, is undefinable. This was an inevitable consequence of the intrinsically irrational character of experience, and Reichenbach did not hesitate to accept it frankly and explicitly.

Kant, in delegating all organizing activity to reason, unhesitatingly claimed that "reason prescribes its laws to nature" and thus incorporated into his philosophy a strongly idealistic factor which became especially conspicuous in his successors, Fichte and Schelling. Nothing is more foreign to Reichenbach than the Romantic philosophy of Kant's successors, yet he did not seem to realize that, by accepting the basic assumption of Kant about the intrinsic surdity of the empirically given data, he in truth subscribed to the Kantian claim that "Reason prescribes its laws to nature." To a certain extent, Reichenbach was dimly aware of this difficulty, for he repeatedly marveled at how the correlation of the rational, mathematical structure with the irrational content of experience is even possible. For if there is no affinity between the rational form and its sensory qualitative

[20] REA, p. 46.

content, by what right are they associated? The peculiarity of the cognitive relation is that, while the conceptual term of this relation is sufficiently defined, it is not so for the second "real" term. The latter, on the contrary, "receives its definition by its correlation with the equations," i.e., with the rational structure. Thus one side of the relation remains completely undefined ("völlig undefiniert").[21] Reichenbach is vaguely aware of a possible danger of the Berkeleyan idealism which he explicitly rejects.[22] But is his answer convincing? His answer remains basically Kantian in its claim that the very definiteness of the correlation is due to the conceptual, i.e. subjective in the Kantian sense, term of the relation:

> Und wir konstatieren die Merkwürdigkeit, daß die definierte Seite die Einzeldinge der undefinierten Seite erst bestimmt, und daß umgekehrt die undefinierte Seite die Ordnung der definierten Seite vorschreibt. *In dieser Wechselseitigkeit der Zuordnung drückt sich die Existenz des Wirklichen aus.*[23] (Italics in the text).

The word "merkwürdig" occurs repeatedly in Reichenbach's passages dealing with the nature of cognitive relations. And certainly there are reasons for his continuous astonishment. Since he rejects any idea of "pre-established harmony" between the rational structure and the empirically given material, and as he was not inclined to draw idealistic consequences from his Kantian premises, it is not surprising that he ends on a note of resignation: This correlation of the conceptual categories with experienced data remains the ultimate unanalyzable fact ("*letzter unanalysierbarer Rest*").[24]

But there was another difficulty, or rather another aspect of the same difficulty. If experience itself is completely amorphous and structureless, it cannot exert any presure by which the structure of our mind would be shaped; for how can the structure of our mind be regarded as a result of something which is completely

---

[21] REA, p. 38.

[22] REA, p. 40.

[23] REA, p. 40.

[24] REA, p. 52. A few pages later Reichenbach denies any pre-established harmony between the conceptual structure and the empirical material (p. 57).

devoid of any structure? It is impossible to speak about the process of adjustment to something which is completely plastic and which borrows all form and consistency from something else. In other words, the biological theory of knowledge is impossible on the Kantian premises; the evolution of the cognitive forms, which are being molded by the pressure of the external reality, is possible only when this objective reality has a consistency and structure of its own. This was the leading idea of Spencer's epistemology and of all those writers who, consciously or unwittingly, followed him. This was also the idea which Reichenbach eventually accepted. It was implicitly present in his book *Relativitätstheorie und Erkenntnis Apriori* in which the idea of the mutability of reason was upheld.[25] More explicit references can be found a decade later in *Atom und Kosmos* and, in particular, in *Ziele und Wege der heutigen Naturphilosophie* where Mach's term, "steady adjustment" (*ständige Anpassung*) of reason to experience, occurs.[26] The most explicit and definite statement occurs in one of his last books, *Rise of Scientific Philosophy*, where the contrast with his original Kantian views is especially obvious:

> The conception of a corporeal substance, similar to the palpable substance shown by the bodies of our daily environment, has been recognized as an extrapolation from our sensual experience. What appeared to the philosophy of rationalism as a requirement of reason —Kant called the concept of substance synthetic a priori—has been revealed as being the product of a conditioning through environment. The experience offered by atomic phenomena makes it necessary to abandon the idea of a corporeal substance and requires a revision of the form of the description by means of which we portray physical reality. With the corporeal substance goes the two-valued character of our language, and even the fundamentals of logic are shown to be the product of an adaptation to the simple environment into which human beings were born.[27]

Compare this passage to the words which may be found on the very first page of Bergson's *Creative Evolution*:

> The history of the evolution of life, incomplete as it yet is, already

---

[25] REA, pp. 69, 74.

[26] ZW, p. 54. Mach used this term in his inaugural address in Prague, October 18, 1883: *On Transformation and Adaptation in Scientific Thought*. Reprinted in *Popular Scientific Lectures*, pp. 219 ff.

[27] RSP, pp. 189-190.

> reveals to us how the intellect has been formed, by an uninterrupted progress, along a line which ascends through the vertebrate series up to man. It shows us in the faculty of understanding an appendage of the faculty of acting, a more and more precise, more and more complex and supple adaptation of the consciousness of living beings to the conditions of existence that are made for them. Hence should result this consequence that our intellect, *in the narrow sense of the word*, is intended to secure the perfect fitting of our body to its environment, to represent the relations of external things among themselves—in short, to think matter. Such will indeed be one of the conclusions of the present essay. We shall see that the human intellect feels at home among inanimate objects, more especially among solids, where our action finds its fulcrum and our industry its tools; that our concepts have been formed on the model of solids; *that our logic is, preeminently, the logic of solids...* (Italics mine)

Compare also Reichenbach's criticism of corporeal substance with the anticipatory remarks made by Bergson more than half a century before *The Rise of Scientific Philosophy* and prior to the main revolutionary changes in physics:

> But the materiality of the atom dissolves more and more under the eyes of the physicist. We have no reason, for instance, for representing the atom to ourselves as solid, rather than as liquid or gaseous, nor for picturing the reciprocal action of atoms by shocks rather than in any other way. Why do we think of a solid atom, and why of shocks? Because solids, being the bodies on which we clearly have the most hold, are those which interest us most in our relations with the external world; and because contact is the only means which appears to be at our disposal in order to make our body act upon other bodies. But very simple experiments show that there is never true contact between two neighbouring bodies; and besides, solidity is far from being an absolutely defined state of matter. Solidity and shock borrow, then, their apparent clearness from the habits and necessities of practical life; *images of this kind throw no light on the inner nature of things.*[28] (Italics mine)

Reichenbach would have been probably deeply shocked, had he realized the affinity of his epistemology with that of Bergson; there are only a few other thinkers who are more disliked or more ignored by the majority of positivists than the author of *Creative Evolution*. Yet the similarity is undeniable: it is one of those unintentional agreements which sometimes occur, ironically enough, between two thinkers who, in other respects, may be

---

[28] *Matter and Memory*, trans. N. M. Paul and W. Scott Palmer, pp. 263-264.

radically different. In both Reichenbach and Bergson we find basically the same biological orientation, in particular in epistemology; the same opposition to the rigid apriorism of Kant—that is, to Kant's idea of the immutability of reason; the same belief that human intellect in its classical form is a result of adaptation to a certain environment; finally, the same conviction that the nature of matter is profoundly different from its superficial macroscopic appearance.

If this is so, then a natural question arises: why are both philosophers, in spite of a very similar epistemological basis, so radically different in other respects? The most important clue to this question may be found in Reichenbach's theory—or rather theories—of time. A detailed analysis of this aspect of his thought, which is beyond the scope of this article, would reveal some interesting affinities with as well as radical differences from Bergson. Within the limits of this paper it can be shown only that Reichenbach failed to draw certain consequences from his biologically oriented epistemology in two important respects. First, his epistemology always retained a certain Kantian tinge which, though outwardly modified by the positivistic and conventionalistic terminology, is clearly discernible even in his last books. Second, Reichenbach's thought was becoming gradually more and more physicalistic, and in this respect it moved steadily away from Kant. As a result, Reichenbach's epistemology shows an uncertain oscillation between conventionalism with some Kantian overtones and physicalism.

3. *The Remaining Ambiguities: Conventionalism and Physicalism.*

Both Reichenbach's conventionalism and physicalism would require a separate and documented study. For our present purpose it will be sufficient to point out briefly their incompatibility with Reichenbach's biological theory of knowledge. From this theory Reichenbach did not hesitate to draw the consequence that the classical concepts such as that of Euclidian space, absolute simultaneity and sharply localizable corpuscules, are, with a great degree of approximation, applicable to the world of daily experience; that is, to the zone of the middle dimensions to which they are adjusted; but their practical usefulness ceases outside this zone; hence the apparently paradoxical character of the microscopic events and of the non-Euclidian curvature of cosmic space. As early as 1920 Reichenbach insisted that "the general theory of relativity declares simply and clearly that the propositions of Euclidian geometry are in respect to physical reality false;" but only a few pages later he claimed that space is nothing but an ordering schema, without objectivity in a physical sense ("nichts physikalisch gegenständlich").[29] In his *Philosophie der Raum-Zeit Lehre* he concedes to the Riemannian geometry the advantage of greater logical simplicity in its application to physical experience; but this simplicity, according to him, is not to be confused with objective truth and does not possess any cognitive value ("Erkenntniswert"). The dispute whether Euclidian or non-Euclidian geometry is "real" or "true" is as meaningless as the question whether the metrical or English system of units is "true."[30] Our answers depend on definitions which are merely conventions, that is, *volitional decisions* which are *neither true nor false*; we select those which lead to a simpler description of reality.[31] In this regard Reichenbach is a *conventionalist* somewhat in the sense that Poincaré was; the important difference between them is that the latter believed that Euclidian geometry will always have the advantage of greater logical simplicity in the description of physical reality while, according to Reichenbach, the very opposite is true.

The source of Reichenbach's conventionalism may be traced to his criticism of Kant in 1920. We recall that he distinguished two meanings in the Kantian term "a priori": a) "timelessly valid"; b) "object constituting" ("Gegenstandkonstituierend"). While he rejected the first meaning, he retained the second. It is undeniable, he says, that experience contains "rational elements"; but these "rational elements" are not, as Kant claimed, constant,

---

[29] REA, pp. 3, 8.

[30] *Philosophie der Raum-Zeit Lehre* (Berlin, 1928), especially pp. 46-47. Hereafter referred to as PRZL.

[31] PRZL, p. 23; RSP, pp. 179-180; also *Experience and Prediction* (1937), pp. 9-10.

but rather *arbitrary*.[32] The contribution of reason consists in introducing not constant, but arbitrary elements into experience. From this it is only a step to Reichenbach's theory of definition and his conventionalism in general. At the same time, he insists that the a priori in the second sense is dependent on experience and should be determined in an inductive way. All such discrepancies merely reflect the basic conflict between his objectivist theory of knowledge, in which emphasis is laid on the gradual adjustment of our cognitive forms to wider and wider segments of reality, and his conventionalism which stresses the allegedly decisive importance of subjective and arbitrary elements in our physical experience. A certain affinity of Reichenbach's conventionalism with Kant's epistemology is unmistakeable. No matter how far he believed he had moved away from Kant, he always shared one basic conviction with him: that sensory experience is completely devoid of any structure and receives its organization from the activity of reason independent from experience. But while this activity of reason for Kant is embodied in continuous functioning of a priori forms and categories, for Reichenbach it has a more volitional character as it manifests itself in arbitrary defining acts of mind.

For the three remaining decades of Reichenbach's life the conflict between objectivist, biologically oriented epistemology and subjectivistically colored conventionalism remained unresolved. From this basic discrepancy stemmed a strange ambiguity which affected his whole philosophy of science. Nearly every specific problem receives from him, implicitly at least, two different solutions. Consider the question of the real structure of space: is it Euclidian or Riemannian? From the point of view of the biological theory of nature, the propositions of Euclid are, physically speaking, false, since they apply only to a limited area of reality and the universe as a whole is Riemannian; however, from the conventionalist point of view, the question is meaningless. Does time possess intrinsic direction? Reichenbach-conventionalist claims that the answer depends on the choice of the coordinating definition; Reichenbach-objectivist apparently be-

[32] REA, p. 85.

lieves that the difference between the future and the past is genuine and is the ground for the objective character of the probability of microphysical events.[33] Are there simultaneous events distant in space? No, says Reichenbach-physicist, at least not in an absolute sense; but according to Reichenbach-conventionalist, if the simultaneity of distant events cannot be empirically verified, it still can be *defined*; consequently, Reichenbach claims, even in the post-relativistic period it is logically possible by conveniently modified definitions to obtain the same simultaneity of distant events *for two different observers moving with different velocities*![34] Similarly ambiguous is Reichenbach's treatment of the problem of material substance. In drawing correctly the consequences of the general theory of relativity, he pointed out as early as 1920 that, if matter is merely a local modification of the non-Euclidian continuum, it is meaningless to speak about individual bodies, since the traditional distinction between the particles and the surrounding space is blurred; the whole concept of thinghood or substance in physics becomes outdated.[35] This critical attitude of Reichenbach toward the classical concept of substance was maintained by him thirty years later, in his *Rise of Scientific Philosophy*;[36] yet, three years before, while conceding that it is meaningless to use the corpuscular language as a state-

---

[33] *Atom und Kosmos*, Ch. 18, especially p. 274. PFQM, p. 17; RSP, pp. 162-165. As early as 1924 Reichenbach correctly observed that determinism in its rigorous form leaves the difference between the past and the future entirely unexplained. For this reason Reichenbach had already—prior to the discovery of the uncertainty principle—upheld the *objective* character of probability laws. "Die Kausalstruktur der Welt und der Unterschied von Vergangenheit und Zukunft," *Sitzungberichte der mathematisch-naturwissenschaftlichen Abteilung der bayerischen Akademie der Wissenschaften in München* (1925), p. 133. On the other hand, Reichenbach remains thoroughly Kantian and conservative in his acceptance of the classical concept of spatio-temporal continuity in spite of very serious doubts which appeared in connection with the quantum theory. In his posthumous book he regards Bertrand Russell's outdated book *Our Knowledge of the External World* (which appeared in 1914) as a modern and, apparently, final vindication of the classical concept of continuity. (*The Direction of Time*, p. 6.)

[34] PRZL, p. 172.

[35] REA, pp. 97-98.

[36] RSP, pp. 189-190. Cf. Note 27.

ment about physical facts, he claims that it is permissible to use it as *definition.*[37] Formally, there is no contradiction, for definitions are neither true nor false and are not statements about empirical reality at all. If so, what is the purpose of introducing them? The definitions introduced by Reichenbach are in fact pervaded by the macroscopic language which, according to his biological theory of knowledge, is utterly unsuitable for describing the interphenomena. To use such a language is, to put it mildly, misleading. It is misleading for Reichenbach himself because, in spite of his verbal declarations, his subconsciousness remains realistic and, though he claims that his definitions are not statements about physical reality at all, he still uses such significant expressions as "exhaustive description" and "corpuscular interpretation." [38] Description of what? Interpretation of what? Not of empirical reality, according to Reichenbach himself. Such discrepancies would remain inexcusable and inexplainable if we did not have a psychological clue to their presence: the persistent conflict in his thought between two incompatible epistemological attitudes.

Only a few concluding remarks about Reichenbach's physicalism. It is occasionally very crude and hardly distinguishable from the primitive materialism of Lamettrie and Büchner.[39] To criticize it would mean restating the objections which have been already raised in the past from Descartes to Lovejoy and Blanshard. But in fairness to him it must be recognized that his physicalism in its more critical formulation approaches the double-aspect theory rather than materialism.[40] We ought not to forget, however, that even the double-aspect theory (or "automaton theory" as it was called by William James) was a natural and

[37] PFQM, p. 29. Reichenbach explicitly compares this problem to that of the simultaneity of distant events.

[38] PFQM, §§ 8, 25.

[39] *Experience and Prediction*, in particular, p. 226 where the mental processes are regarded as being inside the body in the same spatial sense as the physical process within a photoelectric cell. See also RSP, pp. 263-265, 274.

[40] See Reichenbach's criticism of behaviorism in *Experience and Prediction*, pp. 241-243. He blames the behaviorists for trying to eliminate introspection, that is, "the most privileged observer." See also pp. 238-239.

almost inevitable outgrowth of the mechanistic science of the last century which, in the light of Reichenbach's own epistemology, was an illegitimate extrapolation of our intellectual habits beyond the limits of their natural adaptation. Although Reichenbach, at certain times at least, seemed to be aware of the outdated character of the "l'homme-machine" idea, he nevertheless failed to draw any consequence from this awareness. Every interpretation of life other than the physico-chemical one is to him equivalent to supernaturalism.[41]. His theory of consciousness oscillates between the plain materialism of Büchner and more sophisticated epiphenomenalism of Thomas Huxley, both of which belong to past modes of thought.

These critical remarks should not obscure the significance of the most valuable trend in Reichenbach's epistemology: his genetic and biological theory of knowledge. Freed from ambiguities and inconsistencies, this theory leads beyond conventionalism and outdated physicalism toward a more comprehensive view of reality.

[41] RSP, p. 202.

CHAPTER 8

THE SIGNIFICANCE OF PIAGET'S RESEARCHES ON THE PSYCHOGENESIS OF ATOMISM

What is remarkable about the history of atomism is the fact that it can be traced back to the very beginning of human reflection on nature. It is needless to recall the Pythagorean monadism and the names of Leucippus, Democritus and Lucretius: atomism is clearly as old as the first scientific and even proto-scientific explanations of nature. What is even more remarkable is that human thought was able so early and without the aid of microscope and measuring devices to anticipate one of the main findings of modern science. The significance of this anticipation is not diminished by the fact that only the most general features of modern atomism were present in the thought of its ancient ancestors: it was still quite an achievement in the fifth century B.C. to hold the view that space is infinite, that matter is homogeneous and discontinuous in its structure and that all diversity and all changes in nature are reducible to the configurations and displacements of homogeneous, permanent units. The question why and how such a successful anticipation of modern physical science occurred, has been and still is largely ignored; and when it is not ignored, the explanations suggested are hardly satisfactory. These explanations fall roughly into two distinct groups.

The first kind of explanation, apparently still widely accepted, is that the so-called anticipation of modern atomism was a sheer coincidence: Leucippus was just lucky in his guess. This view is held by the majority of professional physicists and by not a small number of philosophers and philosophers of science. It is hardly any answer to the question raised above; furthermore, it is based on the false assumption that there is a radical difference between ancient speculative atomism and the modern empirically based atomistic theories. Such view is historically incorrect; for, as Kurd Lasswitz showed in his classical *Geschichte der Atomistik*, the tradition of ancient atomism survived through the Middle Ages, despite the official rule of the Aristotelian philosophy of nature, to be revived by Gassendi and others in the seventeenth century. But precisely Gassendi's atomism represented an intermediate stage between the Greek speculative form and its modern empirical form which began with John Dalton. In truth, even ancient atomism cannot be called 'purely speculative'; any attentive reader of Lucretius' *De rerum natura* knows about the wealth of empirical observations, no matter how elementary, to which the author refers. On the other hand, philosophical themes were present not only in the atomism of Newton (cf. the concluding pages of his *Optics*), but in that of Dalton as well.

The second answer to the same question was along the Kantian or neo-Kantian lines: the preference for atomistic explanations is an inherent feature of human reason. It is true that Kant in his Transcendental Dialectics regarded atomism as one thesis of the second antinomy which can be proved with equal cogency as its antithesis, i.e. the continuity of matter. But if we look more attentively at Kant's antithesis, we see that it is not as antithetical to the thesis as Kant presents it.[1] The so called 'continuity' of matter is nothing but its infinite divisibility which is entirely compatible with the concept of *material point* which, unlike the material atom, is strictly dimensionless. This was the view which Kant held in the earlier stage of his thought and which is implicitly present also in his later work *The Metaphysical Foundations of Natural Science*; this was also the view of Boscovich before Kant and, after Kant, that of Ampère, Cauchy, Fechner, George Cantor and Russell, at least in his earlier stage. But the material points are as external to each other as the corpuscles of a finite radius; thus the so called 'continuity' of matter is nothing but 'discontinuity infinitely repeated.' In this light the second antinomy is hardly an antinomy at all since both the thesis and antithesis affirm the discontinuity and indivisibility of the first elements of matter; they disagree only about their *number* and their *size*. The thesis postulates the finite number of the atoms of a finite size; the antithesis an infinite number of the dimensionless material points. It was thus logical when the above mentioned Kurd Lasswitz, who besides being a historian of atomism was also a disciple of Kant, made explicit what was merely implicit in Kant: our preference for atomistic explanations is a part of the *a priori* structure of human mind.[2] The views of Emile Meyerson are not basically different, although he, unlike Lasswitz, avoided the Kantian terminology: according to him, the atomistic explanations of change and diversity in the terms of the permanent self-identical units is one of those identifying patterns which, in his view, underlie all rational explanations.

But how does the assumption of the *a priori* character of the atomistic schemes explain their successful application to experience? Orthodox Kantianism claimed that all *a priori* categories are simply *imposed* on the amorphous sensory experience and it is known how obsolete and unsatisfactory this view proved to be. Meyerson's view was more critical; he was well aware that experience is not amorphous and that it resists the identifying patterns which our reason is imposing on it. This surd element in experience was called by him 'l'irrationnel'; this accounts for the fact that the identifying patterns do not apply strictly to experience since neither change nor diversity can be entirely eliminated. But Meyerson still failed to explain an enormously successful application of these identifying schemes, in particular of the conservation laws and of atomism: why sensory experience yields to such extent to these preconceived patterns? The only explanation – and hardly a plausible one – would be to assume that there is a kind of 'pre-established harmony' between the structure of our reason and the objective reality. But it is precisely this 'harmony', this agreement between what Kant called 'form' and 'material' of experience, which requires an explanation and not a mere statement.

There is thus a need for another, more satisfactory theory which, in my view, is provided by that which, for brevity sake, can be called a 'biological theory of knowledge.' The central thesis of this theory (of which there are some varieties) is the claim that the structure of our reason, instead of being ready-made and immutable, is the result of the long process by which our mind gradually adjusted itself to that sector of reality which is biologically important for us. This was nothing but an application of the general theory of evolution to the cognitive functions of man; for these functions, as parts of the psychophysical organism, cannot be exempt from the evolutionary process of which the organism itself is a result. This accounts for the fact that this theory began to appear approximately at the time when Darwin and Wallace proposed their theory of biological evolution. In truth, the first edition of Herbert Spencer's *The Principles of Psychology*, in which the evolutionary and genetic approach to epistemological problems was for the first time consistently applied, appeared in 1855, – four years prior to *The Origin of Species*. Spencer was followed by a number of outstanding and influential psychologists who shared with him his philosophical interest as well as his positivistic orientation. Helmholtz, Mach and Avenarius in Germany, Ribot, Fouillé and, in some aspects of his thought at least, Henri Poincaré in France, – to name at least those who were the most influential.

The biological theory of knowledge corrected Kant's apriorism in the following way: "What is *a priori* for the individual is *a posteriori* for the species" (Spencer) or, in Helmholtz' words, what Kant regarded as *a priori* forms of knowledge corresponds to the inborn psycho-physiological patterns which themselves are – like any other organic features – the results of a long phylogenetic development. There were two main reasons why the interest in this theory began to decline in this century. First, there was a powerful reaction after 1900 against classical positivism and against psychologism in general. Not only the opponents of positivism, like Husserl, but also the neopositivists of the Vienna School, Bertrand Russell and others, especially those interested in the problems of logic and language, dismissed every genetic and evolutionary approach to the problems of epistemology as irrelevant, as being guilty of the 'genetic fallacy'; according to them, the question of the truth of certain beliefs is independent of the question how these beliefs were formed, which biological and psychological processes were involved in their formation. This attitude was forcibly expressed by Wittgenstein in his *Tractatus* (4.1122): "Darwin's theory has no more to do with philosophy than any other hypothesis in natural science."

This reaction obviously went too far; it is difficult to see how epistemology can ignore *any* area of experience, in particular how it can ignore such crucially important facts as those of genetic psychology and evolutionary biology. Fortunately, the pendulum now starts moving in the opposite direction. It was Professor Quine, one of the foremost American logicians, who concedes that the applicability of our logic to reality can be explained only in the evolutionary way. Professor Donald T. Campbell is now detecting the elements of the biological theory of knowledge in the thought of Karl Popper. I, myself, tried to point out the same elements in the thought of two so widely different thinkers – Henri Bergson and Hans Reichenbach.[3] But the most significant contributions to genetic epistemology in the last three decades have been made by Professor Jean Piaget. His researches on the genesis of the concepts of object, of space, of number, and, in particular, his investigations on the psychogenesis of atomism and of the conservation laws shed, in my view, a new light on the question why certain seemingly *a priori* categories of thought apply to the macroscopic level of experience and also why they fail beyond its limits.

Piaget starts from the sound principle that "every response, whether it be an act directed to the outside world or an act internalized as thought, takes the form of an adaptation or, better, of a readaptation." He thus consistently regards intelligence as "a generic form of organization or equilibrium of cognitive structures." In other words:

It (i.e. intelligence) is the most highly developed form of mental adaptation, that is to say, the indispensable instrument for interaction between the subject and the universe when the scope of this interaction goes beyond immediate and momentary contacts to achieve far reaching and stable relations. But, on the other hand, this use of the term precludes our determining where intelligence starts; it is an ultimate goal, and its origins are indistinguishable from those of sensori-motor adaptation in general or even from those of biological adaptation itself.[4]

For our purpose the most important of Piaget's investigations are those dealing with the formation of the notion of permanent object and those dealing with the psychogenesis of atomism and the laws of conservation. It is clear that the latter presuppose the former; for the idea of atom as an invisible permanent object can be grasped only after the notion of permanent object in general is formed. In his book *La construction du réel chez l'enfant*, Piaget[5] showed how the belief in permanent objects gradually develops in a child's mind under the pressure of experience and how the elaboration of the belief in the reality of space takes place concomitantly. There is no place here to go into all details of Piaget's investigations; let us recall that in the first phase the child's universe consists of images which are capable of recognition, but still lack substantial permanence as well as spatial organization; that only in the fourth phase, the child begins to search for the objects that disappeared from the sensory field without taking into account their displacements; that in the fifth phase, the displacements are taken into account, but only as long as they are perceived; and that only in the sixth phase (16–18 mo.) is the belief in the permanence of the objects existing outside the sensory field established, even when their displacements are not perceived. But, as Piaget stressed again in his Columbia lecture *Genetic Epistemology* (Columbia University Press, 1970) the notion of identity is not equivalent to that of quantitative conservation. The identity which the child's mind grasps is of a qualitative kind; it is the ability to recognize the same things in its different positions without realizing that its substance, weight and volume remained the same. This is only natural since certain phenomena seemingly suggest the absence of quantitative conservation. Piaget's monograph *Le développement des quantités physiques chez l'enfant*: *Conservation et atomisme*[6] is extremely instructive in this respect. He observed the intellectual reactions of the children of 6–12 yr old observing the dissolution of sugar in a glass of water. This looks as a sheer annihilation of a piece of sugar which seemingly literally disappears in water – and it is so interpreted by children at first. The next stage, when the attention was drawn to the fact that water remains sweetened, the process of dissolution is interpreted as a sort of qualitative change: water + sugar = sweet water. The child seems to be in the Aristotelian stage when the components in the mixtures and solutions are regarded as *ceasing* to exist actually; the resulting solution is a *new* stuff, homogeneous in all its parts. Only when attention is drawn to the fact that the weight of the glass of water remains *increased* by adding a piece of sugar to it and that the total volume also remains slightly larger, does the child begin to imagine that a piece of sugar was divided into tiny indestructible particles scattered through the water; only then the conservation of the mass and finally that of the volume is firmly grasped. (This takes place in the age 10–12 yr.) There is no place here to go into all the details and all the stages of these experimentation; what was given above is a very simplified sketch. It is exceedingly interesting to see how the insight into the conservation of matter is correlated with the notion of corpuscular structure both in the individual human development and in the collective intellectual development of mankind and how early the permanence of the object, apparently contradicted by sensory appearances, is attributed to tiny material particles.

In the same book Piaget showed how the logico-arithmetical operations develop concomitantly with the operations performed with physical solids; in other words, as Piaget says in his more recent work, the operations executed originally materially, with the solid bodies, become 'internalized', that is, performed in thought.[7] This was already clear in his book *La genèse du nombre chez l'enfant* when he showed how the belief in the elementary commutation law $a+b=b+a$ is an internalization of the operation originally performed with material objects; this belief was impossible as long as the permanence of the objects in the aggregate symbolizing their sum was not explicitly recognized. The reversibility of arithmetical operations is thus based on the reversibility of physical operations. Helmholtz explicitly anticipated Piaget's conclusion when he wrote in *Zahlen und Messen* that the countability of objects presupposed their immutability; during the process of counting "they must neither disappear nor merge nor to be split nor any new one be created."[8] Thus, there seems to be a close correlation between atomism and arithmetization of reality.

Yet, there is one essential difference between Piaget's genetic theory of knowledge and the classical biological epistemology of which Helmholtz was one representative. They both agree that the structure of our reason is a result of the gradual adjustment of our psycho-physical organism to the structure of reality. But Helmholtz believed – like Spencer, like Mach, like even Poincaré – that this process is more or less complete; this confident belief was characteristic of the culminating period of classical physics. If Piaget knows bettel, it is because the twentieth century physics taught us to be more cautious and more modest. Needless to repeat what has been said so many times before about the failure of intuitive corpuscular models on the microscopic scale. Physical atomism has been transformed beyond recognition by its very triumph; for if one thing is certain, it is the fact that not a single feature of the classical concept of particle survived the present revolution in physics. The 'particles' of modern physics are neither immutable, nor permanent, that is, neither indestructible nor uncreatable; their 'motions' cannot be traced along continous trajectories nor can be even localized precisely. In truth, the very usage of the term 'particle' or 'corpuscle' is nothing but a mere inertia of the traditional language. In his article 'The Child and Modern Physics' (*Scientific American*, March, 1957) Piaget made a following pertinent remark:

> Contemporary physicists abandoned some old intuitions about the nature of the physical world. They have, for instance, renounced the concept of the permanence of objects in submicroscopic realm: a particle does not exist unless it can be localized; if it cannot be located at a particular position, it loses its title as an object and must be described in other terms. Now by an extremely curious coincidence it was found that a very young baby acts with regard to objects rather like a physicist.

This sounds paradoxical, but what appears strange becomes completely intelligible within the framework of the new genetic epistemology. There is no reason to be afraid that modern physics invites the physicist to re-descend to the mental, pre-linguistic and pre-operational level of the infant. The babies are hopelessly wrong about our macroscopic environment. But Piaget's investigation of their mentality showed that the structure of human mind is far more flexible than we assumed. Apparently, it does not possess any *a priori* structure; or, if it does, it is certainly not that which Kant had believed he found in analyzing the cognitive functions of man. For what Kant analyzed was not the *a priori* structure of mind but its particular modification which is imposed on it by our continuous intercourse with the solid bodies of our macroscopic experience which eventually was systematized in the Euclid-Newtonian conceptual framework. And it is this structure which proved to be inadequate beyond what Reichenbach called "the realm of the middle dimensions," located between the realm of quanta and the realm of the fleeting galaxies. It fails beyond these limits for the same reason why it triumphs within the realm of the middle dimensions; for it is itself by its own nature the result of the adaptation to this middle region. Our most important and most widely applied category – that of the permanent, identifiable object – is out of place both on the microscopic and megacosmic (extragalactic) scale. In the second volume of his *L'Introduction à l'épistémologie génétique*, Piaget pointed out that we still continue to speak of the universe as a whole, either finite or infinite, without realizing that we are illegitimately transferring our category of object which is a diaphanous replica of the macroscopic solid body beyond the realm of its applicability.[9] He could have added one powerful argument based on the relativity theory: because of the nonexistence of the objective cosmic 'Now', it is meaningless to treat the universe at large as an aggregate of simultaneously existing objects or particles. Our logic seems to be indeed, the logic of solid bodies and the very term 'object' seems to be nothing but a macroscopic prejudice.[10]

But the crisis of the concept of object or 'thing' may go even deeper and imply more than a departure from the Newton-Euclidian corpuscular model of nature. Its full significance is suggested by Reichenbach who after pointing out that "the experience offered by atomic phenomena makes it necessary to abandon the idea of corporeal substance," concludes:

> With the corporeal substance goes two-valued character of our language, and even the fundamentals of logic are shown to be the product of an adaptation to the simple environment into which human beings are born.[11]

Does it mean that not only Euclidian geometry and Newtonian mechanics, but also the two-valued logic is a result of environmental conditioning, more specifically, of the adaptation to the realm of middle dimensions? This view will be probably always resented by logicians, especially those who are oriented toward some kind of Platonism. But, as we have seen, even Professor Quine concedes that the applicability of our logic to experience is the result of natural selection. But he probably assumes with the classical biological theory of knowledge that this adjustment is complete, that is, that what he calls the "immemorial doctrine of ordinary enduring middlesized physical objects" does not require any basic revision and that our two-valued logic truly 'mirrors' the reality in its *whole* extent. But these views are hardly compatible not only with the new genetic epistemology which stresses the incomplete character of our cognitive adjustment, but with the facts of contemporary physics as well.

This is now being more and more recognized. Hilary Putnam raised the question whether logic is also not, after all, an empirical science. David Finkelstein in his recent paper 'Matter, Space and Logic' raised the question whether the present upheaval in physics will not require a similar far reaching *widening* or *generalization* of logic, as revolutionary as the generalization of classical mechanics and of Euclidian geometry was.[12] Within such generalized logic the traditional two-valued logic would still retain its legitimate place, since it would remain applicable to our macroscopic experience. Yet, the traditional logic would lose its privileged, absolutist status since, like the geometry of Euclidean and Newtonian determinism, it would be applicable only to a very large, yet still limited region of our experience.

NOTES

[1] Cf. N. Kemp Smith *A Commentary to Kant's Critique of Pure Reason*, Humanities Press, New York, 1962, pp. 489–91. The correlation between extensionless points and mathematical continuity of space can be seen especially clearly in Bertrand Russell's *Principles of Mathematics*, The Norton Library, 1964.

[2] K. Lasswitz, *Atomistik und Kriticismus*, Braunschweig 1878, in particular Chapter VII, 'Das Apriori in der Physik'.

[3] Cf. W. Van Orman Quine, *Ontological Relativity and Other Essays*, Columbia University Press, New York, 1969, pp. 126–7; Donald T. Campbell, 'Evolutionary Epistemology', to appear in: *The Philosophy of Karl R. Popper*, The Library of Living Philosophers, The Open Court Publishing Company, La Salle, Illinois; M. Capek, 'The Development of Reichenbach's Epistemology', *The Review of Metaphysics* **11** (1957) 42–67; 'La théorie biologique de la connaissance chez Bergson et sa signification actuelle', *Revue de métaphysique et de morale* **44** (1959) 194–211.

[4] Jean Piaget, *Psychology of Intelligence*, Littlefield, Adams & Co., Patterson 1960, p. 7.

[5] *La construction du réel chez l'enfant*, Neuchatel 1937, pp. 11–96

[6] *Le développement des quantités physiques)chez l'enfant: conservation et atomisme*, 2nd ed., Neuchatel 1962, esp. pp. 81–140.
[7] *Ibid.*, p. 279.
[8] *La genèse du nombre chez l'enfant*, Neuchatel 1950; 'Zahlen und Messen', in *Hermann von Helmholtz, Schriften zur Erkenntnistheorie* (ed. and annotated by P. Hertz and M. Schlick), Berlin 1921, p. 82.
[9] *L'Introduction à la épistemologie génétique*, Paris 1951, II, pp. 212–13.
[10] The term used by Ferdinand Gonseth in *Les mathématiques et la réalité*, Paris, 1936, p. 158.
[11] *The Rise of Scientific Philosophy*, University of California Press, Berkeley, 1951, pp. 189–90.
[12] H. Putnam, 'Is Logic Empirical'? in *Boston Studies in the Philosophy of Science*, Vol. V, D. Reidel and Humanities Press, Dordrecht and New York, 1969, pp. 216–41; David Finkelstein, 'Matter, Space and Logic', the same volume, pp. 215. Cf. in the same volume my article 'Ernst Mach's Biological Theory of Knowledge', pp. 400–20. Putnam's radical empiricism in his essay is hardly compatible with his insistence on the timelessness ('tenselessness') of reality, including that of the future, in his article 'Time and Physical Geometry' *The Journal of Philosophy* 64 (1967) 240–7, where he upholds a position very close to the most extreme forms of traditional rationalism.

CHAPTER 9

TOWARD A WIDENING OF THE NOTION OF CAUSALITY

I. THE ORIGINS OF CLASSICAL DETERMINISM

If we wish to speak of the widening of the idea of causality, we must first specify the exact meaning of this concept, the modification of which is now being considered by many contemporary philosophers and scientists. In order to shed light on the classical concept of causality, it is almost impossible to avoid approaching it from the genetic point of view. Without a historical perspective we have only a very limited understanding of the content of the classical concepts by which this philosophic as well as scientific tradition has been constituted. By showing the deep and tenacious roots of our belief in rigorous determinism, we shall better understand certain types of resistance which today are opposed to any attempt at making determinism more flexible.

It is no exaggeration to say that the belief in strict determinism is almost as old as Western thought itself. Without discussing the mythical belief in an impersonal "destiny" to which even the gods were sub-

Translated by Wells F. Chamberlin.

mitted, we find the first precise formulation of determinism in Democritus, when he writes: "All things are determined by necessity, things that have been, things which are, and things which are going to happen." Twenty-two centuries after Democritus, Laplace expressed the same conviction, based on a conception of the universe which does not differ essentially from that of Greek atomism:

> Given for one instant an intelligence which could comprehend all the forces by which nature is animated and the respective situation of the beings who compose it—an intelligence sufficiently vast to submit these data to analysis—it would embrace in the same formula the movements of the greatest bodies of the universe and those of the lightest atom; for it, nothing would be uncertain and the future, as the past, would be present to its eyes.[1]

It is true that there is a very important difference between the determinism of Democritus and that of Laplace. The latter possessed a conceptual apparatus far more complex and flexible than the former. This is only natural; in the interval of time which separated Greek determinism from modern determinism, there occurred two events, which were, moreover, very closely associated with each other: the discovery of infinitesimal calculus and the founding of classical mechanics. The laws of mechanics, especially the law of inertia and of the conservation of the quantity of motion and of energy, were only guessed at by the Greek atomists, and their precise formulation had to await the cosmological revolution of Copernicus and Giordano Bruno. It is still true, however, that Democritus insisted just as vigorously as did Laplace and the modern determinists on the absence of contingency in nature. It is also true that in other respects Democritus anticipated certain aspects of Newtonian physics, for example, the infinity and the homogeneity of space, as well as the qualitative unity of matter, its permanence, and its atomic structure. Thus we see the justification of the Meyerson thesis, according to which, philosophically speaking, the difference between Greek atomism and classical physics is one of degree and not of nature, which means that, given the close connection between the corpuscular models of nature and absolute determinism, the distinction between the "necessity" (ἀνάγκη) of Democritus and the "necessity" of Laplace is also a difference of degree.

Laplace's formula, so frequently quoted, has been expressed many times in more concrete and more colorful language, pointing out clear-

1. Pierre Simon, Marquis de Laplace, *A Philosophical Essay on Probabilities,* trans. F. W. Truscott and F. L. Emory (New York: John Wiley & Sons, 1902), p. 4.

ly that not only inorganic nature but also the most concrete details—and, in appearance, the most contingent details—of human history are only parts of the same network of universal necessity by which all effects are joined to their causes. According to Du Bois-Reymond, Laplacian intelligence would be capable of deducing the most insignificant details as well as the most important events of human history from its huge system of differential equations. It matters little if the events to be deduced belong to the past or to the future. The universal intelligence would know if the sky were clear or cloudy when Pericles embarked at Piraeus to go to Epidaurus; it would also know the exact future date on which the Orthodox cross would be raised over the Mosque of St. Sophia in Constantinople.[2] (We should bear in mind that the date of Du Bois-Reymond's lecture—1877—explains this belief in the inevitability of a Russian conquest of Constantinople.) It is obvious that, from the strictly deterministic point of view, social history is only a particular case of universal physical history. The human body, including the nervous system, is composed of the same elementary particles as inert matter, and, consequently, it obeys the same physical laws; thus Hippolyte Taine was merely consistent when in 1870 he wrote: "In supposing the science to be complete, we should arrive at a mathematical formula enabling us to sum up in some one law the different positions and relations of all the nervous particles."[3]

The idea of inescapable necessity even penetrated literature, where the theme of the inevitability of personal destiny and of all human thoughts and actions became very popular in the naturalistic and in the psychological novel. Let us mention just two examples: Tolstoi and Anatole France. The philosophic epilogue of *War and Peace* is pervaded by the same idea of universal necessity as the philosophic discourses of Dr. Socrates in the *Histoire comique,* when he insists that the whole cosmic past has, as it were, conspired to make M. Chevalier's suicide inevitable. "Even when the solar system was only a pale nebula with a radius a thousand times greater than that of Neptune," says Anatole France, speaking through Dr. Socrates, "the actions of all men, including this particular and tragic one of M. Chevalier, were already inexorably predetermined—for the human mechanism is only a special case of the universal mechanism."

2. E. Du Bois-Reymond, "Über die Grenzen des Naturerkennens," *Wissenschaftliche Vorträge,* ed. J. H. Gore (London, 1896), p. 38.

3. H. Taine, *On Intelligence,* trans. T. D. Hege (New York, 1871), p. 104.

However, it would be a serious mistake to think that rigorous determinism had never been associated with any philosophic system other than the mechanistic and materialistic ones. It is found just as often in the idealistic philosophers. What is seemingly even stranger is the fact that we find a formula just as intransigent as that of Laplace in one of the so-called defenders of human freedom, Immanuel Kant. In a rather little-known passage of his *Critique of Practical Reason,* Kant, long before Laplace, applied Laplacian determinism not only to the human body but also to the human intelligence:

> It may therefore be admitted that if it were possible to have so profound an insight (*so tiefe Einsicht*) into a man's mental character as shown by internal as well as external actions, as to know all its motives, even the smallest, and likewise all the external occasions that can influence them, we could calculate a man's conduct for the future with as great certainty as a lunar or solar eclipse; and nevertheless we may maintain that the man is free.[4]

This passage will seem less paradoxical if we remember that, according to Kant, the category of causality applies to the world of phenomena without any restriction—not only to the "external" phenomena which constitute the contents of our outward perception but also to the introspection, which is placed by Kant on the same phenomenal level as sensory experience. The question has frequently been raised as to what this famous "intelligible freedom," to which Kant alludes in the last words of the text just quoted, could be. For the moment we shall not discuss the question whether freedom is compatible with the denial of time, as Kant himself believed; but we shall return to this point later. Let us merely retain one very important fact: that, as far as the "phenomenal world," including human psychophysical nature, is concerned, Kant was as deterministic as La Mettrie, or any other materialist or mechanist. Even Johann Gottlieb Fichte, certainly one of the most intransigent idealists, did not hesitate to affirm predetermination and the complete predictability of all psychological states.[5] In an entirely consistent fashion, although it appeared to be somewhat disrespectful, Friedrich Paulsen, one of the founders of Neo-Kantism, applied the Laplacian explanation to the mind of his master himself, when he wrote that an

4. Immanuel Kant, *Kant's Critique of Practical Reason and Other Works on the Theory of Ethics,* trans. T. K. Abbott (London: Longmans, Green & Co., 1909), p. 193.

5. J. G. Fichte, *Die Bestimmung des Menschen, Sammtl. Werke* (Berlin, 1943), II, 182–83.

omniscient physiologist would explain . . . the author of the *Critique of Pure Reason* just as he would explain a clock-work. In consequence of this particular arrangement of the brain-cells and of their interconnections with each other and the motor nerves, certain stimuli exciting the retina and the tactile nerves of the fingers had to occasion certain movements, which are in no wise different from those of a writing-automaton or a music-box.[6]

It is quite clear that the doctrine of psychophysiological parallelism, according to which there is no interaction between the consciousness and matter, found another argument in its favor in the Kantian doctrine of causality. If the category of causality is applied to the whole phenomenal world, it must be applied to all motions of matter, including the molecular displacements in the cerebral tissue of Kant himself. Thus human freedom is denied by the Kantians and Neo-Kantians as effectively as by the materialists and the mechanists. The only difference between materialism and Kantism is that for the latter the physical world is only a world of phenomena, while for the former it is a reality in itself, a *Ding an sich*. But interaction between the consciousness and the brain is as radically eliminated by Kant and the Neo-Kantians as by the physiological psychology of the nineteenth century.

Curiously enough, even some of those who are opposed to the parallelist doctrine and who defend a kind of psychophysical interaction nevertheless accept the doctrine of absolute determinism. Hans Driesch, for example, although opposed to mechanistic explanations in biology, has nevertheless stressed that his *vitalism* is not to be confused with indeterminism. Moreover, in a passage in his principal work, *Die Philosophie des Organischen,* he has explicitly stated that the complete knowledge of a certain state of the physical world, added to the complete knowledge of all the states of all the entelechies at the same instant, would make possible for us the integral prediction of any future moment.[7] It is difficult to find a more convincing proof that rigorous determinism is not the exclusive domain of materialism or of naturalism. As we have seen, it can be combined with the idealistic doctrine or even with a vitalist interactionism.

It would be pointless to give more examples. What we have already said is sufficient to show that the doctrine of absolute necessity, which implies the integral predetermination of the future, represents a tend-

6. Friedrich Paulsen, *Introduction to Philosophy,* trans. F. Thilly (New York: Henry Holt & Co., 1912), p. 88.

7. H. Driesch, *Die Philosophie des Organischen,* p. 290.

ency which is present in idealism as well as in naturalism, at least in their classical forms.

A brief survey of the history of philosophy will show that this conclusion is not so paradoxical as it may seem. Rigorous determinism has appeared three times in the history of Western thought: in ancient Greece, in the Middle Ages, and in the science of Galileo and Newton. As we have already stated, it appeared for the first time in the system of Leucippus and of Democritus. By placing the name of Democritus beside that of Laplace, we have already indicated that the modern form of determinism differs only in degree from its classical form. Even if we take into account all the distance which separates the speculative atomism of the Abderite from the experimental atomism of Dalton and from the kinetic theory of gases, the agreement on all essential points obscures the differences of detail and even the difference of method. This difference of method is not so absolute as is often claimed. We must not forget the speculative origins of modern atomism and the influence ancient atomism has had on it. This influence was especially evident in the formative seventeenth-century period; everyone knows the historical bonds between Gassendi's atomism and that of Lucretius. But the influence of Democritus, that is, of the atomist whose system was not spoiled by the curious notion of the undetermined *clinamen,* was no less strong. The name *Democritus reviviscens* which Johannes Chrysostomus Magnenus gave to his book is certainly significant and expresses very well the idea of the return to classical atomism which inspired the physics of the seventeenth century. The global vision of reality is, on all essential points, the same in Greek atomism as in Newtonian physics: the universe is composed of little grains of homogenous matter which move according to strict laws. All diversity of nature is due to differences in *configuration* and in motion. Any *qualitative* transformation is only an appearance produced by the changes in position of particles which always remain the same. Any contingency and any novelty are merely illusions due to our ignorance. Thus it is scarcely an exaggeration to say that the first and the third forms of determinism differ only in details which, however important they may be for the historian of the sciences, are of secondary importance from the philosophical point of view.

In the period which separated Greek atomism and Newtonian mechanics, there appeared a second form of determinism which seemed to be completely different. This was the *theological determinism,* which found its most striking expression in the doctrine of predestination. This form of determinism has certainly been no less rigid than the naturalistic determinism of the Greek and of the modern period. All the concessions—verbal ones, moreover—which have been made by theologians to the notion of human freedom were inspired by motives which were completely foreign to the doctrine itself. Human freedom, in the systems of Augustine, of Thomas, and of the Protestant reformers, is as incompatible with the doctrine of absolute predestination as the *clinamen* of Lucretius was with his mechanistic system. The modern doctrine of absolute necessity is, according to Professor Charles Hartshorne, the result of the "secret alliance" between naturalistic determinism and theological determinism.[8] An assertion of this kind is less surprising when we take into account the common historical origin of these two determinisms. We intend to show that this common source is the philosophy of Parmenides of Elea, whose decisive influence on the development of Western thought is probably without parallel.

The Eleatic origin of Greek atomism is generally recognized. It is known that Leucippus and Democritus, according to Windelband's picturesque expression, "broke Parmenides' sphere into little pieces" which move through empty space according to strict laws. Parmenides' principle of the permanence of Being became the principle of the conservation of matter of the atomists, who, on this point also, anticipated another discovery of modern science. It is true that there are important differences between Democritus and Parmenides. The latter is a monist, while the former was a pluralist. Parmenides denied all change; Democritus admitted at least the reality of change of position. But, despite these differences, there is a profound kinship. Democritus' atom is as permanent, that is, as uncreatable and indestructible, as Parmenidean Being. The quantity of matter which it contains always remains the same. Its essential quality, that is, its *plenitude,* remains as absolute and as immutable as the same quality in the Eleatic Being. If the atomists admitted change, they admitted it in its most innocuous form, that is, in the form of *change of place,* which affects neither the total quantity nor the quality of Being. The change admitted by the atomists is change in the spatial relations of atoms, that is, change which is only half-real. For the void of the atomists, although different from the pure non-

8. Charles Hartshorne, "Contingency and the New Era in Metaphysics," *Journal of Philosophy,* XXIX (1932), 429.

Being of Parmenides, does not have the same degree of reality as matter itself. Consequently, the changing of relations in the void is doubly removed from the *primordial* reality of the substantial *plenum*. Since the time of Democritus, change, as well as multiplicity, is admitted by philosophers; no one, not even Spinoza or Bradley, returned to static monism, as radical and as arrogant as that of the Eleatics. However, the influence of the latter was strong enough to induce most philosophers to regard change and plurality as semireal, that is, as not possessing the same dignity as the underlying Being which remains one and immutable. As Émile Meyerson has shown in his classical works, static monism has remained an ideal model which, although never attained, has inspired philosophic systems as well as scientific explanations.

The continuity of theological determinism with Eleatic philosophy is probably less known and less evident, but it remains no less real. Space does not permit us to give a detailed historical analysis; we shall merely sketch the essential points. What is certain is that the fusion of the idea of Good with that of One, proposed for the first time by Euclid of Megara, and later accepted by Plato and Plotinus, had a profound influence on the formation of Christian theology. In spite of all the differences between Neo-Platonism and the philosophy of Aristotle, the medieval idea of God has the same Eleatic traits. That is why all the eminent Christian philosophers, such as Augustine, Johannes Scotus Erigena, Anselm, and Thomas, identify God with Being, which is One, indivisible, and absolutely immutable—for no change, however insignificant it may be, is compatible with the supreme perfection and incorruptibility of the divine Being. We must not forget that all change, all development, all succession, were regarded by the Christian theologians—as they were, moreover, by the Jewish and Moslem theologians—in a completely Platonistic and Eleatic way, as a corruption unworthy of the absolute perfection of the supreme Being. If we read the first twenty-six questions of the *Summa theologica* of Thomas, we become sufficiently aware of the extent to which the attributes of his God are the attributes of the Eleatic Being. There is no doubt that the religious difficulties in what may be called "theological Eleatism" were very serious. It is almost moving to see Thomas struggling desperately between the biblical idea of a personal and acting God and the Greek idea of a God conceived as an immutable, metaphysical principle and to see him trying to breathe some life and warmth into the cold concept of

Greek metaphysics.[9] In identifying their God with non-temporal Being, the theologians had no other choice than to place his wisdom and his knowledge outside of time. His knowledge must be limited by time; it embraces in a single, indivisible glance the totality of past, present, and future events, which are past, present, and future only for our imperfect human intelligence. Thus omniscience implies foreknowledge, and foreknowledge implies detailed providence, and, consequently, predestination. Within the divine intelligence there is no succession; there is no unrolling of time. All is traced out in advance in the most minute details and cannot be changed. Answering those who ask if prayer for the intervention of the saints can change the eternal decision of God, Thomas says that, if there is a change, it exists only in appearance, because even prayer and the interventions of the saints have been foreseen by the omniscient God, and thus they form a part of total and indivisible predestination. Although this uncompromising doctrine was relaxed in the official semi-Pelagianism of the period which preceded the Reformation, it was taken up again with the same vigor by Luther and especially by Calvin and Zwingli.

The transition from theological determinism to modern naturalistic determinism was not a sudden one. The most important transitional phases were the pantheism of Bruno and, a century later, that of Spinoza. In medieval theology pantheism was only virtual, although several eminent thinkers were coming close to it; but, as long as the duality of the world and of God remained preserved by the very structure of the Aristotelian world, that is, by the duality of the celestial world and the sublunar world, lurking pantheism could not become explicit. But when Giordano Bruno swept away the last sphere of the fixed stars, which was still retained by Copernicus, and when he thus proclaimed the unity of nature in the infinity of cosmic space, the way was open to the explicit and heretical pantheism which would replace the *Deus et Natura* of the Scholastics with the *Deus sive Natura* of Spinoza. We know the profound upheaval which this passage from medieval theism to modern pantheism produced in the sixteenth and seventeenth centuries. But we must not forget that the revolutionary character of modern pantheism was only apparent, because it was virtually present in the thought of theologians before the Renaissance. That explains why the God of Bruno and of Spinoza possessed the same Eleatic traits

9. See *Summa theologica,* Part I, particularly Questions IX, XVIII, XIX, XX, and XXII.

as the God of medieval theology and of Neo-Platonistic philosophy. "The divine Spirit," Bruno writes in his *Summa terminorum metaphysicorum,* "sees all things at once, in a single, simultaneous glance, that is, without distinction between past, present, and future; all things are present for it."[10] As in the preceding philosophical and theological systems, the notion of predestination followed inevitably from that of divine omniscience, but, in the thought of Bruno and of Spinoza, divine predestination was identified with the immanent determinism of nature. This was only natural—for nature and God are but two words for a single cosmic substance. Theological determinism, pantheized in this way, has found itself in natural agreement with the determinism of modern science, the fundamental principles of which were established at the same time. Let us not forget that Spinoza was a contemporary of Newton. After the deistic interlude, which was so unsatisfactory from the philosophical as well as from the religious point of view, God became the impersonal order of nature. Laplace's omniscient mind is only a metaphorical expression for the causal order immanent in nature, but we may also say that it is simply the God of Thomas and of Augustine secularized. Like the God of Christian theology, the "One" of Plotinus and the "Being" of Parmenides, it remains outside of time, outside of change, outside of duration. Let us remember this conclusion, which is of capital importance: the causal order of classical knowledge is a metaphysical entity which is outside of time and which thus implies a radical denial of succession.

II. SUPERFLUITY OF TIME IN THE DETERMINISTIC SCHEMA

Thus, if we accept strict determinism in all its consequences, we are faced with this question: Why do we have the appearance, or, if one prefers, the *illusion,* of time? What is the true place of succession in a strictly determined world? We have already emphasized the fact that no one after Parmenides had had the audacity to deny the reality of time and of change in such a complete and radical manner as the School of Elea had done. A rather curious compromise was generally preferred: becoming, instead of being completely denied, was banished only from the metaphysical realm of the true Being to be lodged modestly in the region of phenomena. In other terms, ultimate reality was placed outside of time while the true Being was almost always re-

10. *Jordani Bruno Nolani opera Latine conscripte* (Florence, 1889), i. 4. c. 14. 32, 33.

garded as static and immutable. It was only its phenomenal aspect—a surface aspect—which was considered as unrolling in time. It matters little if this true Being was the Sphere of Parmenides, the Matter of Democritus, the *Ens realissimum* of the medieval Scholastics, the Substance of Spinoza, the *Ding an sich* of Kant, the Unknowable of Spencer, the Absolute of Bradley, or the impersonal order of nature symbolized by the Universal Intelligence of Laplace—the conclusion always remained the same: time, change, succession, becoming, do not belong to "reality in itself" but to the semireal region of phenomena. Thus the dynamic aspect of reality was merely *reduced in rank*, or *weakened*, instead of being simply eliminated. Although time did not possess as authentic a reality as the immutable ontological background, it nevertheless existed *in some way*, although this manner of existence did not have the same dignity as the underlying Being. However, when one admits the dichotomy of "reality in itself," which is outside of time, and of the "Region of Becoming," in which phenomena succeed each other, he has merely stated the question without even attempting to solve it. Since Plato's time, the following question had been asked: Why is the real cut into two regions, that of the Immediate and Perfect and that of Change and the Imperfect? William James asked it in a concise and precise way in reference to Hegelianism, but his question also concerns all static monisms:

> Why, if one act of knowledge could from one point take in the total perspective, with all mere possibilities abolished, should there ever have been anything more than that act? Why duplicate it by the tedious unrolling, inch by inch, of the foredone reality? No answer seems possible.[11]

Although various explanations of the relation of the temporal and the eternal have been attempted, those who have done it have most often been satisfied with mere words. It has been compared to the relationship of the Perfect to the Imperfect, of the Original to its Copy; Aristotle would quickly have emphasized that such metaphors have no explanatory value and that the theory of the two regions of reality creates metaphysical difficulties instead of solving them. However, this judgment did not stop Aristotle from remaining more Platonic than he wished to, and, consequently, it did not stop philosophers from continuing to split reality in a more or less Platonic manner into two

11. William James, "On Some Hegelisms," *The Will To Believe and Other Essays in Popular Philosophy* (New York: Longmans, Green & Co., 1915), p. 271.

domains without explaining their relationship and, above all, without explaining the superfluity of the temporal. In the Middle Ages, while duality of the world and of God was maintained, the affirmation of the reality of the world involved the reality of time. But, when philosophers began insisting with Giordano Bruno and with Spinoza on the fundamental unity of God and of nature, the status of the temporal was inevitably weakened because the non-temporal eternity of the divine substance inevitably entailed the static eternity of the world. If people avoided this conclusion, it was for the purpose of avoiding the conflict with immediate experience which remains irreducibly temporal. At least it was admitted that temporal experience was real, even though it was illusory. But how can such an illusion be explained? How could static reality of which all the parts exist simultaneously, in a block, be deformed or mutilated into a fragmentary form of temporal development, without ceasing to be immutable? The proposed explanations were only apparent if they were not purely verbal evasions. Thus Spinoza establishes after Bruno the distinction between *Natura naturans* and *Natura naturata,* and he asserts that God, *insofar as he is infinite* (*quatenus infinitus est*), is completely different from what he is, *insofar as* he constitutes human intelligence. William James aptly observed that the main device of Spinoza's philosophy is in the word "*quatenus*":

> Conjunctions, prepositions, and adverbs play indeed the vital part in all philosophies; and in contemporary idealism the words "as" and "quâ" bear the burden of reconciling metaphysical unity with phenomenal diversity. *Quâ absolute* the world is one and perfect, *quâ relative* it is many and faulty, yet it is identically the self-same world—instead of talking of it as many facts, we call it one fact in many aspects.[12]

It is obvious that such a reconciliation of non-temporal reality with its successive and changing appearance is purely verbal; but at least these philosophic prestidigitations, by their very vanity, reveal the impossibility of eliminating succession and change. The temporal character of experience is too authentic and too obstinate to be ignored, and the fact that even static monism in its most varied forms at least recognizes its "phenomenal," that is, its semireal, character without simply denying it, is very significant. It was only natural that scientists and even

12. William James, *A Pluralistic Universe* (New York: Longmans, Green & Co., 1909), p. 47.

philosophers inspired by science, and who, for that reason, were less obsessed with subtle metaphysical problems, did not hesitate to admit the reality of time, frankly and without reservations. However, they also believed, as late as the beginning of this century—and there are many who still believe it even today—that the authentically temporal character of the world is compatible with the most rigorous determinism. Is this true? Are temporality and determinism of the Laplacian type truly compatible? We are now facing the basic question of this article. Upon our answer will depend our attitude toward the general question of determinism and indeterminism, as well as our attitudes toward more special problems, such as that of freedom and of contingency in physical nature—the problem which today is at the center of the controversy over the interpretation of Heisenberg's indetermination principle.

At first glance, the question so stated seems strange and almost devoid of meaning because the answer given to it by common sense is completely clear and negative: there is no incompatibility between succession and strict determinism. From the days of the mythical belief in Destiny to the Newtonian concept of strict causality this answer has not varied. This is only natural. Nothing seems more familiar than the notion of the temporal process the phases of which, although strictly determined, are nevertheless successive. All classical scientific thought, not only in the physical sciences, but also in the biological and social sciences, is based on, or appears to be based on, the idea of the *necessary* connection of successive events. The association between the idea of succession and that of causality is so close and so familiar that, before the French contingentists and especially before Bergson, no one questioned their compatibility. Kant, followed on this point by many others, instead of questioning the compatibility of causal necessity and temporal succession, insisted on their *inseparability;* for him, the only way of saving freedom was to put it outside of time. Even after Bergson people continued to believe the same thing and were surprised if the question was raised. Consider what an American philosopher, Ralph Barton Perry, said in his critique of Bergson: "It is entirely possible to maintain the existential priority of time, and be a vigorous determinist as well." According to Perry, even a strictly determined mechanical system *ages,* although it ages according to a precise law. A simple motion of a material particle, of which all the future positions are predictable with complete accuracy, contradicts, according to Perry, the Bergsonian assertion that temporal evolution and causal necessity are incompatible.[13]

There is no doubt that all the evidence appears to sustain Perry's views and those of his followers. This is especially true if one looks at classical physical science, but it is also true about today's biological and social sciences—for these sciences still remain pervaded by the spirit of classical physics. This is, moreover, only natural. Even among physicists today the question of the strict determination of phenomena is still being debated. Before discussing briefly the changes which have taken place in contemporary physics, we must first expose a serious difficulty which arises for all who claim that the necessary determination of events is compatible with their successive character.

What, then, is the precise meaning of the concept of necessary connection between two successive events? There is agreement on this point: if we affirm that event *b* follows necessarily after event *a,* we are affirming that all the particular traits of the former can be deduced from the latter; supposing our knowledge of a certain event to be complete, there would be no uncertainty even about the most individual and apparently most contingent details of any future event whatsoever. There is no point in quoting Du Bois-Reymond or Anatole France again. This is completely clear in Ralph Barton Perry's example concerning the motion of a material particle; it is obvious there that all the positions as well as the future velocities of the particle in question are predictable. But we know that, according to the determinists, there is in principle no difference between the causal determination of physical events and the necessity of historical events—there are only differences of complexity. It is only their complexity which makes the prediction of events in society so difficult. However, "social physics" does not differ essentially from physics conceived in its original sense. In the one as in the other, the present state implies, without any ambiguity, all future states.

However, by this very assertion, a determinist encounters a difficulty which, in my opinion, is insurmountable. It is known that any logical implication is *ex definitione* non-temporal. It is a commonplace in elementary courses of logic to distinguish logical implication, which is outside of time, from the psychological process of inference by which

13. R. B. Perry, *Present Philosophical Tendencies* (New York: Longmans, Green & Co., 1916), pp. 251–52.

we deduce a conclusion from premises. Although, psychologically speaking, the conclusion is *preceded* by the premises, that is, preceded in the temporal sense, it nevertheless remains true that, logically speaking, there is no succession, no unrolling, in the temporal sense of the word. And let no one be deceived by the ambiguity of the word "flow"; *there is no logical flow in the temporal sense of the word.* If we say that the conclusion "flows" from the premises, we are using this word only in the metaphorical sense. A logical antecedent is not a temporal antecedent; a logical consequence has nothing in common with temporal succession. The premises are not, in the temporal sense, *before* the conclusion, and, in the same way, the conclusion does not *follow* the premises in time. It is more exact to say that the conclusion *pre-exists* in the premises or that it is *contained* in them logically. We *discover* it after the premises in the actual process of human thought, but we do not *create* it by that process itself. The simultaneity of the conclusion with the premises can be illustrated in a convincing way by analyzing a form of classical syllogism: All men are mortal; Socrates is a man; consequently, Socrates is mortal. Or, in symbols: All *M* are *P;* all *S* are *M;* consequently, all *S* are *P*. It is obvious that the expression "consequently" has no temporal meaning. One is easily persuaded of this if he draws the famous Euler's circles, which symbolize the classes, or the logical extensions in question. Not only is class *M* contained in class *P* at the *same time* that class *S* is contained in class *M* but it is easy to see that class *S* is contained at the same time in class *P*. In other terms, the conclusion and the premises are *simultaneous*. The very possibility of symbolizing logical relationships of inclusion by spatial diagrams whose parts are, by their very nature, *juxtaposed,* therefore *simultaneous,* is the reason for this. For there is not a trace of succession in the relationship of inclusion, that is, in the relationship of container and contents. Unquestionably, every conclusion *coexists* in the logical sense with its premises, although it is *thought* and *pronounced* after the premises.

We must not confine our attention to one particular example of the traditional syllogism, for the pre-existence of the conclusion is postulated in every valid reasoning. That is why we say that we *discover* the truth, instead of saying that we *create* it. Just as in the classical syllogism the inclusion of class *S* in class *P* coexists with the two inclusions symbolizing the two premises, so in the solving of a mathematical equation, for example, the "unknown" quantity is *determined in advance* without any ambiguity; thus it is *unknown only to us,* and we discover it in the same way that Columbus discovered America. We say that the solution is simply *waiting for* our discovery, that it exists, so to speak, before our discovery, just as the American continent existed *before* the voyage of Columbus. In the same way, if the future is determined in all its details and without any ambiguity, have we not the right to conclude with Laplace that it is already present and that it is merely waiting to be unveiled to our limited consciousness?

But, if that is true, the same question we have already asked arises again: *Where does the illusion of succession come from?* Why is the future unrolling of universal history not yet unrolled, although it is predetermined in all its details and although the present moment already contains it? If the future history of the universe pre-exists logically in the present, why is it not already here? Why does it require a certain interval of time to become actual, that is, present? Why is there this distance between "it will be" and "it is"? Why does the future require a certain time for its own realization, for its own "becoming present"? Where does this strange time lag come from, a time lag not at all justified by the structure of logical implication, all parts of which are simultaneous? For the average scientist such a question is even more difficult to answer, because for him time is as real as causal necessity; thus he does not permit himself to avoid the difficulty by the traditional expedient of philosophers such as Spinoza, Bradley, McTaggart, and others, who confine succession in the realm of phenomena while excluding it from reality itself.

The incompatibility of causal necessity with the fact of succession was fully emphasized by several French thinkers of the second half of the nineteenth century, such as Jules Lequier, Charles Renouvier, Émile Boutroux, Joseph Delboeuf, and, finally, Henri Bergson. Outside France, it was principally Charles S. Peirce and William James—the latter influenced, at least partially, first by Renouvier and later by Bergson—who insisted on the reality of objective contingency as an essential element of temporal reality. But the intellectual climate of that time was not very favorable to the ideas of this kind. The principle of causality appeared as a simple consequence of the law of conservation of energy (Spencer's law of persistence of force), which in its turn expressed in a new and much more precise way the ancient principle of the indestructibility of substance. This law was considered a sacred dogma, not only by virtue of the empirical evidence in its favor, but also because it was looked upon as a prolongation, and even as a culmination, of the tendencies which had dominated philosophic thought since its beginnings. It is only quite recently that, under the pressure of the new physical discoveries, we have begun to treat the concept of objective contingency with more tolerance. Nevertheless, in Boutroux's time, and even in Bergson's, necessitarian dogmatism, to use Peirce's expression, continuously strengthened by the triumphs of scientific prediction and by the constantly repeated successes of mathematical deduction in the physical sciences, so fascinated minds that almost no one paid any attention to Bergson when he showed that absolute necessity and real succession cannot be reconciled. In his *Creative Evolution,* in a passage which has become classical, Bergson pointed out that the equations of mechanics are concerned only with the extremities of temporal intervals while the intervals themselves are ignored. Even when we talk about them, we scarcely attach any importance to them:

> Common sense, which is occupied with detached objects, and also science, which considers isolated systems, are concerned only with the ends of the intervals and not with the intervals themselves. Therefore the flow of time might assume an infinite rapidity, the entire past, present, and future of material objects or of isolated systems might be spread out all at once in space without there being anything to change either in the formulae of the scientist or even in the language of common sense. The number *t* would always stand for the same thing; it would still count the same number of correspondences between the states of the objects or systems and the points of the line, ready drawn, which would be then the "course of time."[14]

Several pages farther on, after having quoted the famous passage from Laplace, Bergson adds:

> In such a doctrine, time is still spoken of: one pronounces the word, but one does not think of the thing. For time is here deprived of efficacy, and if it *does* nothing, it *is* nothing.[15]

Bergson was probably not entirely right when he affirmed that a determinist pronounces the word "time" without thinking of real succession. The state of mind of an average determinist is certainly more

14. *Creative Evolution,* authorized trans. Arthur Mitchell (New York: Henry Holt & Co., 1913), p. 9.

15. *Ibid.,* pp. 38–39.

complex, and it was more accurately analyzed by Bergson in his first book, where he showed that belief in the necessary connection of events consists in the association of two irreconcilable ideas: that of logical necessity which requires the preformation and even the pre-existence of the future, which ceases to be future by the very reason of its pre-existence, and the idea of the temporal process of which the phases are authentically successive.[16] These two ideas are combined in such a close association that they are almost inseparable, and their incompatibility, their very distinction, is, as it were, submerged by the deceptive feeling of familiarity which is only an effect of habit, of prolonged automatization. After Bergson, philosophers should have shown more mistrust in respect to such deceptive feelings of familiarity. No progress can be made in the solution of the problem of causality except by carrying the logical as well as the psychological analysis as far as possible, effecting a separation of the incompatible elements which are fused into the deceptive unity of instinctive belief or of automatized association. Progress can be made only by questioning all the tacit implications, based upon a confusion of the logical evidence with the psychological feeling of familiarity. The revision of scientific concepts proceeds by such an analysis, by what M. Bachelard calls "the psychoanalysis of knowledge." This could be illustrated by a practically limitless number of examples in the history of the sciences. Naturally, such an effort to break the almost unbreakable associations by which the classical scientific tradition was constituted can only be difficult and even painful. That is why we must never be surprised by the constantly renewed resistance which rises in the mind when it is confronted with a profound revision of the classical concepts. It was precisely resistance of this kind which prevented even the most serious and honest minds from perceiving the fundamental incompatibility between real succession and timeless necessity. Quite probably, even Laplace, Du Bois-Reymond, certainly Tolstoi and Anatole France, believed sincerely in the reality of time, although time had no justification in their view of the universe. For them, the question asked by James and by Bergson, "Why is the future, which must be present, still not present?" did not even arise.

16. *Time and Free Will, an Essay on the Immediate Data of Consciousness,* authorized trans. [of *Essai sur les données immédiates de la conscience*] R. L. Pogson (New York: Macmillan Co., 1913), pp. 212–18.

However, in some cases—and they were rare—the determinist philosophers were aware of this difficulty. Thus Hans Driesch, after having affirmed, in spite of his vitalism, his quite Laplacian belief in the integral predetermination of the universe, asked himself the following significant question: Why does the activity of the timeless entelechies manifest itself in time instead of expressing itself by a single, complex act? Why does it manifest itself in the laborious development of the organism from its egg to its adult form? He frankly admits, "For that question, we have no answer."[17]

Such a question is certainly strange, but a consistent determinist is obliged to ask it. More frequently, the incompatibility between real succession and deductive necessity was only vaguely felt, but this vague feeling at least found its expression in certain particularities of language, invented to hide the incompatibility. The difference between cause and effect is too real to be entirely ignored. There is nothing surprising in the fact that the feeling of this difference is not entirely absent, even in the most uncompromising determinist mind. However, as the determinist insists on the absolute equivalence of cause and effect, unwittingly he faces a dilemma of which he is only half-aware. According to what we have said, it is obviously necessary to choose one of two assertions: *either* real succession with the element of real contingency *or* complete determinism with total absence of succession. Since most frequently the deterministic scientist does not see this dilemma clearly, he tries to retain causal necessity alongside temporal succession, but, as these two ideas are incompatible, he succeeds only in veiling with ingenious verbal formulas the conflict which goes on in the depths of his thinking. What is more, this conflict, as we have already said, suppressed by his conscious thought, manifests itself indirectly by certain particularities of his language. William James showed this in a very clear and precise way in his posthumous book:

Nemo dat quod non habet is the real principle from which the causal philosophy flows; and the proposition *causa aequat effectum* practically sums up the whole of it. . . . But if the maxim holds firm that *quidquid est in effectu debet esseprius aliquo modo in causa,* it follows that the next moment can contain nothing genuinely original, and that the novelty that appears to leak into our lives so unremittingly, must be an illusion, ascribable to the shallowness of the perceptual point of view.

Scholasticism always respected common sense, and in this case escaped the frank

17. Driesch, *loc. cit.*, p. 326.

denial of all genuine novelty by the vague qualification "aliquo modo." This allowed the effect also to differ, *aliquo modo,* from its cause. But conceptual necessities have ruled the situation and have ended, as usual, by driving nature and perception to the wall. A cause and its effect are two numerically discrete concepts, and yet in some inscrutable way the former must "produce" the latter. How can it intelligibly do so, save by already hiding the latter in itself?[18]

And in a footnote on the next page James adds:

The cause becomes a reason, the effect a consequence; and since logical consequence follows only from the same to the same, the older vaguer causation-philosophy develops into the sharp rationalistic dogma that cause and effect are two names for one persistent being, and that if the successive moments of the universe be causally connected, no genuine novelty leaks in.

There is no need to emphasize how that which James calls "the sharp rationalistic dogma" agreed with the energetist conception of reality, in which the cause and its effect were only two energy equivalents, the apparent succession of which masked their underlying identity. Thus, as in the monistic idealisms, ultimate and authentic reality is conceived of as permanent and as always identical with itself, whereas succession belongs only to its phenomenal manifestations. To avoid conflict with our immediate consciousness, which remains irremediably temporal, both physical determinism and idealistic determinism invent ingenious formulas. Instead of denying the reality of time outright, one says that time is only "phenomenal"; instead of saying that the effect is entirely identical with its cause, one says that it is "virtually," or *aliquo modo,* present. Through these verbal concessions, it is possible to avoid the truthless conclusion of Parmenides, which by eliminating succession entirely at the same time eliminates even the superficial difference between cause and effect. Let us say it again: if modern determinism, in its scientific as well as in its idealistic form, hesitates to follow the Eleatic School all the way, it is because the incompatibility of rigorous determinism with the reality of time is at least vaguely sensed.

III. WIDENED CAUSALITY

If we admit that absolute necessity is incompatible with the reality of succession, a single conclusion forces itself upon us. We must abandon the classical concept, that is, the Laplacian or Spinozist concept of causality. Such a conclusion frightens many serious thinkers. They are

18. William James, *Some Problems of Philosophy: A Beginning of an Introduction to Philosophy* (New York: Longmans, Green & Co., 1931), pp. 192–93 and n., p. 194.

frightened because they believe that, with the denial of classical determinism, the intelligible character of the world is forever destroyed. For them the denial of classical causality is equivalent to a "capitulation," even to a "suicide," of reason. Similar apprehensions were expressed when non-Euclidean geometry supplanted the classical geometry of Euclid. Naturally, if one looks upon Euclidean geometry as the only possible geometry, such fears would be justified. In that case, and only in that case, the denial of the fifth postulate of Euclid would result in the ruin of all geometric thought. In an analogous way, if Laplacian causality is the only form of rational coherence which the universe may assume, there would be a reason for fearing that, in eliminating it, we might destroy all possibility of rational explanation. The arguments of Herbert Spencer, John Fiske, Hippolyte Taine, and all the other determinists of the last century against free will were inspired by this facile confusion of the two terms "rational" and "determinist." As William James remarked in 1884 in his essay "The Dilemma of Determinism":

> Nevertheless, many persons talk as if the minutest dose of disconnectedness of one part with another, the smallest modicum of independence, the faintest tremor of ambiguity about the future, for example, would ruin everything, and turn this goodly universe into a sort of insane sand-heap or nulliverse, no universe at all.[19]

Then, two pages farther on, James gives some samples of the argumentation by which the determinists try to reveal the fundamentally irrational and even absurd character of their rivals: "A man's murderer may as probably be his best friend as his worst enemy, a mother be as likely to strangle as to suckle her first-born, and all of us be as ready to jump from fourth-story windows as to go out of front doors, etc."[20] In other words, it is believed that, without strict causality, the world is only the domain of the most capricious chance. More recently we have seen the same mistrust on the part of philosophers in reference to the revision of determinism in contemporary physics. René Berthelot, Léon Brunschvicg, and Hans Driesch, to name only a few,[21] have shown

19. William James, "The Dilemma of Determinism," *The Will To Believe*, pp. 154–55.

20. *Ibid.*, p. 157, n. 1. James adds: "Users of this argument should properly be excluded from debate till they learn what the real question is. . . . Persons really tempted often do murder their best friends, mothers do strangle their first-born, people do jump out of fourth-story windows, etc."

21. H. Driesch, "Naturwissenschaft und Philosophie," *Actes du Congrès International de Philosophie, à Prague* (1934); R. Berthelot, *Bulletin de la Société Française de Philosophie*, Vol. XXXIV, No. 5 (October–December, 1934); L. Brunschvicg, *La Physique du vingtième siècle et la philosophie* (Paris, 1936).

their skepticism concerning the objectivist interpretation of uncertainty relationships. As Jean Louis Destouches has asserted,[22] this resistance was inspired by philosophical motives which are not essentially different from those which were found in Spencer, Taine, and Fiske. It is feared that the rational universe may crumble into a shapeless mass of disjoined and capricious facts.

Let us say immediately that such fears are hardly justified because they are based on the gratuitous supposition that the indetermination now being envisaged is a *complete* and, so to speak, *absolute indetermination*. Now this is not at all the case. Absolute indeterminism is a very rare phenomenon, even with philosophers. It can be found in Epicurus and in Lucretius and, in the modern era, in Renouvier, at least in a certain phase of his philosophy when he was defending the notion of "absolute beginning." But, if we read carefully the works of those who defend the indetermination of the universe in the name of the reality of time, we see that their indeterminism is far from being absolute. The temporalistic philosophers, or, as they are called in English-speaking countries, the "process-philosophers," insist vigorously on the continuity of the past with the present, on the cohesion of the successive phases of becoming. Reread the passages of William James on the stream of consciousness or on the continuity of the perceptual flux; reread Bergson, especially that passage, so infrequently quoted, in *Matter and Memory,* where he affirms that creation is never *creatio ex nihilo* because each present moment is colored by its past; reread Whitehead when he speaks of "causal efficacy" in nature.[23] What, then, is the difference which separates them from the classical determinists? There is only one: when they speak of connection, of continuity, of cohesion of cause and of effect, they affirm that this connection, this continuity, this cohesion, is *temporal* in the true sense of the word, and as such it cannot be the equivalent of static connection, of logical implication; consequently, that it must contain an element of *irreducible novelty,* an *authentic differentiation between cause and effect,* a differentiation which has in it nothing irrational and nothing miraculous because it expresses the distance between the present and

22. Jean-Louis Destouches, *La Physique moderne et la philosophie* (Paris, 1939), pp. 39–40.

23. A. N. Whitehead, *Process and Reality* (Cambridge: Cambridge University Press, 1929), *passim.*

the anterior moment. Briefly, if we venture to use a formula which is perhaps too condensed, we can say that for a modern contingentist time *truly flows* and that the partial indetermination of each temporal moment is only a manifestation of this real flow, whereas, for the classical determinists, time flows, according to Bergson's expression, only because reality demands this sacrifice, "taking advantage of an inadvertence in their logic."[24] We can also say that, for modern contingentism, the *future remains* future, that is, virtual by its own nature, whereas for Spinoza, Laplace, and the others, the future is only a *hidden present.*

In recognizing the virtual character of the future, modern contingentism admits the category of *possibility* which, according to classical determinism, possesses no objective character, being only a manifestation of our ignorance. For Spinoza, for Hegel, and for Laplace, the *real* and the *necessary* are two *synonymous expressions*—for that which is not real is impossible. Consequently, there is no middle ground between the necessary and the impossible. That is why the future, being necessary, must be, for a consistent determinist, *as actual as the present* and as completed as the past. The unlikely and even absurd character of such a consequence has already been fully exposed by Émile Boutroux:

> Is it to be admitted that all possibles are, in their essence, eternally actual; that the present is made up of the past and is big with the future; that the future, instead of being contingent, already exists in the mind of the one supreme purpose or understanding; and that the distinction between being and the possible is but an illusion caused by the interposition of time between our point of view and things in themselves?
>
> This doctrine is not only unwarranted and impossible of proof, it is also unintelligible. To say that each thing is actually all it is capable of being is to say that it unites and reconciles, within itself, contraries, which, from the knowledge we have of them, can exist only by replacing one another. But how can we conceive of these essences as formed of elements that are mutually exclusive?[25]

The logical force of this passage was recognized, at least implicitly, even by Alfred Fouillée, who has always remained a staunch adversary of contingentism. It was probably under the influence of the passage we have just quoted that Fouillée wrote in his critique of contingentism:

24. H. Bergson, *The Creative Mind,* trans. [of *La Pensée et le mouvant*] Mabelle Andison (New York: Philosophical Library, 1946), p. 220.

25. É. Boutroux, *The Contingency of the Laws of Nature,* trans. F. Rothwell (Chicago: Open Court Publishing Co., 1920), pp. 21–22.

> We live in time and we reason in time. Now in time it is *contradictory* to say that the future exists and acts, since, in that case, I am at once living and dead, really living and really dead, my future death being already real, as is my present life. Such a theory means the elimination of all possible thought and of all possible experience, since thought cannot admit the simultaneous actuality of contradictories, and since experience cannot grasp the present and the future simultaneously.[26]

It is obvious that here Fouillée sought to answer the question which the contingentists always ask: "If the future is certain in all details, why is it not already present?" To this question Fouillée answers: "It is the incompatibility of the successive events which prevents the future from being contemporaneous with the present." Succession is thus only a consequence of the law of contradiction. Moreover, the same idea had already been expressed by Leibniz when he defined time as "the order of inconsistent possibilities."[27] But neither Leibniz nor Fouillée was aware that, by such a concession, they were indeed undermining the ground on which their determinism had been built. For the fundamental incompatibility of the successive phases, which they admitted, is precisely *completely contrary to the connection of logical necessity* which, *according to them, joins the successive events.* One of two propositions must hold here: *either* the successive phases of each temporal process are mutually deducible, *or* they are logically incompatible. But it is clear that they cannot be at the same time mutually derivable and incompatible. This impossibility is only another aspect of the fundamental incompatibility of strict determination and real succession.

The fear that the elimination of rigorous causality may destroy all intelligibility of the universe is, let us say again, childish. On the contrary, it is contingentism which makes causality—or rather let us say *causation* (reserving the term "causality" for Laplacian causality)—more intelligible. We have seen that rigorous determinism virtually destroys the temporal character of reality as well as all the difference between cause and effect. But have we not then the right to wonder, along with Boutroux: *"Would this also be a consequent, an effect, a change, if it differed from its antecedent neither in quantity nor in quality?"*[28] By re-establishing the temporal character of causation, we escape the bizarre paradoxes of necessitarian determinism of which the deter-

26. Alfred Fouillée, *La Pensée et les nouvelles écoes anti-intellectualistes* (Paris: Alcan, 1911), p. 140.

27. G. W. Leibniz, *Phil. Schriften*, I, 568.

28. Boutroux, *op. cit.*, p. 29.

minists themselves were often unaware. But in thus restoring the real difference between cause and effect, we are conceding the reality of contingency, or at least of the element of contingency; for the difference between the successive phases of becoming is only another name for the element of contingency, of unpredictability, of radical novelty, which is the very essence of temporal causation.

Let us stress the fact that it is this notion of widened causation which contemporary physicists—or at least most contemporary physicists—are tending to adopt under the impact of recent discoveries. The concept of objective possibility, which was always looked upon as legitimate by the contingentists, comes into the field of science in the form of the concept of *objective probability*. For the classical physicists the concept of probability was only a useful conceptual tool which could be used when the physical events were too complex to be analyzed in detail. However, nothing objective corresponded to this conceptual fiction despite its practical utility. Such an attitude was entirely logical. If there are no real possibilities, there are no real probabilities, either; for, as the German physicist, Weizsäcker, quite recently observed, the concept of probability is only the quantitative form of the concept of possibility. The contingentists were always opposed to this subjectivist interpretation. Let us remember Cournot, let us remember Renouvier, when he insisted in his *Essai de logique générale* that "the equal possibles of Laplace are to be understood in the final analysis as truly indetermined possibles in themselves, as possibles which are rigorously ambiguous." Let us remember James, when he had the courage to maintain as early as 1884 that *"somewhere,* indeterminism says, such possibilities exist, and form a part of truth."[29] Bergson's attitude seemed more ambiguous because he resolutely denied the pre-existence of the future in any form, even in the form of possibility. That is at least the thesis defended in the first two essays of his book, *The Creative Mind* [*La Pensée et le mouvant*]; but, on the other hand, we have to reread pages 204–12 of *Time and Free Will,* in which Bergson, while rejecting the mathematical preformation of the future in the present, still affirms that there is a preformation of another sort, which constitutes our consciousness of time—this is the preformation of the future "in the form of pure possible." Thus we see that the category of the possible has its place in Bergson's thought, which is not surprising, for the

29. O. Hamelin, *Le Système de Renouvier* (Paris: J. Vrin, 1927), p. 147; William James, "The Dilemma of Determinism," *The Will To Believe,* p. 151.

complete elimination of this category is found only in the defenders of integral necessity. Space does not permit us to discuss here the precise meaning of the Bergsonian views on possibility, which, in appearance, were seemingly contradictory. The reader should consult the final chapter of M. Jankélévitch's book on Bergson. On this point, the French and American contingentists anticipated the tendencies of contemporary physics or at least the objective interpretations of uncertainty relationships. Although Reichenbach recently proposed to replace strict causality by probable implication, he was scarcely aware that such a probabilistic interpretation of uncertainty relationships agreed with the conclusions of Cournot, Renouvier, Boutroux, James, Bergson, Peirce, and, more recently, Whitehead. At the same time we see why contingentism can be called a *relative determinism:* the future *is* determined, but only in its general character, never in its actual details. It is this general orientation of each present moment that contemporary physics grasps in the form of probabilistic laws.

In the light of quantum physics we can today answer the objection which Ralph Barton Perry raised against the Bergsonian affirmation of the incompatibility between rigorous necessity and the reality of time. It will be remembered that, according to Professor Perry, the simple fact of mechanical motion establishes irrefutably the compatibility of time with rigorous necessity: a material particle, whose trajectory is entirely determined by the laws of mechanics, nevertheless *moves,* that is, it occupies diverse positions in space *in successive moments.* But this example is obviously borrowed from macroscopic (i.e., from classical) physics. Its plausibility and its apparent clarity are completely deceptive in the light of recent physics. The predictability of the positions of any given macrophysical particle—and we observe only macrophysical particles—is only *approximate* and, as such, remains entirely compatible with the fundamental contingency of underlying microphysical events. The predicted trajectory of a particle, which, in our macrophysical perspective, appears as a precise geometric curve with no transverse thickness, is, in reality, a *thin tube,* a *bundle of possible routes,* which, although very thin, still has transversed dimensions corresponding to the quantic indeterminations of the future positions.[30]

30. See my articles: "The Doctrine of Necessity Re-examined," *Review of Metaphysics,* V, No. 5 (1951), 40–45; "Relativity and the Status of Space," *ibid.*, Vol. IX, No. 2 (1955); and "La Théorie bergsonienne de la matière et la physique moderne," *Revue Philosophique,* Vol. LXXVII (1953).

Thus even in the example considered by Professor Perry, the so-called route of the future is far from being "the only possible route," because it is composed in reality *of the entire field of the possibilities,* which, although very close to each other, still remain distinct. In other words, it is only by virtue of our macroscopic myopia that the field of the diverse possibilities seems to shrink so that it appears finally as a precise infinitely thin line of "the only possible route." There is no need to emphasize that such expressions as "the only possible future route" and "the necessary route of the future" are completely equivalent; classical determinism, by eliminating all the future possibilities save one, in fact eliminated the category of possibility, which was thus reduced to a human and temporary ignorance. In the light of recent physics such an elimination of the concept of possibility is no longer legitimate, although we understand how the character of the macroscopic world, as well as the limitations of our perception, made it inevitable before the time of quantum physics. Nor is there any need to emphasize that the concept of a solid and permanent particle is no longer adequate on the microphysical scale, since solidity itself is only an illusion—a necessary illusion, it is true—of our gross perception. The microscopic reality seems to be composed of *events* rather than of *things.* We may wonder to what extent the Eleatic and atomistic habits of our thinking have been determined by this "logic of solids," which, according to Bergson and Bachelard, is a subconscious foundation of the classical intelligence and which is virtually outlined in the very structure of our macroscopic perception. This is the question which the modern followers of Parmenides and Democritus do not ask themselves.

Not only quantum physics but also relativistic physics confirm the temporal, therefore contingentist, conception of reality. Such an affirmation may appear surprising because it is opposed to the rather widespread presumption according to which the fusion of time with space in the theory of relativity operates in favor of space and that the space-time of Minkowski is a static entity in which the alleged successive phases of cosmic history coexist in their eternal juxtaposition. We do not have space here for a detailed critical analysis of this singular misunderstanding, to which Minkowski himself contributed. Let us merely remember the numerous criticisms made of this erroneous interpretation, from Langevin to Eddington and to Meyerson. Quite justifiably, we can affirm that the fusion of space with time operates, contrary to the easy popular notions, in favor of time and that, instead of the spatialization of time we have rather a temporalization, or at least a dynamization, of space.[31] Let us simply recall the fundamental principle of relativistic dynamics according to which there is an upper limit to the transmission of any causal action: this is the speed of the electromagnetic waves. This is, as Paul Langevin said, the speed limit of causality. Thus there are no instantaneous transmissions in nature; there are only successive connections. In other terms, the theory of relativity has boldly stressed the idea that *the effect is never contemporaneous with its cause* and that causation is always irremediably, and by its very nature, successive. We have already seen that the reality of contingency inevitably follows from the successive character of causation. One may raise an objection by pointing out that contingency is not at all introduced into the theory of relativity. But that is due to the *macrophysical* character of the theory—the microphysical indetermination is, so to speak, masked by the laws of the big numbers on the macrophysical scale, and that is why it has been discovered only on the microphysical scale. But we must not be deceived here: the dynamic and unfinished character of physical reality is as present on the macrophysical scale as it is in the microcosm.

If real novelties exist even in the physical world, there is nothing surprising about finding them in the area of life and of consciousness. Moreover, almost all the objections which have been raised against indetermination on the biological and psychological scale have been inspired by dogmatic belief in physical determinism. It is obvious that the widening of the notion of causality creates a novel situation for the problem of freedom. All the contingentists were aware of it, although they have confused microphysical indetermination with the freedom of living beings. But the discussion of the very complex problem of relationships between contingency and freedom would lie outside the scope of this article.

31. Louis de Broglie, "L'Espace et le temps dans la physique quantique," *Revue de métaphysique et de morale*, LIV (1949), 119–20.

CHAPTER 10

SIMPLE LOCATION AND FRAGMENTATION OF REALITY*

The term "fallacy of simple location" was coined by A. N. Whitehead in 1925 in his book *Science and the Modern World;* the two passages of the book that deal with this problem are worth being quoted in full and may serve as an introduction into our topic.

> The Ionian philosophy asked, What is nature made of? The answer is couched in terms of stuff, or matter, or material—the particular name chosen is indifferent—which has the property of simple location in space and time, or, if you adopt the more modern ideas, in space-time. What I mean by matter, or material, is anything which has this property of *simple location.* By simple location I mean one major characteristic which refers equally to space and time, and other minor characteristics which are diverse as between space and time. The characteristic common both to space and time is that material can be said to be *here* in space and *here* in time, or *here* in space-time, in a perfectly definite sense which does not require for its explanation any reference to other regions of space-time . . . In fact, as soon as you have settled, however you do settle, what you mean by a definite place in space-time, you can adequately state the relation of a particular material body to space-time by saying that it is just there, in that place; and so far as simple location is concerned, there is nothing more to be said on the subject.[1]

It is fairly obvious that the term "simple location" used in this passage is merely a new name for the old concept of *atomicity* of nature. Atomic theory of nature, whether in its classical speculative form associated with such famous names as Democritus, Lucretius

*To be published as a contribution to *The Hartshorne Festschrift, Process and Divinity,* eds. W. R. Reese and Eugene Freeman, scheduled for publication by Open Court in May, 1964.

[1]A. N. Whitehead, *Science and the Modern World* (New York: Macmillan, 1926), pp. 71–72.

and Pierre Gassendi, or in its modern scientific garb, claimed that nature is made of discontinuous solid entities moving through homogeneous Euclidian space. At each particular moment each of these atomic entities occupies a definite portion of space with well-defined boundaries. The shape of these atomic volumes varied through centuries; at first, it was quite irregular because the Greek atomists needed hooks, indentations and various irregularities of surface in order to explain the fact of cohesion; this tendency was also characteristic of the seventeenth century atomism whose main features were consciously borrowed from its ancient ancestor. But in the modern period, especially with the coming of the kinetic theory of gases, the spherical shape was gradually more and more preferred. But no matter what kind of shape was attributed to the ultimate particles of matter, they always were regarded as *particles,* occupying a definite place in space at each particular moment. A series of successive positions of one single particle constituted its trajectory in space; the particle itself was regarded as immutable through time and the only change which was admitted was change of its spatial relations in respect to other particles or in respect to the Newtonian absolute space.

It is clear that Whitehead's term "simple location" is in its meaning equivalent to what John Locke called *principium individuations* (principle of individuality).[2] For the only thing which differentiates one atom from all others is its *spatial location* at a certain particular instant and nothing else. For atomists of all ages agreed that atoms are made of the same qualitatively homogeneous stuff; the idea of the *qualitative unity of matter* is one of the most characteristic features of not only atomism, but of the whole classical science and philosophy which it inspired. The view that atoms are qualitatively different, that is, possessing inherently different qualities, was always a sign of certain immaturity of the atomistic thought whether we find it in Anaxagoras and Empedocles in antiquity or in Galileo's contemporaries like Sennert and Berigard, or finally in John Dalton at the beginning of the last century. The case of John Dalton is especially instructive because, in spite of his authority as one of the founders of modern chemistry,

---

[2] John Locke, *An Essay Concerning Human Understanding,* Bk. II, Chap. XXVII.

there were always scientists, not speaking of philosophers, who hoped that the qualitative differences between chemical elements are only apparent and that they would be eventually resolved into the differences of configuration of some more basic elementary particles. The electron theory to a very considerable extent fulfilled these expectations. But if the basic units of matter are made of the same stuff, then, as already stated, the only differentiating features between such homogeneous particles are *differences in spatial location*. One and the same particle cannot be at the same time in two different places and two different particles cannot occupy one and the same position at the same time; if they do, they cease to be different and their twoness itself disappears. Or, more accurately, to speak of two different particles being simultaneously at one and the same location, means to apply two different names to a single entity. It is clear that in the property of simple location two basic features of classical science are united: the assertion of *impenetrability of matter* as well as *the denial of action at a distance.* Classical physics and, in particular, philosophers inspired by classical physics, were as reluctant to give up the first property of matter as to accept any action at a distance. Space is lacking here to show this in a sufficiently documented way: for our present purpose it must suffice to state generally that every type of physical interaction was always interpreted in terms of solid and immutable entities being transmitted between bodies. From the old Democritus's theory of ἔιζώλα up to Newton's particles of light emitted by luminous sources, the pattern of explanation remained essentially the same: every physical interaction was in the last analysis based on *contact* or *impact.* But whether particles are touching each other as in the entangled hook-like atoms of Greek atomists or whether they are clashing together as in the modern kinetic models of light and gravitation, their individuality remains preserved. They may be *adjacent* one to each other, but they never *merge* together; their impenetrability prevents their fusion while their confinement to certain spatial regions prevents them from acting *where they are not.* Thus the principle of simple location remains unviolated.

It is precisely this principle which Whitehead challenged:

> I also express my conviction that if we desired to obtain a more fundamental expression of the concrete character of natural fact, the

element in this scheme which we should criticize first is the concept of *simple location*. In view therefore of the importance which this idea will assume in these lectures, I will repeat the meaning which I have attached to this phrase. To say that a bit of matter has a *simple location* means that, in expressing its spatio-temporal relations, it is adequate to state that it is there where it is, in a definite finite region of space, and throughout a definite finite duration of time, apart from any essential reference of the relations of that bit of matter to other regions of space and to other durations of time . . . This idea is the very foundation of the seventeenth century scheme of nature. Apart from it, the scheme is incapable of expression. I shall argue that among the primary elements of nature as apprehended in our immediate experience, there is no element whatever which possesses this character of simple location. It does not follow, however, that the science of the seventeenth century was simply wrong. I hold that by a process of constructive abstraction we can arrive at abstractions which are the simply located bits of material, and at other abstractions which are the minds included in the scientific scheme. Accordingly, the real error is an example of what I have termed: The Fallacy of Misplaced Concreteness.[3]

If we read this passage attentively, we see clearly that Whitehead is far from simply denying the reality of atoms. He was certainly aware that the empirical evidence for the atomic structure of matter is overwhelming and that it is impossible today to repeat the mistake of Mach and Ostwald who denied the existence of atoms on spurious epistemological grounds. But at the same time, he is equally aware that the atomicity is only one aspect of nature. The dynamic connection of each individual entity with the rest of the universe is another equally essential aspect of reality which cannot be ignored. If we ignore it, we commit, according to Whitehead, a fallacy of misplaced concreteness. In other words, we reify an abstraction; for, no matter how justified and valid an abstraction may be, it remains an abstraction which should never be confused with the richness and fullness of the concrete fact. Or, as he says himself a few lines after the quoted passage:

The disadvantage of exclusive attention to a group of abstractions, however well founded, is that, by the nature of the case, you have abstracted from the remainder of things. In so far as the excluded

---

[3] A. N. Whitehead, *op. cit.*, pp. 84–85.

things are important in your experience, your modes of thought are not fitted to deal with them.

The question of primary importance is then as follows: is it possible to construct a model in which nature is represented by an arithmetical sum of discontinuous and mutually external entities *or* is the very concept of discrete atomic elements a result of artificial procedures by which we dissect the continuity of reality into discrete units, disregarding all dynamical links which join them together? Or, more simply: is simple location the intrinsic feature of reality or a result of distorting abstraction and conceptualization? From the quoted passages it is clear where Whitehead stands. But in order to understand better his position it is important to realize that no matter how original his criticism of the concept of simple location was, it had its antecedents in a wider movement, originating in the last decades of the last century. This movement, which was present both in physical sciences and psychology, gained momentum around 1900 and Whitehead's philosophy was one of its later phases. The fact that the reaction against the atomistic conception occurred simultaneously, or nearly simultaneously, within such widely different fields as physics and psychology only increases its significance. It was a symptom that the traditional modes of thought had begun to change, and we know that the first indication of the incipient change is the appearance of doubt and criticism.

I

If we respect the historical chronological succession, then we have to consider the reaction against the fallacy of simple location first in *psychology*. This reaction is fairly well known under a more familiar name of *Gestalt-Psychology*. But before we analyze it, it will be useful to remember briefly how much the idea of the atomic entity dominated psychology in the last century. The atom in question was, of course, not a physical atom. The psychology of the last century was far less physicalistic than the American behaviorism and Russian reflexology is today; even physiological psychology recognized the existence of introspective data even though these data were interpreted as simple accompaniments of cerebral processes. But it is significant how these introspective data were treated; in this respect there is hardly any difference between naturalistically oriented psychology or its idealistic or dualistic opponents. The belief which the more idealistically oriented Herbart shared with the positivistic psychology of Wundt, Taine, both Mills, Bain and others was precisely the *atomistic* view of psychological processes. This psychological form of atomism, which is better known under the name of *associationism,* had its roots in the eighteenth century; John Locke and David Hume are its true ancestors. According to this view the basic elements of consciousness are *sensations* and the whole rich variety of psychological life results from various forms in which these basic elements are associated. Thus the difference between perception and thought is merely that of *degree of complexity* and *sensory vivacity;* thought is merely a more complex aggregate of less vivid sensations. Associationism and sensualism were traditionally associated. The name of these basic elements of consciousness greatly varied; but no matter whether they were called *sensations, impressions, Vorstellungen,* or simply "mental states" or "elements", they always retained their quasi-atomistic character. Naturally, being of *psychological* nature, they were not regarded as entities existing in space; but in all other respects their resemblance to physical atoms was striking. Like physical atoms, they were well defined entities, mutually external and permanent in time; their disappearance from the field of consciousness was only apparent because they continued their existence under "the threshold of consciousness" until some accidental cause brought them back to "the surface of consciousness." To be sure, all these various spatial expressions were regarded as metaphors; but the choice of these metaphors only betrayed the close kinship of the physical and mental atoms. Even today our psychological language is imbued by the unconsciously adopted atomistic terminology; we speak about ideas *clashing* together, *sinking* under the level of consciousness, *re-emerging* from the *depth* of subconsciousness; we speak about the various degrees of complexity of our mental processes without realizing that we are borrowing this metaphor from physics and chemistry where the observed variety is a result of different degrees of electronic, or atomic or molecular configuration. A great part of the psychoanalytical language is *associationist,* that is, *atomistic* in its character. Thus although "the movement of ideas" or sensations was supposedly taking place *in time,* not in space, it was only natural to ask with Bergson to what extent the time of the associationistic psychology was not a verbally disguised space of Newtonian mechanics.[4] Both physical atoms and the atomic entities of the associationistic psychology were *substantial entities* moving through a homogeneous container; but while physical atoms were moving in space, psychological atoms were moving through time only. But in spite of verbal differences the imaginative background remained the same. Hume himself who so vigorously attacked the Cartesian concept of substantial soul, conceded explicitly that his "impressions" are substances too, although on a much smaller scale:

> My conclusion . . . is that since all our perceptions are different from each other, and from every thing else in the universe, they are also distinct and separable, and may be considered as separately existent, and have no need of any thing else to support their existence. They are therefore substances, as far as this definition explains a substance.[5]

This passage states clearly the principle of simple location applied to psychology; and because the definition of substance, which Hume applied, is Cartesian, we should designate Descartes rather than Hume as the thinker mainly responsible for introducing the principle of simple location into psychology. For the Cartesian "thinking substance", *substantia cogitans,* or simply *res cogitans* is only a species of substance in general which is defined as "an entity which does not need anything else for its own existence" (*res quae ita existit, ut nulla alia re indigeat ad existendum*). In other words, the main characteristic of substance is its independence from the rest of the world; but this feature, as Hume himself conceded, belongs to his impressions as well. We can thus say without exaggeration that Hume merely cut the Cartesian substantial soul into small, but no less substantial, fragments; in spite of his criticism of substance, he did not basically depart from the substantialism.

There is not enough time to review even briefly all stages of the reaction against associationism in psychology. Let us recall only the

---

[4] H. Bergson, *Time and Free Will,* trans. F. L. Pogson (New York: Harper, 1910), Chap. II, passim, esp. pp. 108–09.

[5] *Treatise on Human Nature,* Bk. I, pt. IV, sec. 5.

most significant intellectual events. Although it is difficult to assign a definite date to the birth of certain movements, historians of psychology regard James Ward's article in the *Encyclopedia Brittanica* in 1886 as the beginning of the new era.[6] This is not entirely accurate; James' brilliant analysis of the atomistic psychology appeared two years before in his article "On Some Omissions of Introspective Psychology" in the British review, *Mind*. Moreover, we do not have to forget that as early as 1874 Franz Brentano protested against the associationistic theory of judgment.[7] Judgement, or, as we say today, after adopting the nominalistic terminology, *proposition*, is, according to Brentano, a primary indivisible act irreducible to a mere aggregation of ideas. This does not diminish the significance of Ward's article which was characterized by one historian of psychology as "a blow from which the associationistic psychology never recovered."[8] Ward showed convincingly that what we call by a rather misleading term "complexity of mind", comes about not through the combination of various elementary units, but rather as a result of a gradual differentiation of a primary unity. After Ward's article the attacks on associationistic psychology followed in rapid succession. In 1889 Bergson published his *Essai sur les donnêes immêdiates de la conscience*, in which the successive continuity of consciousness was stressed and the concept of arithmetical multiplicity rejected as thoroughly inadequate for psychology. If the mental states are "many," they are many in a sense different from that in which physical bodies or arithmetical units are many; they are without sharp boundaries; they pervade each other. One year later William James's *Principles of Psychology* appeared and, independently of Bergson, the continuity of "the stream of consciousness" was stressed again and psychological atomism under the name "mind-stuff theory" critically analyzed. In the same year Ehrenfels writes his article "Über Gestaltqualitäten,"[9] where the term 'Gestalt' appears for the first time. Three years later Wilhelm Dilthey writes

---

[6] "Psychology," *Encyclopedia Britannica*, 11th ed., 22 (1886), 547–604.

[7] R. Müller-Freienfels, *The Evolution of Modern Psychology*, trans. W. Beran Wolfe (New Haven: Yale University Press, 1935), pp. 83–84.

[8] J. C. Flügel, *Hundred Years of Psychology* (New York: Macmillan, 1933), p. 150.

[9] *Zeitschrift für wissenschaftliche Philosophie*, 14 (1890), 249–292.

his "Ideen über eine beschreibende und zergliedernde Psychologie" in which the unitary and comprehensive approach toward psychology is contrasted with the atomistic, associationist, analytical, or, as he says, 'dismembering' approach.[10] Gestalt-Psychology was thus founded; the subsequent work of Wertheimer, Kofka, Köhler is too well known to be dwelt upon.

Two features of this wide anti-atomistic trend in psychology are relevant to our topic. First, its empirical character. The associationistic psychology correctly claimed to be more empirical than the Cartesian theory of substantial soul; Hume observed that no such substantial entity is ever an object of introspective experience; what an introspective glance always discloses is a *particular* impression, *particular* feeling, *particular* idea, but never the bare "Ego" or pure "I". From this there is only a step to the conclusion that Ego is merely a verbal entity, that is, a collective name applied to the totality of concrete impressions, feelings and ideas. It is important to realize that the criticism of associationism was based not on some *a priori* speculative grounds, but on a more refined introspective experience. James significantly called himself a radical empiricist; he, like Bergson, Ward, Dilthey and Gestalt psychologists, pointed out that psychological atoms, whether they are called *impressions, sensations, ideas, representations,* or simply *'mental states'* or *'elements'*, are merely methodological fictions, artificial entities, carved out of the continuity of the stream of consciousness, and that by manipulating these entities we can obtain merely an inadequate and clumsy imitation of concrete reality. If we then concede that the associationistic sensualism of Hume and his followers is more empirical than rationalistic psychologies of Descartes or of Christian Wolf, it nevertheless remains true that Gestalt psychology and the related trends were even more attentive to experience, especially to some of its more elusive and non-sensory aspects. In particular it was Alfred Binet in France and the Würzburg school in Germany which by their investigations of imageless thought convincingly pointed out that the terms "experience" and "sensory experience" are far from being synonymous.

This empirical character of the anti-atomistic trend in psychology increases the significance of its *second* feature which is more

---

10 Müller-Freienfels, *op. cit.*, pp. 98–101.

closely related to our topic. The Gestalt approach in psychology directly challenges the principle of simple location, in particular, its application to time. According to this principle, each particular event is where it is, that is, in one particular moment, but it is *nowhere else*. In other words, its presence is confined to a narrow moment; but once this moment is gone, the corresponding particular event is gone irrevocably and forever. Once passed, it will sink into the abyss of the past, being completely and forever *excluded* from the present, completely *external* to the present. If this were true, then any influence of the past on the present would be completely impossible; consequently, neither causation, nor memory would be ever possible. This is at least what the principle of simple location requires when we apply it to time: the past event is the past event and the present event is the present event and both are mutually external; *the action at a distance in time* is impossible in the same sense in which the action at a distance is impossible in space in virtue of the same principle applied to space. Unfortunately, or, rather fortunately, this is not what experience discloses. Ehrenfels demonstrated it conclusively by analyzing our awareness of *melody*. At the first superficial glance melody appears to be an aggregate of successive individual tones; in this sense our awareness of melody would be built of gradual addition of successive auditory sensations each of which would be temporally external to all others. But if it were so, no awareness of melody would ever arise. Nothing would be present to consciousness except one single tone, living for one single moment; the fact is, however, that not only an immediately preceding moment, but *the whole antecedent musical phrase,* in spite of its pastness, is in an undefinable way present in the present tone which, without losing its musical individuality, acquires a peculiar coloration within its antecedent musical context. Remove this context and the consciousness of melody will disappear, or rather it will be replaced by another temporal *Gestalt,* i.e., by the sensation of the individual tone within the context of the antecedent silence. Deceived by our language and by the fallacy of simple location unconsciously applied, we speak of one and the same tone no matter whether it is embedded in a melody or preceded by a silence. But is it really "one and the same" tone? By no means: it is a *similar* tone, not an identical one. It is true that the differentiating features introduced by different temporal contexts are very elusive, so elusive that it was possible to ignore them for such a long time; this accounts for the plausibility of the associationist psychology. The fact, however, remains that by claiming a complete self-identity and self-sufficience of the individual sensations we can never successfully explain the continuity of our experience, the reality of temporal *Gestalten,* the fact of immediate memory, and that we are thus driven to all the absurdities to which the doctrine of external relations inevitably leads. The conclusions of Zeno of Elea are sufficiently known; the conclusions of Aenesidemus of Knossus are perhaps less known; and those of Herbart, although he lived in the last century, are nearly forgotten: they all agree that in the name of the principle of simple location, motion, change, becoming are impossible; if our experience shows the contrary, so much the worse for experience!

II

It is a peculiar coincidence that approximately at the same time when the notion of the independent simply located entity began to be questioned in psychology, some important changes began to take place in physics, the changes which eventually led to the question raised by Whitehead in his *Science and the Modern World.* Was it really a sheer coincidence? Or was it rather a symptom that the human mind began to become increasingly sensitive to certain inadequacies of traditional modes of thought? For in spite of all the differences between the methods of introspective psychology and methods of experimental and theoretical physics, it is the same human knower equally conditioned by the same influences in both cases. This truth has been shown, I believe convincingly, by Bergson when he pointed out that our logic is basically a *logic of solid macroscopic bodies* and that the very concept of *substance* or of *thinghood,* the Cartesian *res*—whether it is *res cogitans* or *res extensa*—has its roots in the concept of isolated simply located and impenetrable body.[11] It is true that when we speak of a certain thing, we do not necessarily think of a physical body, but also of some abstract entity; but this entity, though diaphanous and dis-

[11] H. Bergson, *Creative Evolution,* trans. A. Mitchell, pp. XIX, 169–173; *Time and Free Will,* pp. 130f.

colored, nevertheless retains some basic features which are characteristic of solid bodies. If we realize this basic principle of Bergson's epistemology, we shall also be less surprised by the otherwise unexplainable parallelism of certain trends in physics and psychology. As it was basically the same fallacy of simple location which colored the development of both physics and psychology in the last centuries, it is hardly surprising that the critical reaction against this fallacy has some analogous features.

First doubts about the adequacy of the corpuscular view of reality began to appear in physics in the last century. As early as in 1844 Faraday in his article "A Speculation Touching Electric Conduction and the Nature of Matter"[12] pointed out that the distinction usually made between solid material particles and forces which, so to speak, emanate from these particles, is artificial; for what will remain of matter if we take away its dynamic manifestations? Even the so-called impenetrability and inertia of material particles cannot be tested in any other way except by interaction with other bodies. And if we consider two other properties of matter, that is, its gravitational and electromagnetic actions which pervade the whole space, we can understand Faraday's conclusion that "matter is not merely mutually penetrable, but each atom extends, so to say, throughout the whole of the solar system, yet always retaining its center of force." The significance of Faraday's view did not escape Bergson who quotes it at least twice, in *Matter and Memory* and in *Creative Evolution;* similarly, Whitehead recalls Faraday's view about the ubiquity of atoms in *The Concept of Nature*[13] in a passage in which his future criticism of the fallacy of simple location is anticipated. Wolfgang Köhler, one of the outstanding representatives of Gestalt psychology, was also clearly aware of the affinity of his own attitude with the anti-atomistic implications of the physical field-theories. He quotes the following passage of James Clerk Maxwell:

> We are accustomed to consider the universe as made up of parts, and mathematicians usually begin by considering a single particle, and conceiving its relation to another particle, and so on. This has

[12] M. Faraday, *Experimental Researches in Electricity,* 2, 293.

[13] *Matière et mémoire,* p. 233; *Creative Evolution,* p. 222; *The Concept of Nature,* p. 146.

generally been supposed the most natural method. To conceive a particle, requires a process of abstraction since all our perceptions are related to extended bodies, so that the idea of *all* that is present in our consciousness is perhaps as primitive an idea as that of any individual thing. Hence there may be a mathematical method in which we proceed from the whole to the parts instead of from the parts to the whole.[14]

But in the times of Faraday and Maxwell there still seemed to be a possibility to save the principle of simple location. Even if we admit that the particles are inseparable from the energetic field which surrounds them and that their very existence without their dynamic relations to the whole field is unthinkable, it was still conceivable that the field itself is *granular* in its structure, that is, made of some ultimate minute particles,—so-called aether-particles. In other words, it was hoped that the mechanical model of the aethereal field could be, in principle at least, constructed, and Maxwell himself tried repeatedly to provide such a model. The basic idea of atomism would thus have been preserved; it is true that the hypothetical aethereal atoms had to be pictured as having an extremely minute radius,—much smaller than the radius of the electron. The failures of all such attempts at mechanical interpretation of aether are sufficiently known and to survey them all would mean to write another history of the concept of the electromagnetic aether in the second half of the nineteenth century.

It may be objected that while corpuscular models of aether were unsuccessful, the corpuscular character of the elementary physical particles is established beyond any doubt; who today seriously doubts the reality of electrons, positrons, neutrons, mesons, etc.? On the other hand, does not the theory of quanta reintroduce the atomicity into the energetic field itself which acquires thus the granular character? In answering the second objection first we shall be less liable to be deceived by the superficial plausibility of the first objection. For it is clear that the quantization of energy does not mean in any sense a return to the corpuscular models of aether; the quantized electro-magnetic field resembles the Huyghens'

---

[14] J. C. Maxwell, *A Treatise on Electricity and Magnetism*, 3rd ed., 2 (Oxford, 1891), 176–179; G. W. Hartmann, *Gestalt Psychology* (New York: Ronald Press, 1935), p. 39.

or Maxwell's models of aether as little as photons resemble the luminous particles postulated by Newton in his emission theory of light. Moreover, it is inaccurate to speak about "atomization" of energy; what is "atomized" is not energy, but its product with time, that is, a dimensionally different quantity which is called *action*. But it is precisely the atomic character of action which forces us to regard the adjective "corpuscular" as a mere metaphor and misleading metaphor at that. There is no place here for reviewing, even briefly, the history of quantum theory. Suffice it to say that the quantification, which was originally applied to the electromagnetic energy only, was later—but still before the advent of wave mechanics—applied to mechanical energy of translation and rotation as well. It was finally extended by Louis de Broglie even to "internal energy," which according to Einstein's equation $E=mc^2$ belongs to every mass, even to the mass at rest. It is sufficiently known how far reaching and revolutionary were the changes brought about by the relativity and quantum theory and especially by wave mechanics in the very foundations of physics; but it is important to stress that especially the traditional concept of particle was transformed beyond recognition.[15] Here is the list of the most important changes to which the classical concept of corpuscle was subjected:

1. The solidity, impenetrability, constancy, indestructibility, uncreatibility—all these attributes which characterized the classical atom from the time of Democritus to that of Lorentz do not belong to the "particles" of contemporary microphysics. We know today that mass is not constant, but a function of its velocity; we know that even the nuclei, in which practically all mass of the atom is concentrated, are under certain conditions lacking impenetrability, being "transparent" to the slowly moving electrons (Ramsauer's effect); we know since Anderson's discovery of positive electrons in 1932 that the microphysical particles are *not* permanent as they can be either created or annihilated. Even if it is true that the corresponding "creation" is not *"creatio ex nihilo"* as a particle in question is a result of "materialization" of electromagnetic radiation into which it may be reconverted in the converse process of

[15] Cf. my book *The Philosophical Impact of Contemporary Physics* (Princeton: Van Nostrand, 1961), in particular Chap. XIV.

"dematerialization," it is quite clear that such entities do not possess the eternal solidity of the classical Lucretian atoms.

2. The boundaries of the microphysical "particles" in regard to their surrounding space are ill defined because their precise localization is impossible in virtue of the principle of indeterminacy (which is an unescapable consequence of the atomicity of action) and even their most fundamental properties such as mass and charge are inconceivable without their dynamic interaction with their environment into which their individuality is, so to speak, fused.

Faraday's view about the "ubiquity" of electric charge has been already referred to. It is not incidental that the conclusion about the relational character of not only charge, but even of mass was reached even before the contemporary revolution in physics. For, as James Clark Maxwell pointed out, the third law of Newton makes the concept of isolated force emanating from an isolated substantial entity–whether from a material particle or from an electric charge–physically meaningless. This is in a strange contrast to the general atomistic orientation of classical physics. We have to remember that Newton defined inertial mass as *vis insita,* that is, literally, as *force residing* within the location occupied by matter and constituting, so to speak, its substantial nucleus which is related *externally* to other particles. The belief in the simple location of sharply defined corpuscular entities could have hardly found more accurate formulation: the essence of material particle is its resistance to acceleration, reacting *hinc et nunc* against the *external* influences of other equally well defined corpuscular entities. This can hardly surprise us if we consider how much Newton and classical physics in general adhered to the atomistic tradition. But the situation becomes different when the third law of Newton–the concomitance and equality of action and reaction–is considered. According to this law every force is accompanied by an opposite force of equal magnitude. Maxwell clearly realized that both forces–action and reaction–are merely two partial and complementary aspects of one and the same dynamic phenomenon which he called *stress*:

> If we take into account the whole phenomenon of the action between the two portions of matter, we call it Stress. This stress,

according to the mode in which it acts, may be described as Attraction, Repulsion, Tension, Pressure, Shearing stress, Torsion, etc. But if . . . we confine our attention to one of the portions of matter, we see, as it were, only one side of the transaction—namely, that which affects the portion of matter under our consideration—and we call this aspect of the phenomenon, with reference to its effect, an External Force acting on that portion of matter, and with reference to its cause we call it the Action of the other portion of matter. The opposite aspect of the stress is called the Reaction on the other portion of matter . . . In commercial affairs the same transaction between two parties is called Buying when we consider one party, Selling when we consider the other, and Trade when we take both parties into consideration.[16]

In other words, to isolate one particle and force from the whole dynamical context is as artificial as to claim that buying may take place without selling. Even the alleged substantial core of matter, the *vis insita* of Newton, is a mere fictitious product of our substantializing habits of thought, a mere abstraction as long as we disregard the whole cosmic surrounding to which the individual inertial mass is related. This is what Ernst Mach saw in his famous criticism of Newton's rotating bucket experiment when he claimed that in the principle of inertia there is "an abbreviated reference to the entire universe" and that "the neglecting of the rest of the world is *impossible*."[17] The centrifugal forces in the rotating vessel arise, according to Mach, not in respect to the fictitious Newtonian void, but in respect to the great stellar masses of the universe. In other words, Mach, foreshadowing Einstein's principle of equivalence, suggested that the so-called inertial forces (of which centrifugal forces constitute one sub-class) instead of emanating from the interior of the particle are due to the gravitational effects of the stellar masses of the universe. Thus the "residing force" of Newton, allegedly confined within the narrow volume of the local particle, is in truth nothing but a mere knot in the web of the universal dynamic interaction in which the remotest regions

---

[16] J. C. Maxwell, *Matter and Motion* (New York: Dover, n.d.), pp. 26–27.

[17] E. Mach, *The Science of Mechanics*, trans. T. J. McCormack, 5th ed. (Chicago: Open Court Publishing Co., 1942), pp. 288–289. [6th ed. (LaSalle: Open Court Publishing Co., 1960), pp. 287–288.]

of the universe take part. J. B. Stallo, who anticipated Mach's criticism of absolute space by several years, rightly claimed that without the rest of the universe it would be meaningless not only to speak of the rotation of an isolated body, but even of a body itself:

> A body cannot survive the system of relations in which alone it has its being; its *presence* or *position* in space is no more possible without such reference to other bodies than its *change of position* or *presence* is possible without such reference. As has been abundantly shown, all properties of a body which constitute the elements of its distinguishable presence in space are in their nature relations *and imply terms beyond body itself*. (The last italics added.) [18]

Thus it appears that what we used to call "particle", instead of being an isolated bit of material simply located in a certain region of space and time, is a product of the whole universe or, to use Leibniz's expression quoted by Bergson and Whitehead, "a mirror of the universe"; conversely, each "particle," instead of being confined within a certain region of space and time, pervades its whole cosmic context from which it cannot be separated except by an artificial process of abstraction:

> In a certain sense, everything is everywhere at all times. For every location involves an aspect of itself in every other location. Thus every spatio-temporal standpoint mirrors the world.[19]

Thus the recent development in physics in its revision of the concept of simple location shows the same opposition toward an artificial fragmentation of reality as *Gestalt psychology* in its criticism of the associationistic "bundle theory." This is another instance of "the parallel development of method in physics and psychology" to which Professor Hartshorne called attention some time ago.[20]

---

[18] J. B. Stallo, *The Concepts and Theories of Modern Physics*, ed. P. W. Bridgman (Cambridge: Harvard University Press, 1960), p. 215.

[19] *Science and the Modern World*, p. 133.

[20] C. Hartshorne, "The Parallel Development of Method in Physics and Psychology," *Philosophy of Science*, 1, 420f.

III

But does not the last quoted passage suggest that the denial of simple location logically leads to the monistic idea of "the bloc universe," that is, to the idea of the cosmic whole which is wholly and indivisibly present in all its "parts"? Strictly speaking, such an universe does not have any autonomous parts because what we call its "parts" are merely partial aspects of one single whole. Individuality and autonomy of these aspects is purely spurious; it is merely a result of our analyzing attention which artificially carves them out of the indivisible unity of a single whole. In truth, the monistic assertion of one single whole and the denial of plurality and individuality is merely one statement in two different forms.

History of both Western and Eastern thought would provide us with practically an unlimited number of examples to illustrate what we just said. There are two assertions which are implicitly present in all monistic systems, although they are not always equally stressed: 1. Each alleged "part" of the universe is present in all other "parts"; 2. In each "part" of the universe all other "parts" are present. These two propositions are clearly not logically independent; you cannot assert one without another. The second proposition was explicitly stated by Spinoza. In his system not only the eternal substance necessarily implies each particular mode, but also even the most insignificant *modus* implies the whole eternal substance; or, in Spinoza's words which reveal the striking and ruthless logic of his sytem, the destruction of a single particle of matter would imply the destruction of the whole material universe.[21] The reason is obvious: according to Spinoza, unwittingly anticipating Mach, the totality of matter is indivisibly present in a single minutest particle. But then, as Professor Hartshorne stressed, the whole distinction between substance and its modes, between *natura naturans* and *natura naturata* disappears and the system of Spinoza becomes undistinguishable from the most extreme monism of the Eleatic type in which all plurality in space as well as in time is dissolved in the undifferentiated timeless unity of *ἓν καί πᾶν*, One and All.[22]

---

[21] Opera, ed. J. van Vloten, 3, epistula IV.

[22] C. Hartshorne, "Contingency and the New Era in Metaphysics," *The Journal of Philosophy*, 29 (1932), 457–458.

Is it not strange that practically the same statement is present in the genuinely pluralistic systems of Leibniz, Bergson and Whitehead? Does this perhaps mean that their pluralism is only apparent? This may be true of Leibniz in spite of the fact that he is regarded as a textbook example of metaphysical pluralism and of the philosophy of discontinuity. If Leibniz says that "monads have no windows," he has also another statement which inspired both Bergson and Whitehead: every monad is a "mirror of the universe." Thus as Harald Höffding pointed out, the question of Leibniz is more complex than the conventional textbooks of history of philosophy assume and his difference from Spinoza is more apparent than real.[23] But what about Bergson and Whitehead? They both assert unambiguously the reality of succession, and to assert succession means always to assert plurality, at least *plurality in time.* It is true that pluralistic aspects in Bergson are somehow obscured by his constant emphasis on continuity; but we have to bear constantly in mind that this continuity is of *dynamic* type, altogether different from the static continuity of Spinoza. The monism of Spinoza as well as the "pluralism" of Leibniz is *timeless;* if we want to speak of the monism of Bergson or Whitehead, we have to join the adjective "dynamic" to the noun "monism." The monism of substance is altogether different from the monism of process; William James was right when he focussed attention on the pluralistic implications of Bergson's dynamism and when he pointed out that the denial of plurality and the denial of time are nearly always correlated.[24]

The concrete meaning of the pluralistic implications of the dynamic view of the world may be seen from the way in which the last quotation from Whitehead must be amended. It is true that "each particular event mirrors the world," but we must not forget that 1. the term "world" must not be taken in the sense of timeless, completed entity; 2. that the act of mirroring *takes time,* that it is itself a time-consuming process. These two qualifications require

---

[23] H. Höffding, *A History of Modern Philosophy,* trans. B. E. Meyer, 1 (New York: Dover, 1955), 353.

[24] W. James, *A Pluralistic Universe,* Chap. VI; "The Dilemma of Determinism," *The Will to Believe and Other Essays in Popular Philosophy,* esp. pp. 150–152.

fuller explanation, which within the limits of this article must be necessarily brief.

1. Each particular event "mirrors the world," but this world is made of two heterogeneous parts: the first part is represented by the causal impact of the past events on the event in question; the second part consists of the virtual future events which may—but not necessarily will—radiate from the same present event. In other words, each present event is pervaded by the causal impact of the past events and it implicitly contains the future events. But the manner in which the past is immanent in the present is altogether different from the way in which the future "pre-exists" in it. Future "pre-exists" only in a general, ambiguous and potential sense; the definiteness and *even the very individuality* of the future events as well as their causal efficacy is lacking as long as they remain future. On the other hand, the past events are by their own nature definite, complete and, in principle at least, causally effective. Future events do not act; the past does. In any genuine affirmation of the reality of process, futurity and potentiality are synonymous terms. This contrast between the definite, settled and causally effective character of the past and the ambiguous, indefinite and causally inefficacious character of the future is nothing but the consequence of the basic asymmetry of every temporal process. This contrast is obscured when the static and abstract term "world" is used; for the "world" which every spatio-temporal event reflects is, so to speak, made of two heterogeneous halves: of the completed past and of the virtual future. On the contrast between the irrevocability of the past and the plasticity of the future the irreversibility of time is based. This was clearly recognized by Aristotle when he limited the omnipotence of God by the following words: "Of this one thing God is deprived—namely to make undone the things that have been made." His disciple St. Thomas stated it even more incisively when he wrote: *Praeterita autem non fuisse contradictionem implicat.*[25] It was the same Aristotle who stressed the unreality and indefiniteness of the future to such an extent that he denied the applicability of the law of the excluded middle to the future

[25] St. Thomas Aquinas, *Summa Theologica*, Q. XXV, art. 4; Aristotle, *Ethica Nicomachea*, Bk. VI, ch. 2. Cf. also Hartshorne's article, "The Reality of The Past, The Unreality of The Future," *Hibbert Journal*, 27 (1939), 246–257.

situations. According to him the proposition: "There will be a seafight tomorrow" is neither true nor false *now;* it is merely possible *now*; but it *will be* either true or false *tomorrow*. According to the Laplacean determinism this proposition is either true or false even when we do not know it. As in René Clair's picture "It happened tomorrow" the future for a determinist has the same status as the past; it *already happened,* although we have an illusion that it did not. The whole difference between the static and dynamic view of reality is illustrated in these two different attitudes toward the future. In the dynamic view there is merely a *semantic* difference between the two following statements: "Each particular event mirrors the virtualities of the future" and "Each particular event does not mirror any specific future event at all."

2. Can we at least say that each spatio-temporal standpoint reflects the *whole present of the universe?* This was a natural belief until the discovery of the finite speed of all causal actions. Because there are no instantaneous interactions in the world, only that which is past can act on each particular event. When we contemplate the starry sky, the light of Polaris, which we perceive, is fifty years old, that of Sirius eight years, that of Neptune four hours, that of the moon one second. Using Leibniz's and Whitehead's expression, we say that only the cosmic past of the universe—never its present—is "mirrored" on the present sky of our planet. This is the meaning of our previous statement that "mirroring" is not an instantaneous process, but that it *takes time.* It would go beyond the scope of this essay to explain how this fact joined to the negative result of Michelson's experiment led to the denial of absolute simultaneity; for our purpose it suffices to say that the temporal relations are now redefined in causal terms: the relation "anterior to" is defined as "acting on" a certain event; the relation "posterior to" is defined as "being acted upon" by the same event; finally, the relation of simultaneity was superseded by that of causal independence. Thus the denial of simple location should be rephrased in the following way: each particular event reflects that part of the universe which acts on it as well as the potentialities of its own future effects; *but it remains causally unrelated to those events which neither act on it nor will be acted upon by it.* In no case can we say that it reflects the *whole* universe; the expression "the whole

universe" is nothing but a remnant of the static, Eleatic and substantialistic terminology.

It is therefore clear that the denial of simple location implies the doctrine of "the bloc universe" only when this doctrine is assumed at the beginning. When Spinoza assumed that the destruction of a tiniest particle would imply the destruction of the *whole* universe; or when Chrysippus illustrated the same idea metaphorically by saying that "a drop of wine will eventually color the *whole* ocean," the concept of the universe as a completed whole was assumed without any discussion. Classical physics seemingly substantiated this assumption: the classical world was rigidly coherent along its temporal dimension because the iron link of necessity joined, and we may even say, *fused* its successive phases to such extent that even their succession, i.e. individuality in time, was in question. But the classical world cohered no less along its *transversal,* i.e., spatial dimension: the advance of time toward the future (as long as it was reluctantly admitted), had, so to speak, one single wavefront on which the single cosmic "Now," that is, all simultaneous events were carried together to the future. Whether moving or not, the classical world was one single huge bloc; the denial of simple location within such a universe necessarily entailed the spatial and temporal omnipresence of the smallest atom and, eventually, the *implicit identity* of each atom and the universe. ("The identity of Minimum and Maximum," as Cusanus and Bruno used to say.) The universe of modern physics is of a different type. There are definite indications that the cohesion of its successive phases is of a different kind from the rigid necessitarian bond of the Laplacean physics and Spinoza's metaphysics. The universe no longer seems to be a single unbending fact rigidly cohering along its temporal dimension. But the relativity theory loosened also its coherence along its transversal, i.e., spatial dimension when it replaced the relation of simultaneity by that of causal independence. It is possible that both types of loosening are interrelated. James saw this as early as 1884 when he linked the assertion of plurality with that of contingency. Fifty years later Whitehead, too, linked "contemporary independence" to the contingency character of the world.[26] In doing this Whitehead virtually modified his original claim that "each spatio-temporal standpoint mirrors the *whole* universe." In the dynamic universe—or rather "multiverse"—in which individual causal lines, partially separated by the gaps of causal independence, are being continually prolonged in the direction of the future, the concept of the absolutely coherent universe —coherent either in the temporal or spatial dimension—loses its meaning. In other words, *the principle of absence* limits to a certain extent the mutual immanence implied in the denial of simple location. In the language of James whose profound significance appears today even in a clearer light than fifty years ago:

> The monistic principle implies that nothing that is can in any way whatever be absent from anything else that is. The pluralistic principle, on the other hand, is quite compatible with some things being absent from operations in which other things are singly or collectively engaged.[27]

This recognition of the real absences in the world does not, of course, imply a return to the traditional atomism. The previously stated criticism of the artificial fragmentation of reality remains valid; but we have to be careful not to jump to the opposite extreme in believing that there are no articulations and no individualities in the universe. Reality is neither an undifferentiated and completed whole nor is it a bundle of externally related entities; it is a polyphonous process, never complete, never entirely one and never entirely many.

Mach's statement that in the principle of inertia there is "an abbreviated reference to the whole universe" should be now amended by replacing the word "the whole universe" by the words "that part of the universe which is causally affecting the material 'particle' here and now." When I perceive Sirius on the sky, my retina, my nervous system and my mind are affected by the event which is eight lightyears away from me; using Leibniz's language, I can say that my consciousness "mirrors" an extensive spatio-temporal re-

---

[26] W. James, "Dilemma of Determinism," *op. cit.*, pp. 150–151, 181; A. N. Whitehead, *Adventures of Ideas* (New York: Macmillan, 1947), p. 255.

[27] *Some Problems of Philosophy*, p. 144. The term "the principle of absence" was coined by Jean Wahl, *Les philosophes pluralistes d'Angleterre et d'Amérique* (Paris, 1920), p. 126.

gion affecting my particular present event. But the very same event *will remain forever independent* from the photon which will reach my eye tomorrow or after a week, a month, a year. The light from Sirius tomorrow will not reach my present moment, but its future causal successors. All future causal effects of Sirius will remain *forever absent* from this particular present moment. This is the meaning of *the principle of absence* which necessarily limits the denial of simple location; and from the illustration just given it is clear that this principle follows logically from the very fact that the universe is *always* incomplete, that no interaction can take place instantaneously and, *a fortiori,* that no causal action can move "backward in time." But to say that the universe is everlastingly incomplete and that every process exhibits temporal asymmetry is one and the same statement in two different forms.

CHAPTER 11

PARTICLES OR EVENTS?

I believe I should start with a kind of opening statement which will make the purpose of this paper clear and its presentation easier to follow. In the first place, it is not going to be a paper on philosophy or methodology of science – at least not in its usual, orthodox sense – but rather a philosophical comment on one particularly significant trend in twentieth-century physics. You may call it an essay in 'philosophy of nature', if we understand the term properly. I am fully aware of how unpopular and discredited this term has become; it is now rare to find institutions which still offer courses in 'philosophy of nature'. It really takes courage to do so and I commend my colleague Robert Cohen for having introduced courses of this kind in the Boston University curriculum. It is not difficult to trace the causes of this unpopularity and I have analyzed them in some of my previous writings. In the first place, the term itself is a translation of the German *Naturphilosophie* coined by the German idealists in the post-Kantian period, and a lingering disappointment with their speculative and arbitrary constructions comes immediately to mind as soon as the word is mentioned. In truth, we could hardly find another period in which the contrast between sterile and *a priori* speculations such as those of Schelling and Hegel and the genuine progress in the empirical sciences were more striking; we have only to consider the development of geology, biology, chemistry and of the physics of electricity and magnetism during the same period. Second, even if we understand 'philosophy of nature' in a more acceptable and less pretentious way as an attempt to synthesize various scientific fields, that is, as 'completely unified knowledge' in the sense that Herbert Spencer in the second half of the last century defined philosophy in general, some grave doubts remain. When, after all, would scientific knowledge be *fully* unified? Spencer's name itself reminds us of how premature and ambitious his attempt at a 'complete integration of knowledge' was; all he achieved was a codification and integration of nineteenth-century scientific knowledge and only in a rough and approximate sense. Isn't the same thing likely to happen to anybody who would try to synthesize the scientific knowledge so enormously increased and diversified, would not such an attempt be even more unrealistic and more pretentious now?

Thus it is natural to restrict the philosophy of science to a mere methodology and to veto all questions having an even remotely metaphysical ring, such as the questions raised by philosophy of nature undoubtedly have. About fifty years ago, Moritz Schlick still had the courage to name one of his books *Philosophie der Natur*, despite the fact that he belonged to the generally anti-metaphysically oriented Vienna Circle. Today, even terms such as 'the nature of the universe' bring a contemptuous smile to the lips of some scientists and the majority of philosophers of science.

Nevertheless I still believe that philosophy of nature is a legitimate enterprise provided we are careful to redefine its task and its limits. There is no question that there are certain definite trends in the sciences, for example, in present-day physics, which remain either undiscerned or ignored when one confines oneself to a certain narrow field of specialization or when interest is restricted to questions of methodology only. Such trends can be discerned only within a wider context – more specifically, when a broader historical perspective is adopted. How can any trend be discerned without considering its contrasting historical backdrop? How can the inadequacy of the concept 'particle', about which I want to speak, be discerned without first bringing into focus as sharply as possible *all* its essential features and their relations to other classical concepts? It may be argued that no trend can be established beyond doubt since there have been many so-called 'trends' which proved to be reversible, in the sense that they were eventually replaced by trends in an opposite direction. I suppose that today when claims are made that there is no progress being made at all in scientific knowledge, such a view is probably very fashionable.

Such platitudinous generalities about the alleged reversibility of any trend are possible only when all the evidence for the persistence of some trends is disregarded – and such evidence is indeed massive. Furthermore, even if we still regard such evidence as circumstantial, it is greatly strengthened by epistemological considerations which can hardly be ignored. In the context of the problem I am going to discuss, it is not enough to show all the growing evidence for the inadequacy of the concept 'particle' on the microphysical level; one must also show how the psychological origin of this concept, which developed under the pressure of limited macroscopic experience, makes its applicability to the 'microcosmos' exceedingly improbable.

I. THE FRUITFULNESS OF CORPUSCULAR EXPLANATIONS IN CLASSICAL PHYSICS

The inadequacy of the concept of particle, or, at least its inapplicability beyond the limits of our macroscopic experience had been suspected long ago, although mainly on epistemological, logical or metaphysical grounds. To mention just two outstanding examples: Leibniz in the eighteenth and Mach in the nineteenth century, both philosopher-scientists, were severe critics of atomism; one may say that the whole of their thought was pervaded, and in a sense inspired, by a deep distrust of the concept of particle on which the corpuscular-kinetic models of nature and, more generally, the whole mechanistic view of the universe were based. But the fruitfulness and success of these models in explaining various concrete phenomena were such that this criticism was largely ignored, at least by physicists. In fact, even the critics themselves could not remain blind to the triumphs of the mechanistic, i.e., the corpuscular-kinetic models. To use the same examples, Leibniz, while rejecting the concept of atom (i.e., of an indivisible material particle) in his metaphysics, at least favored mechanistic Cartesian models in his physics, and he even accused Newton of smuggling an occult quality under the name of 'attracting force' back into nature. Matter was for him a "well-founded phenomenon" (*phaenomenon bene fundatum*) which, while being constituted of immaterial, non-mechanical entities (monads), nevertheless appears to us in a way which can be described by mechanistic models. The case of Ernst Mach, one of the severest critics of what he called "mechanistic mythology", is even more instructive. He, followed in this respect by Ostwald, did his best to substitute more abstract energetic explanations for corpuscular models; but at the end of their lives, neither of these men could ignore the overwhelming empirical evidence in favour of the reality of atoms. In retrospect the whole revolution against atomism and mechanism at the end of the last century, characterized by the names Mach, Stallo, Ostwald and Duhem, was largely premature, since contrary to the views of these men, the fruitfulness of corpuscular-kinetic explanations at that time had not yet been exhausted. Only with the coming of the relativity theory and, even more conspicuously, with the discoveries of the wave nature of matter, did the basic inadequacy of the concept of particle become obvious and undeniable.

How serious the present crisis surrounding this concept is, will become clearer when we realize how spectacular the previous success of the corpuscular-kinetic explanations of various phenomena had been. My time for even a short historical digression is severely limited; but I would like to review at least briefly the main phases of the development of atomism and what appeared as its ultimate triumph.

It is generally known that the atomistic tradition, i.e., the view that matter consists of ultimate, indivisible particles, can be traced to Leucippus and Democritus in the fifth century B.C.; it is less generally known that it had not entirely disappeared even in the Middle Ages, though it was driven underground by the pressure of the medieval establishment which accepted Aristotelian physics and cosmology; finally, it re-emerged victorious during the cosmological revolution of the sixteenth and seventeenth centuries and its revival coincided with the foundation of modern classical science. The continuity between Democritus, Gassendi, Dalton and even Lorentz is obvious to any unprejudiced person who is acquainted with the history of ideas and is doubted only by those who are not. (I am thinking in particular of such historians of ideas as Cyril Bailey, Kurt Lasswitz, Emile Meyerson and Federigo Enriques.[1] It is usually claimed, in particular by some rank-and-file physicists, that there are essential differences between the ancient atomism which was allegedly purely speculative and the modern one which is based on extensive experimental verifications.[2] This objection, plausible as it may appear, overlooks the fact that Greek atomism was born out of the reaction against the metaphysics of Parmenides and Zeno of Elea, the metaphysics which Benjamin Farrington appropriately characterized as a "reaction against experiential science".[3] In rejecting this reaction, the atomists returned to concrete sensory experience, though not without retaining a large portion of the conceptual apparatus of the Eleatic school. I believe it was Windelband who said picturesquely that "Democritus smashed the Parmenidean sphere of Being into tiny fragments and scattered them through empty space". In other words, the atom of Democritus – and this remained true of the atoms of all the periods up to the end of the nineteenth century – retained all the attributes of the Eleatic Being: it was immutable, that is, indestructible, uncreatable, and indivisible; each atom was 'one' in the sense that it filled the volume it occupied fully and continuously, in a homogeneous and undifferentiated way. There was, of course, one fundamental difference: the atomists recognized the reality of the void between the atoms in order to account for our undeniable experience of diversity and change, – the experience which Eleatic metaphysics was unable to explain and which it simply and arrogantly denied.

It is precisely this feature which made classical atomism implicitly modern, in the sense that it made its future impressive empirical verification possible. There were two basic themes common to atomism in all its forms and phases which have not changed through the centuries: that *all diversity of nature is reducible to the differences in configuration of the basic homogeneous particles*; and, second, that *all apparently qualitative changes are reducible to various displacements of the same basic units*. In this respect classical atomism was far superior to the physics of Aristotle, which upheld real qualitative differences between the four – or rather five – heterogeneous elements and which, in regarding, for instance, the process of evaporation as a real transformation of one element – water – into another – air – also believed in real qualitative changes in nature. The modern concept of matter as a homogeneous stuff differentiated only by the quantitative differences between the ultimate particles – that is, their size, shape, position and motions – was fully anticipated by Democritus while it remained completely foreign to Aristotle's qualitative physics. There was a similar contrast between the infinite space of the atomists, homogeneous in all its parts and isotropic, and the naive spherical universe of Aristotle, differentiated into the celestial and sublunar realms and into the heterogeneous concentric zones of "natural places". There is no question now which of these views was superior and which was closer to the spirit and even to the letter of Newtonian physics.

It is true that it took a considerable time before a more correct view prevailed. As I mentioned, the atomistic view of matter was almost eliminated in the Middle Ages when it was regarded as synonymous with atheism; it survived outside of, and in opposition to, the medieval establishment which made Aristotle's cosmology its own. No wonder the atomists were at that time persecuted or silenced. Thus in 1348 Nicolas d'Autrecourt was forced by order of the University of Paris to recant the following 'errors': that in nature there is no coming into being nor any real annihilation, but merely changes in position; that light consists in local motion which is propagated at a finite velocity; that in nature there is no real generation nor any real destruction, but merely congregation and disaggregation of atoms, etc. Note that all these allegedly erroneous opinions proved to be correct – but only in the century of Newton and Robert Boyle,[4] that is, only after Lucretius's view of infinite space was rehabilitated by Giordano Bruno, and atomism in general by Pierre Gassendi and others. This started the process Dijksterhuis called "mechanization of the world picture" which continued triumphantly and practically without interruption until the end of the last century.

Thus the main difference between classical Greek atomism and modern atomism was one of degree; the former was based on far more limited experience than the latter, that is, on the sensory experience of the naked eye (and naked touch) unaided by telescope, microscope and other devices by which the field of our perception has been enormously extended. Yet this limited experience was analyzed by the early atomists so attentively and with such a finesse that their main conclusions about the nature of the physical world were identical to the conclusions of modern atomists. Anybody reading Lucretius attentively and not merely for aesthetic or philological pleasure is struck by the wealth of empirical facts he dealt with and by the acuteness with which he analyzed them. Only inattentive and superficial readers with a limited knowledge of the physical sciences can still maintain the fiction that ancient atomism was "purely speculative". In truth, some of its anticipations were remarkably precise and astonishingly specific: such as Lucretius's view of hidden molecular motion, imperceptible to our senses, eighteen centuries before its actual discovery by Brown and its correct interpretation by Ramsey,[5] or Democritus' view that empirical differences in macroscopic bodies are due to the differences in shape (σχῆμα), arrangement (τάξις), and position (θέσις) of the atoms[6] – this represents the same general approach to the observed diversity of matter as that adopted by modern structural chemistry in its explanation of isomerism, polymerism, polymorphism; in its interpretation of the diversity of chemical elements in terms of different electronic configurations, etc.

It may be objected that such anticipations are too general to be significant, or even to be called genuine anticipations. It is true that they were consequences of certain general principles which both ancient and modern atomism share. But the list of these general principles shows how important they are: the homogeneity and infinity of space; the conservation and unity of matter; the reduction of all empirical diversity to the differences in configuration and motion; the reduction of all changes to motion. Add to this the fact that ancient atomists came remarkably close to stating the laws of inertia and of the conservation of momentum when they asserted that the motion of atoms is as eternal as the atoms themselves and that each of them continues moving along a straight path until it rebounds from other atoms.[7] If this may appear vague and unsatisfactory to the modern mind, let us compare it to the opposite view of Aristotle according to which every motion requires a mover, i.e. a moving force to keep a body moving – a proposition which vitiated his whole physics and a large part of his metaphysics.

What happened during the period 1600–1900 is too well known to be dwelt upon: there occurred a gradual but decisive empirical verification of all important general insights of classical atomism: the homogeneity and infinity of space, the unity and constancy of matter, the mechanization of the world picture with all its corollaries. All continuous fluids such as phlogiston, caloric and the electric fluid disappeared from physics and even the last one which remained – aether – was sometimes interpreted in a corpuscular, and always in a mechanistic fashion.[8] By the end of last century, the mechanization of the world picture seemed nearly complete; even the electromagnetic theory of matter which at first seemed to be a rival of the mechanistic view, proved to be its ally with the coming of electron theory. For even electrons and protons were, after all, particles and although their motions were ruled by the laws of mechanics combined with the laws of electromagnetism, the hope that the latter laws could be interpreted mechanically by some appropriate model of the electromagnetic aether persisted until the advent of relativity and quantum theory. But in its initial phase, electron theory fulfilled one of the most cherished dreams of the atomistic-kinetic view – the reduction of all qualitative differences to those of configuration: it explained the diversity of chemical elements by the differences in the number of nuclear particles and the corresponding number of orbital electrons. The Democritean "alphabet of Being" seemed finally to be within our grasp.

II. THE CRISIS FOR CORPUSCULAR MODELS

Having said all this, I hope I am beyond any possible reproach of being unfair to atomism or, more generally, to the corpuscular-kinetic view of nature. Yet, it is precisely the very applicability of this view outside our macroscopic experience which is now in doubt; and it is in doubt because the concept of an indivisible, permanent corpuscle moving along a continuous trajectory through space and identifiable at successive instants in time seems to be utterly inadequate on the microphysical scale. At least the circumstantial evidence pointing in this direction is overwhelming and still increasing.

For we all know what happened after 1900 to that impressive looking edifice of classical physics which had appeared nearly complete: not only did its roof collapse, but even its very foundations had to be rebuilt. All fundamental classical concepts had to be either given up or profoundly revised: space, time, matter, motion, causality. It is very difficult to deal separately with each of these concepts since a change of one is related to the changes in the others. This follows from the fact that all these concepts were related in a particular way and in most instances their very definitions contained references to other concepts. What we called the corpuscular-kinetic model of nature had been a precise, conceptual network of the concepts listed above, related in definite ways. It is, for instance, obvious that in the concept of a material particle moving through space along a continuous path, persisting through time and obeying certain laws of dynamics, reference is being made to the other concepts just mentioned; thus the changes of these other concepts do inevitably affect the concept of particle and vice versa. It is important to keep in mind that the crisis for the corpuscular-kinetic models means not only a crisis for the concept of particle; it involves a set of all the correlated changes in all the other constituent concepts. Thus it is impossible to discuss the transformation of the concept of matter and, one might even say, the *elimination* of the concept of particle without referring to all the other correlated conceptual changes. Only such an approach can help us to place a new concept of matter in a proper perspective and to grasp the full meaning of the twentieth-century revolution in physics. Nothing in my opinion is more dangerous to a full understanding of this topic, than a discussion of the change in each individual concept, in isolation from the changes in the whole conceptual framework. Yet, I am afraid this is what frequently happens, as I shall later illustrate, using some concrete examples.

As I said before, the electron theory of matter in its first phase seemed to represent the successful culmination of centuries of effort to interpret physical reality in a corpuscular-kinetic fashion; yet, it is interesting that the first signs of the inadequacy of the classical concept of corpuscle began to appear at the same time. The atom was no longer regarded as indivisible, but as complex; but this in itself was hardly a threat to the concept of corpuscle, since its constituent intra-atomic parts, whether those composing the nucleus, or orbiting around it, were still regarded as *particles*; they were merely the old solid atoms of Dalton, but on a much smaller scale. They were regarded as tiny spheres, possessing a certain mass and consequently being *indestructible* – a simple consequence of the law of conservation of matter; they seemed to be the true ultimate units of matter. But there were some complications which, from the rigorously mechanistic point of view, appeared as serious flaws. First, they were viewed as the basic units of electricity: the electrons with a negative and the protons with an equal positive charge. Now there were two questions involved here, about the relationship between the mass of the particles and their charge, and about the dual character of these charges. Neither of these questions could be answered satisfactorily. Were the mass and the charge two irreducible attributes of the particle? This idea hardly appealed to mechanistic taste, which was always idiosyncratically reluctant to concede any irreducible diversity or even duality in nature. Or was mass merely a manifestation of electric charge? This latter view was fashionable among some physicists at the turn of the century, and as late as 1914, Jean Perrin suggested in his now classical book *Les atomes* that perhaps the whole inertial mass of the electron may be of electromagnetic origin.[9] This view was suggested by the increase in the inertial mass of the electron which follows from Einstein's relativistic mechanics, but was originally interpreted in the terms of classical electromagnetic theory as a result of the reaction of the electromagnetic field to the motion of the electron. From this point of view, the electrons themselves and the elementary particles in general "are nothing but condensations of the electromagnetic field" (Einstein).[10]

But would this mean the reduction of matter to electromagnetism? Einstein continues the passage quoted above as follows: "... our conception of the cosmos recognizes two realities which are conceptually quite independent of each other even though they may be causally connected, namely gravitational aether and the electromagnetic field...". This would mean that not all manifestations of matter are reducible to electromagnetism. It is true that the reference to aether would seem to give another chance to the mechanist provided he could construct a successful mechanistic model of the medium which would account for the transmission of electromagnetic waves and also, it is to be hoped, of gravitational interactions, and in which the elementary 'particles would merely be local structural complications. Such was the hope of all mechanistic theories of aether from that of Huygens up to William Thomson. This would in fact amount to an operation in just the opposite direction, namely the reduction of the whole of physical reality to the mechanics of the aetherial medium. If the structure of this medium were grain-like, as Huygens originally suggested and as a number of physicists still believed at the turn of the century, the concept of ultimate corpuscular units might still be saved, but their dimensions would be incomparably smaller than those of the electron; thus Osborne Reynolds (1903) estimated a corpuscular radius equal to the order of the 10^{-18}cm.[11]

But such hopes were already obsolete at that time, and even more so by the time Einstein gave his lecture about aether. Even prior to the special theory of relativity, the inadequacy of the corpuscular-kinetic models manifested itself in the repeatedly frustrated efforts to construct a satisfactory mechanical model of aether. The final blow to such hopes was the negative result of Michelson's experiment which divested aether of even the most elementary kinematic properties; it had to be neither at rest nor in motion, otherwise there would be inconsistence with the fact of the constant velocity of electromagnetic waves. For all practical purposes the idea of aether wad dead and if Einstein was still willing to retain the original word, its meaning was só thoroughly different from the original one that an entirely different term should have been invented for this purpose. Today only a few people still speak of aether and hardly any of 'aether particles'.

But the term 'material particle' is still very much alive, even though, as I am going to argue, it is no less inappropriate than the term 'particle of aether'. Let us return to the original planetary model of the atom which at first appeared to be another triumph of the corpuscular-kinetic view of physical reality. Without the help of aether the relationship between the mass of the particles and their charge as well as the duality of charges remained mysterious. Fifty years ago it was already clear to Hermann Weyl[12] that to imagine the charge as 'sticking' to the rigid electrons would be nothing but a grotesque naïveté. But this was not the only difficulty. Another was *an apparent lack of proportionality between mass and volume* – a proportionality which was one of the cornerstones of classical atomism. Thus although the proton is nearly two thousand times heavier than the electron, its radius is of the same order of magnitude, i.e., 10^{-13} cm; and this is apparently true of other elementary particles. The third difficulty appeared when Niels Bohr introduced his quantification of electronic orbits in 1913; it then became abundantly clear that the alleged analogy between a macroscopic planetary system and the atom was altogether deceptive. The so-called 'forbidden zones' between discrete electronic orbits clearly did not have any macroscopic counterpart. Even more seriously, they seemed to contradict both the homogeneity of space and the continuity of trajectories on the microphysical scale. Russell's 'axiom of free mobility' which characterizes not only Euclidean space, but also all homogeneous spaces (i.e., those with constant curvature) ceased to be applicable to microphysical space. Related to this was the difficulty of applying a classical spatio-temporal analysis to the so-called 'quantum jumps'. If the electron were really a corpuscle, then its passage from one orbit to another should be along a *continuous* path, no matter how short, from a point in one orbit to a point in another orbit. But from the beginning it was if not obvious, then at least very probable that such a transition from one orbit to another should be regarded as an *indivisible jump* within which neither spatial nor temporal subintervals could be discerned. Thus even in the early phase of the development of quantum theory, doubts began to emerge about the classical continuity of space, time and motion; Poincaré before 1912, followed by Whitehead in 1920 considered the possibility of an 'atom of time' or a 'quantum of time'; this was soon followed by a host of speculations about the discrete nature of space, time and motion.[13] But without spatio-temporal continuity of its path, the identity of the particle, i.e., its identifiability at different points, in space and at successive instants of time is impossible to maintain and the concept of corpuscle itself loses it meaning.

There was another large group of facts, discovered and interpreted by relativity theory which suggested an equally profound revision of the concept of particle. The special theory fused together two concepts traditionally distinct – mass and energy. In classical physics, mass and energy had always remained distinct and one could exist without the other: thus a particle at rest was devoid of energy while the radiation energy was regarded as massless. Even when a particle was in motion, its mass remained unaffected by the kinetic energy associated with it. It is different in relativistic mechanics: an increase in kinetic energy involves an *increase in mass* – in other words, the mass of a particle in motion is greater than that when it is at rest; this fact was amply confirmed on the microphysical scale, for bodies whose velocities were significantly close to the velocity of light. Similarly, allegedly 'disembodied' electromagnetic radiation has a certain inertial mass, though small, and thus exerts a certain pressure – a fact verified even prior to the advent of the special theory, by Lebeděv in 1900. The concept of mass was thus generalized, but at the same time was clearly divested of its original intuitive (i.e., corpuscular) connotation.

Other consequences of the special theory or, more specifically, of the relativistic 'fusion' of mass and energy, point in the same direction. Einstein's equation $E = mc^2$ means that *every* increase or decrease in energy involves an increase or decrease in the corresponding inertial mass. Thus the total mass of a material aggregate is no longer equal to the sum total of the masses of the particles of which it is composed, as was true in classical physics, and as is still approximately true in our daily macroscopic experience. It is either decreased or increased as energy is either absorbed or released in the process of aggregation. Thus, strictly speaking, there is a loss of mass when one mole of CO_2 is formed, since 94.052 calories of energy are liberated; while the absorption of 21.600 calories in the formation of NO results in an increase in mass. It is easy to see that in such reactions, as in every macroscopic chemical reaction, the calculated mass effects, whether positive or negative, are too minute to be experimentally detected. This is why they escaped detection even in the accurate and repeated experiments of Hans Landolt at the turn of this century, the results of which were hailed as a definitive confirmation of the law of conservation of matter (or more accurately, of mass). But what Landolt proved was only that there are no relative variations in weight to the order of 10^{-6}, while the relativistic variations in mass in ordinary chemical reactions amount to less than 10^{-13} of the total mass involved.

But the situation is different when one considers aggregate formations on a nuclear scale of magnitude. One of the most well-known instances is the mass effect resulting from the formation of the nucleus of a helium atom, consisting of two protons and two neutrons; while the mass sum of all the components is 4.03302 atomic mass units (a.m.u.), the mass of the compound nucleus is less – only 4.00280 a.m.u.. In other words, approximately 0.03 atomic mass units have 'disappeared' or, more exactly, have been converted into the binding energy of the nucleus. Similar mass decrements have been found for other elements; these increase with atomic number up to 0.238 a.m.u. for uranium. The frightful technological application of the energy released by the so-called 'annihilation' of mass is generally known. The opposite process, i.e., the 'materialization' of energy takes place in so-called endergic reactions where energy is absorbed instead of being released; it is especially striking in the reactions between elementary particles of very high energy when some of their kinetic energy is converted into the rest mass of a new particle. The creation of a π-meson by the interaction of two high energy protons is an example of this kind. The first instance of this process was observed in 1932 with the discovery of the positive electron by Anderson; in this case it was the energy of high frequency radiation interacting with a heavy nucleus which was for the most part 'materialized' into the rest mass of two oppositely charged electrons while the excess of the original energy survived in the form of their kinetic energy. The opposite process of 'dematerialization' was discovered at about the same time; in fact, the reason why the positive electron was discovered so late was its extremely short life; its dematerialization occurs after 10^{-8} sec when it encounters a normal, i.e., negative electron, and they both disappear in a puff of high frequency radiation.

The observed variability of the mass of the elementary particles as well as their creation and annihilation are the most serious threats to the applicability of the concepts of corpuscle on the microphysical scale. For is it meaningful to apply the term 'corpuscle' – the most salient traditional features of which were immutability and everlastingness – to mesons, some of which 'last' only 10^{-16} sec? Is not the term 'event' a more appropriate name for such evanescent entities? This question becomes even more pressing when we realize that materialization and dematerialization of particles is not an exception, but rather a rule. Even such a 'solid' particle as the neutron decays in twelve minutes, while the more massive hyperons disintegrate in one hundred millionth of a second. At first glance these processes look very different from those of creation and annihilation of particles; the 'disintegration' and 'decay' are perfectly meaningful within the corpuscular-kinetic scheme – all we have to assume is that the so-called unstable particles are really *composite* and not indivisible, and that radioactive decay is nothing but a drifting apart of the constituent particles which were originally closely packed together. But as Niels Bohr observed as early as 1939,[14] to assume that beta-particles *pre-exist* in the nucleus, from which they are then ejected, is as naive as to believe that photons pre-exist in the atom prior to their emission. On the contrary, according to Bohr, instead of *expulsion* of the electrons from the nucleus, we should speak of their *creation*; they are created in a sense similar to that in which photons are created during their emission. The classical idea of the nucleus being made up of smaller, juxtaposed and closely packed subparticles, although appealing to our pictorial imagination, simply fails to represent the true microphysical situation.

This can be shown quite convincingly by analyzing the radioactive decay of the neutron. If a neutron were really a close combination of a proton and a negative electron, and if its disintegration could be regarded as a separation of these two constituent parts, then the opposite process – the transformation of a proton into a neutron and a positive electron would be impossible. Yet this process *does* take place. Now this remains completely unintelligible within the classical corpuscular-kinetic scheme: either the neutron is an aggregate which disintegrates into its constituents – the proton and the negative electron – *or* the proton is complex and the neutron and the positive electron are its parts. But we cannot have it both ways since this would not make any sense: a particle which would be a fragment of another particle cannot contain the same particle as its own part. Here we are witnessing a complete failure of a strictly corpuscular or *configurational* model of the nucleus. Neither proton nor neutron are *configurations* in the classical sense of being composed of the more basic, closely packed, juxtaposed units, *actually* existing prior to their separation. As Otto von Frisch observed, the situation is even worse when we consider so-called 'strange particles'. "The κ-meson can change into two pions or into three pions or into a pion, a muon, and a neutrino, or in several other ways. If we assume that κ-meson is composed of, say, three pions, then we cannot understand the other modes of break-up." He then logically concludes that "the very idea of compositeness must be left behind if we want to understand the subatomic particles." Heisenberg arrived at the same conclusion.[15]

There are, of course, ways to make these paradoxical results appear intellectually more palatable. For instance, it is true that the materialization of the particles is not *creatio ex nihilo*, creation out of nothing, nor is their 'dematerialization' their absolute annihilation; the conservation laws of mass-energy are not formally violated and one may be allowed to say that the mass of a couple of the created electrons 'virtually' *pre-existed* in the mass-energy of the radiation from which they originated. Thus Hans Reichenbach exaggerated when he called these phenomena "causal anomalies".[16] But, as I mentioned above, this formal preservation of the constancy of mass is obtained by the generalization of the concept of mass itself, that is, by merging it with the concept of energy – but only at the expense of divesting it entirely of its original, i.e., *corpuscular* connotation. For it is obvious that such paradoxical fusion of mass and energy implies consequences which appear grotesque and absurd as long as we retain the corpuscular-kinetic framework. In such a framework there is no place for genuine 'virtualities' or potentialities' since its conceptual components are precisely the bits of homogeneous matter assumed to exist *actually*, i.e., *actually occupying* various positions in *actually* existing space. To speak of 'virtual particles' as present-day meson physics does is hardly anything more than a concession to old intellectual and linguistic habits. As Emile Meyerson observed, the notion of a 'potential' or 'virtual' state is a linguistic device to preserve the identity and uninterrupted continuity of the object in time, by assuming that it *somehow* continues to exist even if it *apparently* disappears.[17] Thus its 'potential' or 'virtual' existence during the intervals of its unobservability guarantees its persistence of identity in time. Only in this way can the human intellect (in its classical form, we must add today) eliminate the emergence of genuine novelties and reduce all changes to a mere reshuffling of permanently existing particles. Thus in most instances, especially in the classical era, the term 'virtual' is merely a cover-name for 'actual', and a potential entity is understood as a hidden actuality. Hence the persistent hopes to interpret potential energy as the *kinetic* energy of *actually* moving, invisible particles – hopes which can be traced from Christiaan Huygens to Herbert Spencer, that is, to the very end of the 'classical era'.[18] Fortunately, hardly any physicist today maintains the 'virtual' pre-existence of the mass of the pair of electrons in the sense of their *actual, corpuscular* pre-existence; in fact, the adjective 'virtual' which physicists join to the noun 'particle' is intended as an explicit warning not to take the word 'particle' in a literal sense. But in such a situation, would it not be better to drop the word 'particle' or 'corpuscle' altogether, precisely because it is so loaded with misleading associations which no qualifying adjective can successfully eliminate?

III. POPPER-LANDÉ'S DEFENSE OF THE REALITY OF PARTICLES

I could go on accumulating other examples of the inapplicability of the strictly corpuscular concept, but new examples would not be basically different in their import from those already mentioned. Furthermore, at this stage of my exposition I can expect the following, rather impatient remark from either a physicist or a philosopher of science: "All you are trying to prove is that the term 'corpuscle' or 'particle' cannot now be taken as having its original, literal meaning. But everybody knows this since everybody is fully aware of the fact that corpuscular character is only *one aspect* of physical reality and that there is another aspect, equally essential and complementary to it, corresponding to the *wave* nature of matter. The inadequacy of the classical concept of particle is due to the fact that the wave aspect must also be taken into account – and that everybody knows."

Well, not everybody. There are serious and outstanding thinkers who still claim that the wave aspect is merely secondary and, so to speak, apparent and thus unrelated to the fundamental physical reality which consists of particles. This is, for instance, the view of Karl Popper and Alfred Landé who claim that Heisenberg's uncertainty relations have been habitually misinterpreted since an attentive analysis of them will disclose that they do not impose any definite limit on the precision of *simultaneous* determination of the position and the momentum of a particle and thus are compatible with the reality of particles.[19] This view is diametrically opposed to the view of such widely different thinkers as Arthur Eddington, Philipp Frank and Max von Laue who argue that there is an *objective limit* to the precision of our measurements and that the impossibility of measuring exact position and exact momentum in conjunction can mean only one thing: that such a conjunction simply *does not exist* in nature.[20] Since such a simultaneous conjunction of position and momentum is nothing but a particle itself, the denial of the reality of such a conjunction is equivalent to the denial of the reality of particles.

This would follow as a direct consequence of an objectivistic interpretation of Heisenberg's principle, the correct name of which, in such a case, should be *Indeterminacy* rather than *Uncertainty*, principle. If this view is correct – as I have tried to show elsewhere[21] – then the principle of indeterminacy would be the final *coup de grâce* to the concept of particle or rather to its applicability, the many inadequacies of which I have tried to point out.

Now it would be unfair to discuss Popper and Lande's argument in the limited space which remains available for me; this really requires a separate paper. But allow me to make some modest and, I hope, relevant remarks. Popper is justifiably proud to be, as far as his view is concerned, in good company with such men as Einstein, Louis de Broglie, Max Born, Landé and Bohm; but it is only too clear that this company is rather heterogeneous. Thus Einstein and de Broglie's rejection of the usual interpretation of Heisenberg's principle was largely inspired by their commitment to determinism; to some extent, though not entirely, this is also true of David Bohm, but certainly not of Landé to whose views Popper's are closest. They both assert the physical reality of particles together with indeterminism. This sounds strange only if we forget that according to both Popper and Landé, not only quantum physics, but classical physics as well must be regarded as indeterministic or at least not rigorously deterministic – a rather paradoxical view and historically incorrect at that. But this is less important in the context of our discussion than the fact that both Popper and Landé are apparently not ready to face all the consequences of their affirmation of the reality of particles.

If particles are truly physically real, i.e., if they have at each instant an *exact* position and an *exact* momentum, then they should move along continuous trajectories both outside and inside the atom, and even inside the nucleus. This would mean a return to the old, naive planetary model of the atom; instead of energy levels, we would have electrons moving continuously in circular or elliptical orbits around the nucleus; it would mean that an electron would literally jump from one orbit to another and during this jump would move continuously – no matter how quickly – through the intervening zone between the orbits. Finally, it would mean – if Popper and Landé really mean what they say – than even within the nucleus the particles would move continuously, possessing at every instant an exact position and an exact momentum, no matter how quickly their positions and velocities might change. Few, if any of the defenders of corpuscular models go so far, even though it would only be consistent for them to do so. Take, for instance, the spin of the particles. The only meaningful way to integrate it into a corpuscular framework is to interpret it as a rotation of tiny spheres. But as has been shown long ago, this would imply rotating velocities exceeding the velocity of light, i.e., incompatible with relativistic mechanics.[22] On the other hand, to give up this naive interpretation borrowed from the mechanics of macroscopic bodies, while retaining the corpuscular models, is hardly satisfactory; it results in a schizophrenic mixture of incompatible epistemological attitudes, half way between Kelvin and Dirac: an abstract, non-intuitive property is incongruously grafted on to the pictorial image of a tiny sphere. Those are not the only difficulties. A far more serious – in fact insurmountable – difficulty is to interpret the undulatory character of matter in terms of consistent corpuscular models; but more about this later.

Why then are the majority of rank-and-file physicists together with some philosophers of science so strongly committed to the reality of particles? Because they are sincerely convinced that the empirical evidence for their existence is overwhelming and irrefutable. They point out that the discrete tracks in a Wilson Cloud Chamber as well as discrete impacts on the spinthariscope screen or on a photographic plate cannot be interpreted in any other way. Landé himself regards these facts as 'directly supporting' the objective reality of particles. Yet it has been pointed out long ago that this is not necessarily so.[23] What looks like a 'continuous corpuscular trajectory' in the Wilson Chamber under a magnifying glass appears as a discontinuous series of water drops separated by irregular intervals, arranged along almost straight lines in the case of fast-moving and heavy particles, while the 'tracks' of slow-moving electrons are sinuous or broken lines. Each droplet has condensed around an ion and thus what appears to us as the 'continuous track of a particle' is nothing but a discontinuous series of water drops reflecting the light. But this is not all. Each fast-moving electron successively strikes different molecules of gas; it ionizes each of them by tearing away a peripheral electron from it, so that the rest of the molecule acquires a positive charge. (For the sake of argument, I am deliberately using the original corpuscular language of Bohr's model.) Are we now sure that it is the *same* electron which continues along this track? Or is it an electron which was driven from the ionized molecule? Obviously, no answer is possible. In other words, the allegedly 'direct' empirical evidence for the existence of particles guarantees neither the identity of the particle at different points along its alleged trajectory nor the continuity of its path – the two features which are logically inseparable and without which no consistent corpuscular model is possible.

This clearly shows how in thinking about the 'microcosmos' we are easily misguided by false macroscopic analogies. When we see periodical flashes of a firefly in the darkness, we assume that the same firefly continues to exist between its flashes, and we can easily verify this; but, as Professor Margenau[24] asked, are we justified in using such a 'firefly model' in microphysics? Unlike a macroscopic object, an electron's motion cannot be followed without being lost from our sight; Mill's definition of matter as the 'permanent possibility of sensations', applicable to a firefly and the macroscopic objects generally, fails when we try to apply it to micro-objects. But if the trajectories of the particles are not continuous in the sense that the particles have positions and velocities (if we disregard for a moment Heisenberg's uncertainty relations) only at certain instants and not between, we give up their claim to existence in everything but name. It would be an abuse of language to apply the name 'particle' to a string of spatially and temporally disconnected events.

We have just mentioned Heisenberg's uncertainty principle; it is obvious that to deny it or at least to deny its objectivistic interpretation is *essential* for those who, like Popper, retain the reality of particles. For if we accept it, we concede that not only can no particle be observed twice (as Schrödinger correctly observed[25] and as follows from the analysis above), but that it cannot even be observed *once* since there is no such thing as a conjunction of exact velocity and exact position, in nature. It is true that if we regard the principle as a mere limit to our technique of observation (and, according to Popper, a limit which is not unsurpassable), then there may still be hope for the existence of hidden corpuscular structure at the sub-quantum level. Whether such hopes are borne out is another matter; the circumstantial evidence against the divisibility of the quantum of action is so far overwhelming. But it is interesting that not all those who share such hopes accept a corpuscularian philosophy of the Popper-Landé kind. Thus David Bohm doubts that the inner structure at the sub-quantum level "is made of still smaller particles":

> For the field theory describes all motion in the quantum-mechanical domain in terms of 'creation' and 'destruction' of elementary particles. Thus, if an electron is scattered from one direction of motion into another, this is described as the 'destruction' of the original electron, and the 'creation' of another electron moving in the new direction. Hence, there is, in this theory, no particle which permanently retains a fixed identity as a particle. Indeed, if one looks more deeply into the field representation of the motion even for a free particle one discovers that its motion is described mathematically as a destruction of the particle at a given point and its creation at a closely neighbouring point. Thus the motion is analysed as a series of creations and destructions, whose net effect is to continually displace the particle in space.[26]

Bohm concludes that if the notion of a permanent particle fails on the quantum level, *a fortiori* it is inapplicable on the hypothetical sub-quantum level: if there are such micro-microparticles, they are "always forming and dissolving" so that their precise positions and momenta would have little significance. I think that still to call such quickly dissolving structures 'particles' is an excessive concession to traditional linguistic habits.

In a similar way, J.-P. Vigier defines his hypothetical particle as "an average excitation of a chaotic subquantum-mechanical level of matter; similar in a sense to a sound wave in the chaos of molecular agitation".[27] Philosophically, Bohm and Vigier's views belong to the same category as those of Schrödinger and Einstein: the 'particles' are merely temporary perturbations in the continuity of the spatio-temporal field – a tradition whose roots go back to William Thomson, aether theories and eventually to Cartesian physics. Although classical atomism was radically opposed to Cartesianism and its posterity as far as the status of particles was concerned, it shared with Cartesianism two fundamental theses: it accepted the continuity of space and time (only matter is discrete, not space, nor time) and, most importantly, rigorous determinism.

Only on this last point is there a significant and rather strange difference between classical atomism and its new Popper-Landé version. As already mentioned, their commitment to the reality of particles does not prevent them from being indeterminists – if their indeterminism is genuine. Only once in Western intellectual history has such a combination of corpuscularism and indeterminism appeared. Its only (still famous) instance was the post-Aristotelian atomism of Epicurus and Lucretius; but it was generally recognized that in their system the idea of indeterminacy was incongruously grafted onto a mechanistic scheme. From Cicero to Bergson, philosophers have dismissed and ridiculed it; when Gassendi tried to rehabilitate Epicurus's atomism, it was atomism without the spontaneous swerve of the atoms. When some late nineteenth-century thinkers favored the objective status of indeterminacy in the physical world, they tried to dissociate it from its original corpuscular-kinetic framework: such was the case with Renouvier's neomonadism, Boutroux's contingentism and Peirce, Bergson and Whitehead's panpsychism. Very few historians found kind words for Lucretius's indeterministic mechanism: it is true that Emile Meyerson, always looking for historical antecedents, regarded the Lucretian *clinamen* as a remote ancestor of quantum indeterminacy and that Dirac probably unintentionally paid a compliment to the author of *De rerum natura* when he spoke, either jokingly or seriously, of *liberum arbitrium naturae.*[28] But there is hardly any question that when an attempt is made to combine mechanism and indeterminism, something similar to Epicurean atomism must result. For the only way to smuggle indeterminacy into a corpuscular-kinetic scheme is to endow the particles with the capacity to *choose* their future positions and future velocities from a certain range of *possible* positions and *possible* velocities. Popper's theory of particles endowed with 'propensities', when placed in the wider perspective of the history of ideas, appears as a new, highly sophisticated version of the Lucretian doctrine of *clinamen*. It is hardly surprising that it faces the same difficulty: how to integrate a non-mechanistic element of 'propensity' or 'potentiality' (Popper uses both terms)[29] into a corpuscular, i.e., mechanistic scheme.

Besides this major philosophical difficulty there is another which only took on a more concrete form as a result of twentieth-century physics: how do we integrate the undulatory character of matter with the corpuscular model? According to both Popper and Landé, the primary reality, which consists of *particles* and their 'wave-like character', is nothing but the way they are distributed in space. There is thus no duality of particles and waves; this duality was already removed by Max Born, and Landé only regrets Born's occasional reversion to dualist language,[30] language which also prevails in the majority of the textbooks. But then why are the particles distributed in a wave-like fashion? Where does the spatial periodicity of their interference rings come from? Both Popper and Landé reject the attempts to explain this periodicity in terms of some unitary wave theory, whether one speaks of Schrödinger's original theory of particles as high crests of real, substantial waves, or of the later theories of de Broglie, Bohm and Vigier. All such theories are deterministic as well as anticorpuscularian, in the sense that particles are regarded as mere temporary products of the continuous field whose vibratory disturbances are fully describable in a classical deterministic fashion. Such views are obviously diametrically opposed to Popper-Landé's insistence on the substantial reality of particles as well on the irreducible character of the statistical laws. But this leaves the basic question mentioned above unanswered: why do electronic impacts form interference rings, instead of being distributed in a classical, random, non-periodical fashion?

One can feel considerable sympathy for the Popper-Landé criticism of the orthodox interpretation of quantum theory, with its positivistic, phenomenalistic and even subjectivistic overtones. Popper rejects Jeans's view that the probability waves are 'waves of knowledge', and rightly criticizes the oscillation between subjectivistic and objectivistic interpretation of probability which is so characteristic of the Copenhagen school. His real merit is that he stresses that physical indeterminism, i.e., the primary and irreducible character of the statistical laws, does *not* necessarily imply epistemological idealism; he accepts Einstein's epistemological realism without accepting his determinism. This is why he insists that the waves of propensities are "physically real", being "objective relational properties of the physical world." (Landé is more radical than Popper since he insists on the *exclusive* reality of particles and deprives probability of any physical, 'kickable' status.[31] But since he also rejects de Broglie and Bohm's theory of pilot waves, the physical status of his propensity waves remains obscure. The obscurity is only increased by his claim that quantum theory is no more mysterious than any other game of chance and that "all apparent mysteries would also involve thrown dice, or tossed penny *exactly* as they do electrons".[32] (His italics.) Such comparison of the behavior of the electron with that of a macroscopic object is epistemologically untenable, but for both Popper and Landé it follows from their equally untenable denial of any basic distinction between classical and contemporary physics. Their tacit assumption is analogous to that of Einstein: in the same way that Einstein dogmatically assumed that epistemological realism is possible only on a strictly *deterministic* basis, Popper and Landé believe that it is possible only on a *corpuscular* basis. But does the attitude of epistemological idealism necessarily correspond to a return – no matter how disguised and sophisticated – to Epicurus and Lucretius?

IV. A CASE FOR EVENTS

There is another alternative open to us besides those mentioned above: that the basic physical entities are *neither particles nor waves* in their usual macroscopic sense, but *imageless events* whose behavior, when it is communicated to us by very complex artificial arrangements, reminds us alternately of the behavior of macroscopic particles or periodical disturbances in elastic media. Enough has been said about the inadequacy of the concept of particle; the intuitive image of wave as a periodical disturbance in the three-dimensional aetherial medium proved to be inadequate for the same epistemological reasons. Thus we can afford to be fairly brief in our survey of the main difficulties with the concept of aetherial medium.

The first signs of its inadequacy were the enormous difficulties encountered in trying to construct a consistent and empirically satisfactory model. These difficulties had become insurmountable after Michelson's experiment which deprived aether of its last intuitive properties: aether was deprived even of the most fundamental kinematic properties, for it could not be either at rest or in motion. Such difficulties increased enormously with Planck's discovery of the quantum character of radiation which showed that the propagation of light cannot be exhaustively described in terms of continuously spreading waves. Thus a highly unsatisfactory 'wave-particle dualism' was introduced into physics – unsatisfactory as long as 'particles' and 'waves' were understood in their original, intuitive, macroscopic sense. It is generally known that quantum theory was in no sense a simple return to the old corpuscular theory of light; nor could the 'waves of matter' (*les ondes matériel-les, Materiewellen*) discovered by de Broglie, be interpreted in a realistic sense as periodical disturbances in a hypothetical sub-aether, as Schrödinger originally believed. Intuitive models of waves are thus as unsatisfactory as intuitive models of particles and it is rather interesting that de Broglie's discovery of the vibratory nature of matter represented another '*coup de grâce*' for the traditional pictorial image of corpuscle. For no effort of imagination can integrate into one single self-consistent model the image of particle and that of vibration.

The last statement hardly needs any extensive comment: a particle *can vibrate*, but it cannot *consist of vibrations*. If we say that it does consist of vibrations, being, as was originally believed, that region of the 'sub-aetherial field' where the amplitudes of the vibrations are higher than in its neighbourhood, then it ceases to be a particle in the original sense because it loses its substantiality. Conceptually, within the classical intuitive framework, 'particle' and 'wave' remain irreducibly different and can be associated only in an external way. This is why physicists speak of the de Broglie wave length being '*associated with*' a particle according to the formula $\lambda = h/mv$. Strangely enough, an equally fundamental equation $mc^2 = hf$ (m = mass, c = velocity of light, h = Planck's constant, f = frequency) is hardly ever commented upon. What is the meaning of this intrinsic vibrational frequency f, once we give up the intuitive and naive tendency to interpret it as a locomotory oscillation of some imaginary 'sub-aetherial' corpuscles?

On this point, the view of the particle as a *string of imageless events* seems to me the only one that is free of the epistemological crudities of visual mechanistic models, in particular of the intolerable conflict between the two classical images. It presupposes a radical revision of classical habits of thought and as such it cannot be welcomed by those who insist on the 'Cartesian clarity' of physical models. In particular, it presupposes a negation of the infinite divisibility of time as envisioned by Whitehead and, before him, by Poincaré and Bergson. It takes into account the two most philosophically significant innovations of relativity – *the conjunction of space with time* and *the elimination of the distinction between time-space and its physical content.* Thus to say that particles are 'successions of spatio-temporal pulsations' is not a mere figure of speech; as Whitehead pointed out shortly after de Broglie's discovery, "when we translate this notion into the abstractions of physics, it at once becomes the technical notion of 'vibration'."[33] The significance of f in the equation above is clear: it designates *a number of elementary events constituting the duration of a particle per unit time.* These events are precisely the spatio-temporal minima whose extents vary according to the nature of particles, although, in the light of present evidence, it is never significantly below the limits postulated by the chronon and hodon theories.

A physicist will probably object that the view proposed above is too general and qualitative to be satisfactory. Indeed, it is; it is not a model in the usual sense. You may even call it a philosophical guess. But the task of a philosopher of nature is not to prescribe directives too specific to the development of science; such attempts have always ended catastrophically, as the fate of the German *Naturphilosophien*, in particular those of Schelling and Hegel, shows. His task is to discern the *direction* in which the sciences are moving; in this particular case, to take into a synthetic account the cumulative evidence of the inadequacies of the corpuscular, and more generally, mechanistic models. By focussing our attention on conceptual blind alleys, a philosopher of nature can open the way for new channels of thought which appear implausible only because of our adherence to traditional modes of thinking. Thus the implausibility of the view that matter consists of vibrations or – less intuitively – of a succession of exceedingly short events is due to the fact that in all classical models, the particle, besides filling a certain tiny volume of space, also fills continuously and without interruption any interval of its duration, no matter how small. In Whitehead's words, "if material has existed during any period, it has equally been in existence during any portion of that period. In other words, dividing the time does not divide the material".[34] In a more ordinary language, the particles persist uninterruptedly through time. (I do not see any evidence that either Popper or Landé depart from this classical view.) In the space-time diagram such particles are represented by narrow four-dimensional *tubes* of infinite length because of their permanence and indestructibility. It is this model which exerts an inhibitory influence on our minds and which makes the vibratory model of matter so difficult to accept.

But the situation changes when, again in Whitehead's words, "we consent to apply to the apparently steady undifferentiated endurance of matter the same principles as those now accepted for sound and light". He is careful to add that such a vibration "is not the vibratory locomotion: it is vibration of organic deformation". It is not a vibration of *something* like the vibration of the particles of air in sound or that of some mythical aethereal or sub-aethereal particle. For this reason, instead of using Whitehead's highly metaphorical neologism, I prefer to speak of a 'succession of events' or a "succession of spatio-temporal deformations' 'or 'pulsations'; I am aware that these are also neologisms, but they are less tainted by misleading associations. (To every person acquainted with physics, the word 'vibration' inevitably suggests a spatial periodical displacement, of *something* around a position of equilibrium – and thus steers his imagination back to the discredited notion of some substantial particles.) But in this context we immediately see one concrete advantage of our model which at first appears so unsatisfactorily general. As long as the particles are viewed as substantial entities, persisting continuously through time, their interference and diffraction patterns remain unintelligible; highly artificial assumptions must be made to explain why they are scattered *only in certain specific directions* to simulate the wave-like behavior. On the other hand, if the electrons and micro-particles in general are intermittent 'entities' or events, their interference becomes in principle possible. Two particles, because of their permanence, can never merge together nor cancel one another; two events can. Such cancellation clearly occurs in the minima of interference rings, and, more spectacularly, as Otto von Frisch pointed out, in a 'corpuscular version' of Michelson's experiment, where two electron beams *cancel each other.*[35] This does not mean that the electron *consists* of waves, but only that it has intrinsic vibratory, i.e., *periodical* structure. Now the essence of any interference is the presence of the *periodicity in time* which produces the *periodicity in space*, i.e., the alternating regions of maxima and minima.

But the most decisive argument for the view stated above is an *epistemological one*: namely the argument based on the *genetic* or *biological theory of knowledge*. I concede that such an argument can interest the rank-and-file physicists only indirectly, but its cogency is recognized by some philosophically-minded physicists and some philosophers of science. Only the barest outline can be given here; besides this, I refer the reader to my previous articles on Mach, Bergson and Reichenbach and to the whole first part of my book on Bergson. Briefly stated, this theory is as follows: the cognitive functions of the human mind are not static entities, but they are results of evolution; in their present form, they are the result of a long adaptive process by which the human mind adjusted itself to external reality. According to the older biological theory of knowledge, proposed by Herbert Spencer and upheld by Helmholtz, Mach and even Poincaré, this evolutionary process was completed, at least in its essential features; human intelligence in its classical form was regarded as the *final* and *culminating* phase of this adaptive process in the sense that the traditional two-valued logic, Euclidean geometry and finally classical mechanics were believed to represent a true, adequate and in their general features *complete* representation of reality. (I suspect that both Popper and Quine are fairly close to this view.) It is interesting that this older version of biological epistemology agreed in one important aspect with Kant. For Kant, any future departure from the Newtonian-Euclidean picture of reality was logically excluded by the rigid and unchanging *a priori* structure of the human cognitive apparatus; for evolutionary positivists of the last century it was excluded because the Newtonian-Euclidean character of our intelligence is the final adaptation to objective reality; any departure from it would be a step backwards, an epistemological regression.

But such dogmatism is now untenable: the whole astonishing and paradoxical character of twentieth century physics shows that our so-called 'categories, and 'forms of intuition' fail both below and above the limits of our macroscopic experience precisely because they are adapted only to what Reichenbach called "the zone of the middle dimensions". In other words, the adaptation of the human mind to objective reality is *not* complete; this explains both the triumphs of classical physics *inside* the zone of the middle dimensions and its failures when we try to extrapolate it *beyond* its limits. Consequently, contrary to the author of the *Critique of Pure Reason*, which was an epistemological justification of traditional logic, of Euclidean geometry and Newtonian mechanism, our categories are *not a priori* nor are they universally valid; they fit with remarkable precision the zone of the middle dimensions, but generally fail outside it. Ultimately they are of empirical origin, and like all other empirical ideas they have only limited applicability.

Now one of the most venerated traditional categories is that of substance. In the context of our discussion only that of *material* substance is relevant. It is certainly striking how little this concept has changed through the centuries, at least if we disregard the Aristotelian intermezzo which lasted so long only because of factors foreign to philosophy and science. But in physics and the philosophy of nature, the concept of material substance remained essentially the same. Hume was basically right when he reduced it to a stable conjunction of sensory qualities, but he failed to explain why these qualities were restricted to some qualities of sight and touch – while the other, so-called secondary qualities, were eliminated so early. In other words, using his own terminology, why the material susbtance, the atom of Democritus, Newton and Laplace was reduced to a simultaneous conjunction of the basic geometrical and mechanical properties, that is, of mass, space occupancy and motion; or, in the language of analytical mechanics, to that of position and momentum. He was apparently little concerned about this problem and this is why modern classical physicists and philosophers of science generally followed Democritus and Locke rather than him, especially when all empirical evidence then suggested the notion of permanent substance. And not only empirical evidence: why was the law of conservation of matter anticipated so early, more than two millennia before its verification by Lavoisier? Kant was so impressed by it that he regarded substance as an *a priori* category *imposed* on our experience. The point of view of the genetic theory of knowledge is different: the notion of a material particle quantitatively constant and persisting through time is formed by the pressure of macroscopic experience and, as Piaget's research has shown, I think, decisively, in childhood. The notion of a permanent object, persisting through time, is formed much earlier, in truth before the end of the eighteenth month, if I remember correctly. I had a chance to say more about the significance of Piaget's research about three years ago, here at Boston;[36] now, I would like to stress or rather re-stress this: neither the concept of atom, nor that of particle, nor even that of permanent object are *a priori* categories of mind, but the results of our adjustment to a limited range of macroscopic experience; to extrapolate them beyond these limits has led, and if we persist in so doing, will lead to repeated failures. Hence the inadequacy of the concept of a substantial particle on the microphysical scale;[37] this is why the term 'event' is far more appropriate – it is free of misleading corpuscular associations. What we still call 'particles' are either individual events, quite often of extremely short duration or a string of events succeeding each other through sometimes long intervals of time, but always of finite duration. If I regard 'process', 'event', 'change' as basic categories, it is not because of any subjective preference, but because they are far more comprehensive and less restricted in their applications than such terms as 'objects' or 'particles'.

SUMMARY

(a) Classical physics regarded elementary particles as very minute solid bodies with constant mass, shape (usually spherical) and volume, persisting through time and identifiable in the successive positions of their continuous trajectories.

(b) Not a single one of these features remains unquestioned by contemporary physics. The constancy of their mass, shape and volume disappeared with the special theory of relativity; their permanence with the discovery of the creation and annihilation of electrons; and the continuity of their trajectories with quantum theory.

(c) Even the most abstract residue of this concept – the conjunction of definite position with a definite momentum – is challenged by wave mechanics, in particular by Heisenberg's principle of indeterminacy.

(d) Popper-Landé's defense of particles does not take into account the strange consequences to which consistent applications of corpuscular models lead: a return to the original planetary model of the atom, a billiard ball model of the nucleus, etc.

(e) Finally, on epistemological grounds, it is exceedingly implausible that the notion of particle, made in the image of a solid macroscopic body would apply to the region so remote from our daily experience. The individuality on the microcosmic level is very probably the individuality of an *event* rather than that of a thing.

NOTES

1 Cyril Bailey, *The Greek Atomists and Epicurus* (Oxford: Oxford Univ. Press, 1928); Kurt Lasswitz, *Geschichte der Atomistik vom Mittelalter bis Newton* (Hamburg and Leipzig: Voss, 1890); Emile Meyerson, *De l'explication dans les sciences* (Paris: Payot, 1921); Federigo Enriques, *Le dottrine di Democrito d'Abdera* (Bologna: Zanichelli, 1948).

[2] Cf. an effective refutation of this view in Meyerson, *op. cit.*, II, pp. 320–321, and 356.
[3] Benjamin Farrington, *The Greek Science* (Harmondsworth: Penguin Books, 1944), Ch. 4.
[4] K. Lasswitz, *op. cit.*, I, pp. 257–258.
[5] Lucretius, *De rerum natura*, II, vv. 309–333; Bailey, *op. cit.*, p. 332.
[6] Bailey, *op. cit.*, p. 80.
[7] Enriques, *op. cit.*, Ch. III, 'Il principio d'inerzia', pp. 57–91.
[8] Hans Witte, *Über den gegenwartigen Stand der Frage nach einer mechanischen Erklärung der elektrischen Erscheinungen* (Berlin: Ebering, 1906); P. Drude, 'Über die Fernwirkungen', *Annalen der Physik* 62 (1897), pp. XXV–XLIX (on numerous models of gravitation); finally, E. T. Whittaker, *History of the Theories of Aether and Electricity. The Classical Theories* (New York: Philosophical Library, 1951), and Kenneth Schaffner, *Nineteenth-Century Aether Theories* (New York: Pergamon Press, 1972).
[9] Jean Perrin, *Les atomes* (Paris: Alcan, 1914), p. 253.
[10] A. Einstein, 'Relativity and the Ether', in *Essays in Science* (New York: Philosophical Library, 1934), p. 110.
[11] Cf. White, *op cit.*, pp. 216–219; Osborne Reynolds, *The Sub-Mechanics of the Universe* (Cambridge: Cambridge Univ. Press, 1903), p. 1.
[12] H. Weyl, *Was ist Materie?* (Berlin: Springer, 1924), p. 18.
[13] Cf. the bibliographical references in M. Čapek, *The Philosophical Impact of Contemporary Physics*, new paperback ed. (Princeton: Van Nostrand, 1969), p. 242.
[14] Niels Bohr, *Quantum d'action et noyaux atomiques, Actualités scientifiques et industrielles*, No. 807 (Paris, Hermann, 1939), p. 12; Robley D. Evans, *The Atomic Nucleus* (New York: McGraw-Hill, 1955), pp. 30–31.
[15] Otto R. Frisch, *Atomic Physics Today* (New York: Basic Books, 1961), pp. 132, 186, and 192; W. Heisenberg, 'The Nature of Elementary Particles', *Physics Today* 29 (1976), 32–39, "words such as 'divide' or 'consist of' have to a large extent lost their meaning". Hence Heisenberg's skeptical attitude toward the quark hypothesis (*ibid*, p. 39). (This was probably his last article.)
[16] H. Reichenbach, *The Direction of Time* (Berkeley: Univ. of California Press, 1956), p. 265.
[17] E. Meyerson, *De l'explication dans les sciences* (Paris: Payot, 1921), I, Ch. X.
[18] On this point, see J. B. Stallo, *The Concepts and Theories of Modern Physics*, Ch. VI: 'The Proposition That All Potential Energy Is in Reality Kinetic' (Cambridge, Mass.: Harvard University Press, 1960). On Huygens' kinetic model of potential energy, cf. K. Lasswitz, *Geschichte der Atomistik vom Mittelalter bis Newton* (Hamburg and Leipzig: Voss, 1890), II, p. 373. Spencer's view is stated in his *First Principles*, 4th edition, Appendix (New York: Appleton, 1896), pp. 598–599.
[19] K. Popper, *The Logic of Scientific Discovery* (London: Hutchinson, 1959), pp. 215 ff.; A. Landé, *From Dualism to Unity in Quantum Physics* (Cambridge: Cambridge University Press, 1960).
[20] A. Eddington, *The Nature of the Physical World* (New York: Macmillan, 1933), p. 225; P. Frank, *Philosophy of Science: The Link Between Science and Philosophy* (Englewood Cliffs, N.J.: Prentice-Hall, 1957), pp. 215, 230; Max von Laue, 'Über Heisenbergs Ungenauigkeitsbeziehungen und ihre erkenntnistheoretische Bedeutung', *Naturwissenschaften* 22 (1934), 439–441.
[21] M. Čapek, *The Philosophical Impact of Contemporary Physics* (Princeton: Van Nostrand, 1969), Ch. XVI.

[22] H. Margenau, *The Nature of Physical Reality* (New York: McGraw-Hill Co., 1950), p. 313.
[23] E. Bauer, 'Rapports entre la physique actuelle et la philosophie', in *L'Evolution de la physique et la philosophie*, Quatrième Semaine Internationale de Synthèse (Paris: Alcan, 1935), pp. 31–33.
[24] H. Margenau, 'Advantages and Disadvantages of Various Interpretations of Quantum Theory', *Physics Today* 7, no. 10 (1954), 6–13.
[25] E. Schrödinger, *What Is Life and Other Scientific Essays* (Garden City: Doubleday, 1958), p. 175.
[26] D. Bohm, 'Explanation by Hidden Variables at a Sub-Quantum Level', in *Observation and Interpretation*, edited by S. Körner (London: Buttersworth Scientific Publications, 1957), p. 35.
[27] J.-P. Vigier, 'The Concept of Probability in the Frame of the Probabilistic and Causal Interpretation of Quantum Mechanics', *ibid.*, p. 76.
[28] P. A. M. Dirac, *The Principles of Quantum Mechanics* (Oxford: Clarendon Press, 1930), p. 4.
[29] K. Popper, 'The Propensity Interpretation of the Calculus of Probability and Quantum Theory', *Observation and Interpretation*, pp. 65–71.
[30] A. Landé, *op. cit.*, pp. 78–79.
[31] K. Popper, *The Logic of Scientific Discovery*, p. 221; 'The Propensity Interpretation . . . ', p. 69; Landé, *op. cit.*, p. 76.
[32] Popper, 'The Propensity Interpretation . . . ', p. 68.
[33] A. N. Whitehead, *Science and the Modern World* (New York: Macmillan, 1926), p. 193. There is every indication to support the view that Whitehead was not only fully aware of the discovery of the vibratory nature of matter, but even anticipated it. The contrary view of Robert Palter and Abner Shimony is not supported by the texts. Cf. A. Shimony, 'Quantum Physics and the Philosophy of Whitehead', in *Boston Studies in the Philosophy of Science*, Vol. 2 (Dordrecht, Holland: D. Reidel, 1965), p. 307; R. Palter, *Whitehead's Philosophy of Science* (Chicago: University of Chicago Press, 1960), p. 218. On the relation of Bergson's view of matter to that of Whitehead, cf. both my books: *The Philosophical Impact of Contemporary Physics* (Princeton: Van Nostrand, 1969), pp. 368–369, 375 and 391 and *Bergson and Modern Physics, Boston Studies in the Philosophy of Science*, Vol. 7 (Dordrecht, Holland: D. Reidel, 1971), Part III, Ch. 14.
[34] Whitehead, *op. cit.*, p. 73.
[35] Otto von Frisch, *op. cit.*, p. 90.
[36] Jean Piaget, *Le développement des quantités physiques chez l'enfant: conservation et atomisme* (Neuchâtel: Delachaux et Niestlé, 1941); M. Čapek, 'The Significance of Piaget's Research on the Psychogenesis of Atomism', in *Boston Studies in the Philosophy of Science*, Vol. 8 (Dordrecht, Holland: D, Reidel, 1971), pp. 446–455.
[37] The non-substantial character of particles was explicitly stressed by A. March, *Die physikalische Erkenntnis und ihre Grenzen* (Braunschweig: Vieweg, 1960), pp. 58–62 and 95–97. W. Yourgrau, although generally favorable to the Popper-Landé view, concedes that concepts like 'sameness' or 'individuality' do not apply to micro-particles. See his 'On the Reality of Elementary Particles', in *The Critical Approach to Science and Philosophy*, ed. by M. Bunge (New York: The Free Press of Glencoe, 1964), p. 369.

PART III

THE STATUS OF TIME IN THE RELATIVISTIC PHYSICS

CHAPTER 12

THE END OF THE LAPLACIAN ILLUSION

THE PRINCIPLE OF INDETERMINACY AND ITS CONFLICTING INTERPRETATIONS

CHAPTER XV has shed new light on the impossibility of sharp localization of microphysical "particles." According to the prevailing interpretation of wave mechanics, no electron, strictly speaking, exists *at* a given point; there is only a certain probability of finding it there. This probability is the only "thing" which is accessible to measurement. The principle of uncertainty, formulated by Heisenberg,[1] forbids a simultaneous knowledge of the position and velocity of any elementary particle. There is no place here for the explanation of this now well-known principle, whose exposition may be found in any elementary textbook of modern atomic physics. Suffice it to say that any increase of accuracy in the measurement of position is accompanied by a growing haziness in the measured value of velocity and vice versa. This follows immediately from the relation

$$\Delta p \cdot \Delta x \geqq h \qquad \text{(see Chapter XIII)}$$

The haziness in our knowledge of a present state inevitably entails a limitation of our knowledge of future states. Because the present state of any particle is given by the correlation of the precise values of its position and momentum, it is obvious that its future positions and velocities can be predicted only with a certain probability, never with classical Laplacian certainty. We have only to remember that in the classical Laplacian model of the universe "the world at a given instant" was definable as a huge instantaneous configuration of elementary particles, each possessing besides its definite mass also a sharply defined position and velocity; a "state of the world" thus defined virtually contained all past and future configurations and velocities because any event in world history was in principle deducible from any sharply defined cosmic state. According to the principle of uncertainty both the concept of precise position and that of sharply defined velocity lose their meanings; consequently, the concept of the "state of the world at an instant" loses its definiteness too. This concept has already been thoroughly discredited by the relativistic criticism of the simultaneity of distant events; Heisenberg's principle brought merely a final *coup de grâce.*

Not only does this principle exclude the possibility of a precise localization in space, but in its second formulation

$$\Delta E \cdot \Delta t \geqq h \qquad \text{(see Chapter XIII)}$$

it ruins the possibility of a sharp *localization in time;* for any particle localizable at a mathematical instant would be literally without any definite value of energy. Thus the concept of energy at a definite instant has lost its legitimacy in the same degree as that of definite position associated with definite velocity.

After all that has been said about the crisis of classical concepts of particle and motion and about the inadequacy of the concepts of instantaneous position and instantaneous velocity, it is only natural to expect that the traditional concept of *rotary motion* will be modified in an equally profound and revolutionary way. After all, the concept of rotary motion is merely a special case of the concept of motion in general and the transformation of the latter entails inevitably the transformation of the former. This is brought into full focus by the third and the fourth forms of the principle of indeterminacy.

In the third form of the principle the conjugated variables are angular momentum A and angular position γ; the products of the corresponding uncertainties must then be

$$\Delta A \cdot \Delta\gamma \geqq h$$

The complete absence of uncertainty about angular momentum would leave the angular position of an electron on its orbit completely undetermined; a result which is hardly surprising in view of what had been already said about the meaninglessness of its precise localization. From the point of view of wave mechanics, which regards orbital electrons as "standing waves" this is only natural; an electron is, so to speak, "omnipresent" all along its orbit. As the maximum uncertainty of the angular position is correlated with the precise value of the angular momentum, and as the former cannot be larger than 360°, that is, 2π, it follows that the value of the latter must be $h/(2\pi)$; for only then $(2\pi h)/(2\pi) = h$. From this, as F. A. Lindemann pointed out more than a quarter of a century ago,[2] Bohr's condition for the first quantum orbit follows immediately.

The term "orbit" is thus completely misleading. Can we speak at least of its *radius* or, using the undulatory language, of the distance of the electronic "standing wave" from the nucleus? The classical formula for the radii of quantum orbits of the hydrogen atom

$$r = \frac{h^2 n^2}{4\pi^2 m_e e^2}$$

seemingly indicates that we can. For in the expression above, all symbols apparently stand for well-defined quantities: $h =$ Planck's constant, $n =$ quantum number, $m_e =$ the mass of electron, $e =$ its charge. But F. A. Lindemann pointed out that the conjugated coordinate to the square of the charge is the reciprocal velocity $\frac{1}{v}$, in other words, the time taken to pass through a unit distance. "The more accurately we can state the velocity, the more accurately we know the charge. Conversely, inaccuracy in our knowledge of the velocity can be interpreted as inaccuracy in our knowledge of the charge."[3] As the first form of Heisenberg's principle excludes the accuracy in our knowledge of the velocity, the accurate determination of charge is equally excluded; this in turn would bring an inaccuracy into our knowledge of the radius as well. This follows also from the second form of Heisenberg's principle in which the conjugated variables are energy and time.

In the fourth form of the principle of indeterminacy the conjugated variables are the moment of inertia I and the angular velocity ω:

$$\Delta I \cdot \Delta\omega \geqq h$$

Now, if we adhere to the strictly corpuscular image of the electron, the precise value of its momentum of inertia will leave its angular velocity completely undetermined. This conclusion is inevitable as long as the quantities which enter into the formula for the moment of inertia of the electron, i.e., its mass and radius, are regarded as sharply defined values. Thus the very insistence on corpuscular representation leads to a contradiction; the electronic sphere would have a definite radius, definite size, definite mass—but no angular velocity, not even a zero velocity! (For even zero velocity represents a definite value.) This means that we have to be on guard against accepting the alluring image of the electron as a rotating charged sphere, the image which the term "spin" inevitably brings to mind. This image is incompatible not only with the indeterminacy principle, but with the relativity theory as well: for a point on the circumference of the alleged electronic sphere should rotate with a speed 300 times faster than that of light! [3a]

Forms three and four of the principle of indeterminacy are less frequently referred to than the first two. Even textbooks of atomic physics not infrequently fail to treat them explicitly, while in philosophical discussions hardly any mention of them is ever made.[4] Yet, their significance must not be underestimated. They definitely destroy two alluring classical images which survive in microphysics: that of a *rotating solid* and that of a *well-defined continuous orbit* on which a planet-electron, rotating on its axis, revolves around a nucleus. At the beginning of this century physicists and laymen marveled at the "unity of nature in space," especially at the striking similarity between our solar system and the microsystems of planets-electrons. Part I showed how appealing this analogy between macrocosm and microcosm was, how it dominated the imagination of Pascal, Leibniz, and Huygens in the seventeenth century, and how it re-emerged at the beginning of this century in the Gulliverian speculations of Fournier d'Albe and others. The name of A. Righi's book, published in 1921, *Comets and Electrons*, was certainly significant.[5]

We have already pointed out how fallacious the analogy between the atom and our solar system was; the analysis of the last two forms of the uncertainty principle only confirms our earlier conclusion.

The indeterminacy principle sheds an unexpected light on Nernst's discovery, which had been made prior to the rise of wave mechanics: that microphysical particles retain some energy even at absolute zero temperature. Translated into the classical terminology, this means that the particles in question are *still vibrating and rotating* even when, according to the kinetic theory of matter, they should be completely at rest. From the classical point of view, according to which the concept of material body is logically separable from its motion, this discovery was difficult to understand; no reason can be given for the impossibility of completely stationary particles. Not only is there nothing contradictory in conceiving a body without motion, but even according to the principle of relativity *any* moving body can be made stationary by an appropriate choice of the frame of reference. (There is here another limitation of the relativity theory which is due to its macroscopic character: only on *the macroscopic level* is it possible to eliminate any motion by an appropriate choice of the system of reference; similarly only on *the macroscopic level* is it possible to speak very approximately about the continuity of time-space and the infinitely thin world lines.)

But so-called "zero energy" will appear less paradoxical if we bear in mind the results of our discussion in Chapter XV concerning the radical transformation of the concept of motion.

If motion cannot be separated from a "thing moved" and vice versa, then the existence of the so-called "zero energy" is merely another striking confirmation of the inseparability of two concepts which are fused into a single dynamic notion. Needless to say, this synthesis is, psychologically speaking, an extremely difficult operation; all our classical mental habits revolt against it. In the words of Gaston Bachelard:

> Underlying this physical synthesis is a metaphysical synthesis of "thing" and "motion." It corresponds to a synthetic judgment extremely difficult to formulate, since it is vehemently opposed to the analytic habits of ordinary experience, which unhesitatingly divides phenomena into two realms—static phenomena (things) and dynamic phenomena (motions). We must restore to the phenomenon all its linkages and get rid of our notion of *rest*. In microphysics it is absurd to assume matter at rest, since it exists for us only as energy and sends us no message except by radiation.[6]

The psychological difficulties are lessened if we realize that the traditional distinction between matter and motion owes its spurious convincingness to its sensory origin, to the conditioning of our mind by the "zone of the middle dimensions." Outside this zone the distinction loses even its approximate validity. Once we realize this, we shall be in a better position to understand that "zero energy," like wave-mechanical vibrations, is not, properly speaking, a kinetic energy in the traditional macroscopic sense at all. It is not comparable to Brownian motion because it has nothing in common with the thermal vibrations. As Lindemann emphasized, it is only another aspect of the inseparability of the geometrical and dynamic coordinates which is the very essence of the principle of indeterminacy.[7] According to Bachelard's expression, it is "primary mobility, preceding, so to speak, the thermal mobility" ("*mobilité première, antécédente à la mobilité thermique*").[8] Such a statement sounds completely unintelligible and even meaningless within the classical corpuscular-kinetic framework; but in view of what has been said in Chapter XV about the concept of *change* replacing that of displacement, and the concept of *event* replacing that of particle, it loses its paradoxical character, even though its positive meaning does not possess the same deceptive Cartesian clarity as the classical concepts. We shall return to its positive meaning at the end of this chapter.

From what has been said it is clear that physicists hardly exaggerate when they speak of the *crisis of determinism* in contemporary physics. Yet, in the language of those physicists who insist on the inadequacy of classical determinism there are certain features which inevitably produce misunderstandings and confusions, and which eventually weaken their own basic claim. When these physicists speak about the "uncertain position" of the electron or about its "uncertain velocity"; or when they speak about the "probability of the occurrence of an electron" at a given position; or when they speak about the "impossibility of determining simultaneously both the position and momentum of a particle," then not only laymen, but even philosophers or philosophically minded physicists are almost inevitably misled. They are naturally inclined to understand the terms "uncertainty" and "probability" in their usual subjective sense. In such a sense the indeterminacy of the microphysical events is of purely subjective origin, being nothing but the human uncertainty resulting from our technical incapacity to find all determining circumstances of an observed phenomenon. Such an indeterminacy of the velocity and position of microphysical particles would exist in our own knowledge only and not in the nature of things; the microphysical reality *in itself* (Kant would say *an sich*) would be strictly determined, while all haziness, all uncertainty, all indeterminacy would be a result of the intervention of a physicist who by his act of observation modifies the conditions of the observed phenomenon. This view was clearly stated a few years after the formulation of Heisenberg's principle by the French philosopher Léon Brunschwicg:

> It remains to understand that this [Heisenberg's principle] by no means implies the breakup of determinism;

> it merely means that at the present stage of our experimental technique we cannot be satisfied any longer with a simple-minded and dogmatic form of determinism which is interested in reality without being interested in knowledge. . . . The uncertainty relations merely mean that *the determinism of the observed phenomenon* is in itself nothing but an abstraction because it is inseparable from *the determinism by which the act of observation is ruled.*[9] [Italics in the text.]

According to this interpretation the whole situation, to which the instruments of an observer belong as much as does the observed phenomenon itself, is subject to the same rigorous determinism as that of classical physics. If it were so, then the statistical probability of microphysical events would be of the same kind as the statistical laws in the classical kinetic theory of gases. When this theory was proposed in the second half of the last century, nobody doubted that the individual gas molecules move according to the strict laws of Newtonian mechanics, even though in experimentation and calculation physicists were able to deal only with huge molecular aggregates. For the Laplacian mind, which, as we have seen, was merely a picturesque expression for the objective and impersonal order of nature and which would not be bound to laborious experimental and observational procedures, no uncertainty, no indeterminacy would exist; the velocities and positions of the microparticles as well as the values of their energy and their temporal dates would appear in all their definiteness—unequivocal and sharp, though tremendously complex and humanly inaccessible. In this view nature itself has no "fuzzy edge"; all fuzziness, all ambiguity, all indefiniteness resides, not in the nature of things, but in the mind of an observer. Microphysical indeterminacy is due to the disturbing influence of observation; but the compound determinism of the observed phenomenon and the process of observation remains intact.

Thus instead of being threatened by Heisenberg's relations of uncertainty, classical determinism would *explain* their existence. Herbert Spencer, were he alive, would certainly be pleased by such an attitude; for he certainly would regard the existence of the uncertainty relations as another exemplification of his cherished "law of persistence of force," to wit, the law of the conservation of energy applied to the process of observation itself. For if the quantity of energy remains always and rigorously constant, not even its tiniest fraction can ever disappear and therefore its influence cannot be disregarded. But precisely such tiny fractions of energy are present in the form of photons which in the gamma-ray microscope "disturb" the object observed. The conventional explanation of the indeterminacy by the Compton effect has a more Spencerian ring than the authors of the textbooks of physics realize. Even Heisenberg himself, as Karl R. Popper wittily observed, "tries to give a causal explanation why causal explanations are impossible." [10]

Contemporary physicists generally do not share this deterministic view. With a few exceptions, to which, paradoxically enough, both Einstein and Planck belonged, physicists today reject the assumption of the hidden strictly determined processes which would underlie the apparent contingency of the observed phenomena. In their view, the observed statistical laws of macrophysics are not mere surface phenomena, ultimately reducible to classical causal models; on the contrary, the statistical laws are regarded as ultimate and irreducible features constituting the objective physical reality. This attitude is largely due to the positivism prevailing among contemporary physicists, who insist on a consistent elimination of *all* unobservable factors. Einstein explicitly acknowledged his intellectual debt to Hume and Mach; his rejection of the absolute frame of reference was motivated by repeated failures to detect its existence by means of mechanical, optical, or electromagnetic experiments. There is no question but that a similar positivistic motive is conspicuously present in the minds of the physicists dealing with the problem of determinism in quantum mechanics: they consider the empirically unattainable determinism as useless as the hypothetical aether which escapes all means of empirical detection.

It is interesting, however, to see Einstein hesitating to apply

Occam's razor to classical determinism, although he had previously boldly used this razor in eliminating the empirically unverifiable motionless aether. This merely shows that a pure and undiluted positivism in an extremely rare phenomenon; as we shall see, Einstein's reluctance to depart from determinism was due to certain unconscious or semiconscious metaphysical predilections. This explains why Einstein, who was a fairly consistent phenomenalist in his theory of relativity, was much less so in his attitude to quantum phenomena, although at present classical determinism is as unverifiable as absolute space or absolute motion.

It is even more instructive to observe those physicists who, adhering more consistently than Einstein to the rule of parsimony, regard as futile all hopes to explain the uncertainty relations by hidden unobservable causal models. Yet, in spite of all positivistic and agnostic declarations to the contrary, their imagination in most cases remains incurably realistic and they naturally retain an urge to *imagine* in some way the allegedly unobservable background of phenomena. When Heisenberg, for instance, begins the explanation of his principle by his famous *Gedankenexperiment* [experiment in thought], in which a microparticle is "illuminated" by the high-frequency radiation by which its momentum is modified,[11] does this mean that, contrary to his own conscious convictions, he still visually imagines the elementary particles as individual corpuscular entities basically similar to macroscopic bodies of our daily experience? It is probable that Heisenberg and all philosophically minded physicists would claim that, if any imaginative visual content is present in their mind when they use the words "particle" or "corpuscle," it is an irrelevant and accessory psychological phenomenon accompanying the abstract thought which is expressible in precise mathematical symbolism. They would insist that it is the thought which really counts; the concomitant images are completely irrelevant.

I think that this is a serious mistake which stems from an unfortunate separation of psychology and epistemology. It is simply not true that what we conceive abstractly is always independent of the concomitant unconscious or semiconscious imaginative content. A detailed analysis would show that often the convincing power of abstract ideas has its roots in the semiconscious imaginative fringes surrounding them. Instances of this kind have already been mentioned in Part I; let us recall briefly one of them—the classical concept of spatiotemporal continuity. Underlying this concept is our subjective incapacity to stop *imagining* smaller subintervals within any interval of space and time, no matter how minute it may be. It is impossible to separate epistemology from psychology for the simple reason that no sharp boundary separates "reason" from "imagination."

Even those who persistently insist on the independence of their epistemology from psychological factors do accept, either knowingly or unwittingly, a certain theory of mind, that is, a certain *psychology* convenient for their purposes. But they seem to be less aware that this psychology is rather outdated now, since it is based on the artificial separation of "clear" and "pure" thought from the allegedly "confused" sensory content. This division of mind into independent faculties is one of the main characteristics of seventeenth- and eighteenth-century rationalism and it is very definitely present in the Kantian separation of *Sinnlichkeit* and *Vernunft*. But already Locke and Leibniz had begun to question the existence of the sharp boundaries allegedly separating reason and imagination: Locke in moving the the boundary up in order to include reason in sensory perception and its derivatives, Leibniz in moving the boundary down in order to transform sensory perception into a rudimentary form of reason. It is not necessary to commit oneself to a particular epistemology in order to realize that neglected imaginative elements not only accompany but also tinge and influence even the apparently abstract operations of thought, sometimes in a decisive way. Epistemology, and the epistemology of modern physics in particular, would profit enormously from a sort of "psychoanalysis of knowledge" in Gaston Bachelard's sense[12] which would unmask the inhibiting influence of our Euclidean and Newtonian subconscious in the minds of those physicists who sincerely believe themselves to be entirely free from them.

Such subconscious or semiconscious visualization is certainly responsible for why a picture of an individual particle spontaneously emerges in our mind as soon as we pronounce a word like "electron," "proton," "meson," etc. From such an image it is impossible to separate its constitutive properties—its position at a certain time and its motion or lack of motion within a certain frame of reference. The very concept of particle consists of the association of its position, its mass, and its motion; or, more technically, of its position and momentum. In truth, these two features—position and momentum—are the only residuum which abstract dynamics retained from the original crude and sensory image of Lucretius and Gassendi. As Louis de Broglie observed not very long ago,[13] in speaking of microscopic particles we have the tendency to imagine very tiny grains of sand; naturally, we are aware of the inadequacy of such imagery, and following the ancient atomistic tradition, we eliminate from such a picture all sensory characteristics, all "secondary qualities," until eventually nothing but the basic geometric and mechanical properties remain. When finally the concept of corpuscle is reduced to the conjunction of its position and momentum, we have reached *the farthest limit* we can go; one more step will inevitably lead to the elimination of the concept itself.

Psychologically as well as logically it is entirely impossible to divorce the concept of velocity or that of position from the concept of particle without destroying the concept of particle itself. Within a certain frame of reference a particle can have only *one* definite position and *one* definite velocity; a particle without any definite velocity or without definite momentum, a so-called "unsharp particle" ("la particule imprécise") is nothing but a pseudo-idea, a mere combination of words, a sheer *flatus vocis.* But this pseudo-idea is a result of the compromise between the positivistic maxim, which excludes all unobservables, and our almost irresistible tendency to visualize. The same thing is true of the concept of "probability of occurrence of a particle at certain place." Again in this wording the visualizing tendencies of imagination, which manifest themselves in the usage of such terms as "place," "occurrence," "particle," clash with the abstract positivistic principle which demands the exclusion of all unobservables.

From this *impasse* there seem to be only two ways out: either to yield to the natural proclivity of our imagination—needless to say the imagination already purified in the sense of classical physics—and admit that microphysical particles have a precise location in space and are endowed with a well-defined quantity of motion; or to press the revision of our concepts to its logical conclusion and give up the idea of individual particles entirely, like all other sensory or pseudo-sensory notions which physics borrowed from the world of macroscopic experience. The first way would imply the rehabilitation of the classical determinism which would extend to all microphysical events; then the alleged "uncertainties" or "indeterminacies" would be devoid of any objective significance, being merely gaps, perhaps even temporary gaps, in our knowledge. Logically there is hardly any middle ground between these two solutions. It is epistemologically intolerable, at least in the long run, to retain a Newtonian and Democritean subconscious which is entirely incompatible with our conscious convictions and verbal utterances. Yet, this uncertain oscillation between subconscious Newtonianism and its conscious rejection seems to be a characteristic attitude on the part of a great number of physicists today, in particular those who insist on the irreducibility of the statistical laws to some hidden causal micromechanisms.

This is merely another way of saying that the second way is psychologically incomparably *more difficult* than the first one. Classical determinism was, as we have seen, regularly associated with the corpuscular-kinetic models of reality; if we reject determinism and still speak about corpuscles, we evidently use an inadequate language. Hence such monstrosities as a particle without any definite position or velocity. Nearly a quarter of a century ago a dissatisfaction with this situation was clearly stated by Schrödinger:

> The quantum mechanics of today commits the error of maintaining concepts of the classical mechanics of points —energy, impulse, place, etc.—at the cost of denying to

> a system *in a precisely determined state* any definite values for these magnitudes. This shows how inadequate these concepts are. *The concepts themselves must be given up, not their sharp definability.* Attempts are made to avoid the monstrosities of ill-defined concepts by carrying out hundreds of mental experiments to show clearly that the magnitudes in question under the circumstances cannot in principle be accurately measured. [Italics added.] [14]

Not only the concept of corpuscle, but even *the more general concept of precise spatiotemporal location* becomes inadequate (Chapter XIII). Indeed the whole conceptual framework of classical physics is thoroughly incompatible with the claim that the contingency of microphysical events is of an objective nature, instead of being merely a result of our technical limitations or of the disturbing effect by which an observed phenomenon is modified in the very act of observation. But this constitutes a very serious handicap for those who defend this claim, that is, for those who do not share the view of Léon Brunschwicg, quoted above, nor the hopes of Albert Einstein and, more recently, of Louis de Broglie, David Bohm, and Jean Pierre Vigier, according to which the uncertainty relations will be eventually reducible to a causal model, basically not different from the causal models of classical physics.[15]

The defenders of the deterministic interpretation have the tremendous advantage of having the ready-made language with the corresponding conceptual framework strengthened by three centuries of scientific tradition and twenty-five centuries of philosophical tradition; for, let us not forget, philosophical determinism is twenty-five centuries old. It would be unfair, of course, to claim that the deterministic interpretations lack the refinement and complexity of their rivals; but, as we shall see, in their basic features and by their basic motives, they are, often consciously, a return to the past. For this reason the way of deterministic interpretation will, for a certain time at least, appear more attractive and more natural, while the opposite interpretation strikes us as adventurous and baffling. This is hardly surprising; physical indeterminists have neither an adequate language nor a corresponding conceptual framework at their disposal; whenever they try to express their views, they naturally become entangled in artificial mental experiments which are necessarily misleading because they involve macroscopic experimental arrangements and the use of the *corpuscular* (i.e., macroscopic) language, utterly unsuitable for the description of microphysical events.

But in spite of this attractiveness, the deterministic interpretation cannot be trusted now. Before summing up the important epistemological reasons which make a rehabilitation of classical determinism extremely improbable, we have first to consider the weight of physical facts pointing toward the same conclusion.

In the first place, the true significance of the uncertainty principle was and still is being obscured by the considerations in which the role of observation is unduly stressed. Not only a layman, but even a philosopher interested in physics, not infrequently even an average physicist himself, as long as he is not sufficiently acquainted with the logical structure of wave mechanics, has the impression that the microphysical uncertainty is nothing but the *uncertainty of measurement* resulting from the intervention of an observer and his observing instruments. The usual textbook presentation of Heisenberg's principle leads almost inevitably to such a view.

Too much insistence on the act of observation and the presence of the observer brought a misleading note of epistemological idealism into discussions revolving around the principle of indeterminacy. This note was welcomed with an almost equal satisfaction, although for very different philosophical reasons, by idealistically oriented physicists and by neopositivists. This will become less paradoxical if we remember that neopositivists, in particular in Germany and Austria, have never lost their Kantian tinge, especially in their epistemology; the dividing line between positivists and Neo-Kantians is sometimes quite dim, and from phenomenalism to idealism is just one step. After all, the ancestors of contemporary idealists and positivists—Berkeley and Hume—were, in their epistemology at least, not so far apart. An analogous attitude was adopted by the same two groups toward the theory of relativity. The words of Alfred North Whitehead criticizing this attitude may be applied *mutatis mutandis* to present neopositivistic and idealistic interpretations of the principle of indeterminacy:

> There has been a tendency to give an extreme subjectivist interpretation to this new doctrine. I mean that the relativity of space and time has been constructed as though it were dependent on the choice of the observer. It is perfectly legitimate to bring in the observer, if he facilitates explanations. But it is the observer's body that we want, and not his mind. Even this body is only useful as an example of a very familiar form of apparatus. On the whole, it is better to concentrate attention on Michelson's interferometer, and to leave Michelson's body and Michelson's mind out of the picture.[16]

This passage may be almost *verbatim* repeated, if we replace Michelson by Heisenberg and if we substitute "Heisenberg's gamma-ray microscope" for "Michelson's interferometer," and "indeterminacy" for "relativity of space and time." We can go further and say that even the presence of the instrument, whether Michelson's interferometer or Heisenberg's gamma-ray microscope, serves in either case to exhibit some particular effects of the objective features of nature, whether relativistic binding of space and time or microphysical indeterminacy.

This is not intended to discount the correctness of Heisenberg's *Gedankenexperiment;* its only shortcoming is to confine our attention to a special case only. For the influence of the physical apparatus on the observed phenomenon is only a *special instance of physical interaction in general whose character is everywhere and always dominated by Planck's constant h;* the uncertainty principle, that is, the impossibility of punctual localization in space and time, follows logically and inescapably from the indivisibility of the atoms of action. Planck's atom of action is a universal constant, independent of any observer and relativistically invariant; hardly anybody claims that its appearance in empirical data is brought up by an intervention of the observer. If we regard action as a fundamental physical reality, then, on closer reflection, it will become obvious that the reality of sharply localized corpuscles, endowed with precise instantaneous values of momentum and energy, must be denied. We have seen that such a conclusion was reached by Schrödinger in 1934; it was, implicitly at least, formulated by Eddington a few years earlier:

> The suggestion is that an association of exact position with exact momentum can never be discovered by us *because there is no such thing* in *Nature.* [Italics in the original.] [17]

But we have seen that the association of position with momentum is practically the only remnant which survived in abstract dynamics from the original sensory and crude notion of corpuscle. If even this last remnant is questioned, is there anything at all which is left of the concept of corpuscle? A more recent statement of Philip Frank cogently criticizes the intrusion of corpuscular notions into various expositions of the principle of indeterminacy:

> Very frequently, in popular presentations which are occasionally written by scientists, the laws governing the atomic objects are formulated in a misleading way. Some authors have said that according to the contemporary laws of motion for atomic particles the position and velocity of a particle cannot be measured at the same instant. If we measure the coordinate (position), we "destroy" the possibility of measuring the momentum, and *vice versa.* This formulation is misleading because it gives the impression that before the measurement there was a "particle" that possessed both "position" and "velocity," and that the "measurement of its position" destroyed the possibility of "measuring its momentum." As a matter of fact, the atomic object itself cannot be described by the terms "position" or "veloc-

> ity." Obviously, what does not "exist" cannot be "destroyed." [18]
>
> Since the assumption that an atomic object behaves like a "real particle" is incompatible with the observable facts of atomic physics, neither position nor momentum is possessed by the object; *they cannot be measured because they do not exist.* For the same reason, the possibility of measuring them cannot be destroyed because it has never existed. [Italics added.] [19]

The only objectionable term in the passage quoted is "atomic object." The term "object" has subtle corpuscular overtones which may be misleading, if not for Frank himself, at least for his readers. The term "atomic event" is free of such misleading associations; we refer to what has been stated in Chapter XV. The extent of the contemporary crisis in physics cannot be fully grasped if we do not realize that the very concept of *substance* or *thinghood* has become questionable. The term "object" misleadingly suggests the permanence of a certain thing—more specifically, a constant conjunction of observable properties. But we already know that the alleged microphysical "objects" are not permanent; furthermore, *not even within a single instant* can they be characterized as conjunctions of positon and velocity. The atomic structure of action, in which very probably the pulsational character of time-space manifests itself, makes it impossible. Once more: for microphysical indeterminacy we do not have the disturbing effect of observation to blame, but Planck's constant of action itself.

It is always possible to hope that future physical discoveries will disprove the atomicity of action, that it will be shown that the indivisibility of action is spurious and is due exclusively to present limitations of our technique of observation. Has it not already happened several times in the history of science that what was regarded as a definite and ultimate limit of nature has been eventually shown to be only a *temporary* limit reached by our investigation? For instance, has not the allegedly indivisible atom of Dalton eventually been divided into its component parts? Can we be absolutely sure that no fractions of action smaller than *h* exist in nature?

Such an attitude, though theoretically permissible, can hardly be called fruitful, and it is extremely implausible that the expectations which it arouses will ever be fulfilled. To say nothing of the epistemological reasons which will be discussed later, the evidence for atomicity of action is both qualitatively and quantitatively incomparably more convincing than the limited empirical evidence of the last century for the alleged indivisibility of Dalton's atom. It is hardly probable that the mass of evidence accumulated by this time—frequently along independent and divergent lines of investigation—against the further divisibility of action is either a coincidence or a result of our present technical limitations. Moreover, the majority of those who hope to explain the indeterminacy principle as a result of our experimental and observational limitations do not deny the existence of the constant *h*. Apparently they do not realize their inconsistency. To recognize the existence of *h* and to deny objective indeterminacy is as little consistent as to recognize Newton's law and to deny the validity of Kepler's laws, which are its mathematical consequences.

THE CONTINGENCY OF MICROPHYSICAL EVENTS

A short glance at *radioactive disintegration*, that important group of microphysical phenomena, will show that (1) indeterminacy relations are independent of the interaction between the observed fact and the observer and (2) it is extremely doubtful whether they will ever be reduced to some classical corpuscular-kinetic model. The most paradoxical feature of natural radioactive decay is not that it is entirely independent of external conditions like temperature or chemical binding, although it is true that this feature appeared, at the time of the discovery, very surprising. It definitely indicated that radioactive transformation instead of being an ordinary chemical process, in which the atoms of various elements preserve their identity, is an *intra-*

atomic change in which the very constitution of atoms is changed. No matter how astonishing this discovery was after nearly a century of firm belief in the immutability of the chemical atom, it was comparatively easy to imagine a mechanism of the radioactive transformation without basically departing from the classical habits of thought. All that was necessary was to postulate that atoms, instead of being simple and undifferentiated units, are themselves complex; this, as we have seen, was already anticipated by Robert Boyle.[20] Then it is understandable that when one or more constituent parts are ejected from the interior of the atom, its physical and chemical properties will be modified. Even the mechanism of disintegration was easily conceivable, at least in principle; the fact that radioactive explosions occur spontaneously in the heaviest elements, that is, in the elements whose nuclei have a very large number of constituent parts, seemingly suggests that the ejection of alpha particles as well as of beta electrons is *caused* by *apparently random* fluctuations of the internal kinetic energy of other constituent particles.

In this sense the radioactive emission would not be different in principle from the process of evaporation, in which, even under a relatively low temperature, a few individual molecules escape from the main body of liquid. Indeed Isaac Newton did not imagine the mechanism of luminous emission in any radically different way; according to him, the violent agitation of aether particles ejects the corpuscles of light from the inside of material bodies.[21] In this way the correlation between heat and light was qualitatively explained; for Newton, following in this respect his fellow countryman Francis Bacon, regarded heat as motion of minute particles.

But the truly baffling feature of the radioactive phenomena was the discovery that the nuclear alpha particles are emitted from the nucleus *even when they do not have a sufficient velocity to climb over the surrounding potential barrier.* According to the laws of classical mechanics no particle can penetrate a zone in which the difference between its total energy and its potential energy would be negative. The potential barrier surrounding the nucleus represents such a "forbidden zone," where

$$E - V < 0$$

($E =$ the total energy, $V =$ potential energy of a particle). Within the classical framework this was merely a truistic statement that the kinetic energy of any particle can never become negative as this would imply the absurd consequence that its velocity would become imaginary! Yet, what was a sheer physical impossibility for classical physics is regarded only as an *exceedingly small probability* by wave mechanics. No matter how minute these probabilities are, they are realized in the radioactive emissions described above. The inadequacy of classical mechanics could not have been more clearly demonstrated. On the other hand, wave mechanics not only explains facts inconceivable within the classical framework, but leads to correct quantitative derivations of various decay constants which were previously accepted as bare empirical data.

The concept of impossibility and of causal necessity were closely correlated in classical thought, physical as well as philosophical; indeed they were complementary aspects of one and the same ontological claim. According to this claim, whatever happens *must* happen and whatever does not happen *cannot* happen; there is no middle ground between impossibility and necessity and, consequently, possibility or contingency is a mere symptom of human ignorance as Spinoza and Democritus had already stressed.[22] The substitution of the concept of small probability for that of impossibility implied the concomitant substitution of the concept of large probability for causal necessity. These two correlated substitutions measure the departure of wave mechanics from the rigid necessitarianism of classical thought. The concept of objective possibility, so resolutely rejected by classical determinism, re-emerges in contemporary physics in the form of objective probability.[23]

This is especially striking in the concrete case of the radioactive emission of alpha particles. Not only is it *possible* for a particle with insufficient kinetic energy to pass through the potential wall, but also it is *not necessary* for a particle with sufficient kinetic energy to surmount the same barrier. According to classical mechanics the particles of the first kind were *necessarily* reflected from the potential wall while the particles of the second type *necessarily* surmounted it. It is not so in wave mechanics: particles of both kinds *may be either* reflected *or transmitted* by the nuclear potential wall, although the corresponding probabilities are not equal. From this point of view radioactive explosions are regarded as *contingent events* whose irreducible chance character manifests the basic indeterminacy of microphysical occurrences.

This follows also from the way in which Heisenberg's principle of indeterminacy is applied to this particular group of phenomena. While in classical physics the surmounting or crossing of the potential barrier was a manifest absurdity, Heisenberg's principle, applied here, provides us with a certain explanatory insight, even though the paradox, for the reasons which will be stated later, does not entirely disappear. Both forms of the uncertainty principle lead to a plausible explanation of what is called the "tunnel effect." Using the first form of the principle

$$\Delta p \cdot \Delta x \geqq h$$

we see that if the uncertainty concerning the momentum is sufficiently small, the uncertainty about the position of a particle is so large that it can be located on either side of the potential wall. It can be shown that such a case occurs when the width of the potential barrier is of the same order as De Broglie wave length. But if the particle is present on the outer side of the barrier, then, according to physicists, "we should have to regard it as having successfully passed through the barrier." [24] If we apply the second form of Heisenberg's principle

$$\Delta E \cdot \Delta t \geqq h$$

then, for a sufficiently small amount of time, the uncertainty of the energy is so large that there is a finite probability that the energy E of the particle can exceed the height of the potential wall.

From either presentation it can be seen clearly that the uncertainty principle does not depend in any way on the accuracy of measurement.[25] There is no intervention of an observer in the case of radioactive explosions which occur spontaneously and independently of any extranuclear factors. The function of the human observer here is passive; it is reduced to counting particles in a certain interval of time and measuring their energy *after* their emergence from the nucleus. It is true that even this apparently passive role means an intervention in the observed physical process; thus the counting of the emitted particles is impossible without using a spinthariscope or Geiger counter, and to these observational procedures everything which Heisenberg stated in his original formulation of the principle applies. But this intervention does obviously occur *after* the event, that is, *after* the radioactive explosion has taken place. Thus the uncertainty of the radioactive disintegrations is independent of the limitations of human experimental technique; the term "indeterminacy" or "contingency" is far more appropriate and much less misleading than "uncertainty."

The phenomena of radioactivity are not the only ones which support this objectivistic interpretation of the principle of indeterminacy. The facts of radiation belong to the same category. In the terms of Bohr's model the emission of photons occurs when an electron returns from an outer orbit to an inner one; then the energy of the emitted light-quantum is equal to the difference of energy between the higher energy level and the ground-state level. Bearing in mind the inadequacy of the corpuscular model of the atom, it is more appropriate to say that the emission of light quanta occurs when the atom returns from its "excited" (i.e., energetically richer) state to its "ground" (i.e., normal) state. It is thus evident that the presence of an "excited" state is a necessary condition for the emission of light; an atom must first absorb some excess energy in order to radiate it in the form of a photon. But this absorption of energy is *not* the *cause* of the subsequent emission, though it is a required *condition* for it; a spontaneous return of the atom into its normal state with a concomitant emission of radiation has no less a statistical character than a radioactive explosion. (This analogy between luminous emission and radioactive disintegration was noticed, interestingly enough, by Einstein himself, who nevertheless remained an adversary of microphysical indeterminism.)[26]

Neither Bohr's model nor the later wave-mechanical model provides for the causal explanation of "quantum jumps" which correspond to the emission of radiation. In the days of Bohr's model physicists used to say that an electron "spontaneously" returns to its lower orbit, without specifying the cause for it. Although the total energetic balance sheet was in satisfactory agreement with the principle of conservation of energy, *the mechanism of the emission* remained entirely unclear. Wave mechanics merely brought into focus the undetermined and statistical character of luminous emission as well as of radioactive explosions. As early as in 1929 Pascual Jordan[27] spoke of the "radioactive character" of the luminous emission when he pointed out that the existence of so-called "metastable states" established the undetermined character of the emission of light. For in metastable states an atom *persists* in its excited state much longer than in most cases; while in most cases the life of excited atoms is not longer than 10^{-8} second, in the metastable states it may last 10^{-2} second or even 1 second. There seems to be no immediate necessity for the return of an excited atom to its ground level, or, if the corpuscular language of Bohr's original model is used, for the return of an electron to its original inner orbit. The analogy between the various lifetimes of radioactive nuclei and the different lifetimes of "excited atoms" is quite conspicuous, and Jordan's suspicion about the "radioactive character" of luminous emission is quite justified.

On the other hand, opponents of microphysical indeterminism can always claim that the last word has not yet been spoken and that the possibility of a deterministic explanation still persists. In the case of radioactive explosions it seems to be especially plausible to believe in the existence of hidden causes which time and again bring about the ejection of alpha particles from the nuclei. As already stated, the very fact that spontaneous radioactivity occurs only in the very complex nuclei of heavy atoms seems to justify the belief that each particular emission is caused by a momentary fluctuation of the internal kinetic energy of the constituent parts. Thus the so-called "tunnel effect" would not be, properly speaking, a tunnel effect at all; a particle would not cross the "forbidden zone" with insufficient kinetic energy, but it would *climb* over it, having, by a happy coincidence, a sufficient amount of energy for doing so. The terms used by contemporary nuclear physicists—like "heating of the nucleus" or "evaporation of nucleons from the nucleus"—are pregnant with classical kinetic and, consequently, deterministic associations. This is the basis of all hopes of locating on the subquantum level the "hidden variables" determining the quantum phenomena. From this point of view the "random fluctuations" of the energy of alpha particles, which lead either to their emission or their reflection on the potential wall, are only *apparently* random; they are analogous to the apparent irregularities of Brownian motion. As a contemporary opponent of microphysical indeterminism says:

> Hitherto in physics (as in other fields) when one had met with an irregular statistical fluctuation in the behavior of the individual members of an aggregate, one assumed that these irregular fluctuations also had causes, which were however as yet unknown, but which might in time be discovered. Thus in the case of the Brownian motion, the postulate was made that the visible irregular motions of spore particles originated in a deeper but as yet invisible level of atomic motion. Hence, *all* the factors determining the irregular changes in the Brownian motion were not assumed to exist at the level of the Brownian motion itself, but rather, most of them were assumed to exist at the level of atomic motions. Therefore, if we study the level of Brownian motion itself, we can expect to treat, in general, only the statistical regularities, but for a study of the precise details of the motion, this level will not be complete. Similarly, one might suppose that in its present state of development, the quantum theory is also not complete enough

> to treat all the precise details of the motions of individual electron, light-quanta, etc. To treat such details, we should have to go to some as yet unknown deeper level, which has the same relationship to the atomic level as the atomic level has to that of Brownian motion.[28]

What increases the plausibility of this approach is, apart from the general philosophical attitude which is unconsciously or semiconsciously present, the peculiar language of even those who support the indeterministic interpretation. We have seen that the terms "corpuscle," "energy," "velocity," and "positions" abound in texts concerning the applications of the principle of indeterminacy to radioactive phenomena, although in virtue of the same principle these terms lose their meaning. Thus we can find the following passage in one textbook of nuclear physics:

> *A particle cannot be localized more closely than its de Broglie wave length divided by 2π.* In the present case, if the barrier width a is comparable with or less than $\lambda/2\pi$, we cannot say whether a particle whose momentum is $p = h/\lambda$ is found on the right side of the barrier or on the left side. But if the particle is found on the right side of the barrier we should have to regard it as having successfully passed through the barrier.[29]

Contrary to the intention of the author, the impression of this passage is misleading in several ways. The words "we cannot say . . ." convey the impression that the uncertainty in the question is merely subjective; while the final sentence presupposes (*a*) that the particles pre-existed in the nucleus and (*b*) that it moved along a continuous trajectory through the "forbidden zone."

Assumption (*a*) is questionable. As Schrödinger observed, no particle can be observed more than once;[30] what we actually observe is two different *events* which we connect by the image of a corpuscle persisting through time. But this, as Lindemann observed, is "a mere concession to our habits, not to say infirmities."[31] It will be needless to repeat what has already been said about the nonpermanency and the eventlike character of microphysical "particles." In the particular case of radioactive emission, the particles *inside* the nucleus cannot be observed in principle without breaking the nucleus apart; its existence inside of the nucleus is only *assumed* in order to satisfy our substantialistic mode of thinking. Moreover, in the case of the emission of beta electrons, it can be *proved* that it is impossible to assume their pre-existence within the nucleus.[32] According to Niels Bohr, beta particles are *created* in the process of emission, just as photons are created in the act of their own emission. Although the pre-existence of alpha particles in the nucleus cannot be as cogently denied as that of beta electrons, it becomes extremely doubtful once we realize the general inadequacy of corpuscular models in microphysics.

Assumption (*b*) is as already shown (p. 309) plainly impossible; to postulate the spatiotemporal continuity of the trajectory through the "forbidden zone" amounts to the assertion of imaginary velocity. This does not make any sense unless we claim that we are interested only in mathematical symbolism and not in its pictorial interpretation. But this claim would mean giving up all attempts to construct any corpuscular model, and consequently also giving up the belief in the pre-existence of the particles within the nucleus. For the continuity of the trajectory guarantees, so to speak, the identity of the particle through time; we cannot have one without the other.

But opponents of microphysical indeterminism would certainly protest against such an identification of their position with the corpuscular-kinetic scheme of nature. They would probably stress the fact that in spite of the original historical association with atomism, physical determinism is a far broader concept and that its ultimate destiny is not bound up with that of the concept of corpuscle. They would probably point out that already in the last century the bonds between determinism and the corpuscular view were loosening, when field theories, which were as much deterministic as classical atomism, began to penetrate, first the domains of optics, electricity, and magnetism, and eventually the whole realm of physics.

But appearances notwithstanding, the opposition between corpuscular and field theories of matter was superficial rather than basic, and both belonged to the same classical tradition. We have seen that although the fluid theories of matter from Descartes to William Thomson apparently dissolved the contours of atoms in the universal continuity of the cosmic medium, the question whether the aether or the field itself is continuous or atomic remained open; also that by the most consistent minds the second alternative was implicitly preferred. Thus atomism reappeared on a smaller scale of magnitude as the "aether particles" replaced electrons and protons. But even when the radius of "aether particles" was reduced to zero and thus aether was made continuous in the mathematical sense of the word, the belief in the basic discontinuity of nature was rather obscured than given up. For mathematical continuity, being merely a different term for infinite divisibility, is, as such widely different thinkers as Poincaré, Bergson, Weyl, and Cassirer recognized,[33] a *disguised discontinuity;* the concept of a discrete element is retained even when its size is reduced to zero.

From this point of view we may claim that the dilemma "continuity versus atomicity" never really existed in classical physics: philosophically speaking, Descartes, Leibniz, Boscovich, Kant, and Faraday were as atomistic as Gassendi, Dalton, and Lorentz. From Pythagoras' monads and Democritus' atoms to the pointlike centers of force and the pointlike elements of field the route was long and devious; but basically it was one and the same route, which led to the view of nature as composed of mutually external units, even if these units were supposed to exist in infinite number within the smallest intervals of space and time. This concept of spatiotemporal "continuity," or if we prefer a less misleading word, of infinite divisibility of nature in space and time, is the *very basis* of classical determinism and —what is even more important—remains the basis of the neo-deterministic interpretations of contemporary microphysics.

This will become clear when we consider briefly the two most famous opponents of physical indeterminism: Albert Einstein and Louis de Broglie. The case of De Broglie is especially striking because his present deterministic position is a result of recent conversion, or rather reconversion to his original view, which he had given up twenty-five years before. Einstein, on the other hand, maintained his attitude of distrust toward contemporary physical contingentism consistently through all his life. The view of reality that both Einstein and De Broglie prefer belongs to the *Cartesian tradition* in physics and philosophy of science. L. de Broglie himself concedes it explicitly and sees the most promising feature of his theory in a return to Cartesian clarity (*clarté cartésienne*).[34] Let us recall briefly the most important features of the Cartesian tradition in physics and philosophy of science as it may be traced from Descartes to William Thomson. It is characterized by the persistent tendency to eliminate or at least to blur any sharp distinction between material bodies and the space which surrounds them. According to Descartes, matter, being identical with space, must share with space its mathematical continuity; for this reason there can be no atoms. On the other hand, space, being identified with matter, must share with matter its impenetrability; for this reason the void is impossible. This led Descartes and his followers to regard what we call "material particles" as mere complications of the all-pervading aethereal liquid. Needless to repeat what has already been said in the first part of the book and to list all the difficulties which such an ambitious enterprise faced. Let us only recall that the basic difficulty of reconciling the existence of the plenum with that of motion was never overcome by either Descartes or his followers. They were forced to concede explicitly or in a disguised way (cf. Chapter VII) the existence of the void, in an unintentional agreement with the rival atomic hypothesis.[35]

The basic reason why Descartes and his school could not succeed in their enterprise is now clear: they knew only the Euclidean form of space, which is by its own nature homogeneous and rigid, and for this reason cannot be fused with its physical content which is diversified and changing. In this respect Einstein was in a more favorable position because he had more flexible intellectual tools at his disposal. His non-Euclidean continuum, which possesses curvature different in different places and changing from one instant to another, simulates so well the properties of the diversified and changing physical content that it cannot be differentiated from it. In this sense the general theory of relativity may be regarded as a continuation of the Cartesian tendency to fuse space and physical reality into a single entity. To be sure, Einstein's general theory of relativity is Cartesian only in a very special sense; for not only are the Euclidean character of space and its homogeneity given up, but space is fused with time; thus it is more correct to speak of the *physicalization* or *dynamization* of space than of the geometrization of matter and time.[36]

There is, however, another feature of Einstein's general theory of relativity which is more distinctly Cartesian and therefore more distinctly traditional. This is the assumption of the mathematical continuity of space, time, and matter. From this follows another typically Cartesian assertion: that the spatio-temporal continuum in virtue of its infinite divisibility cannot contain any real individualities, any *indivisibilia*, in the etymological sense of the word. Einstein was too cautious to deny the existence of physical particles. He condemned Ernst Mach for opposition to atomism.[37] But from the times of his general theory of relativity he was inclined to interpret the existence of the particles as mere structural complications arising within the continuity of space-time. In other words, he regarded "particles" as mere anomalies of the continuous physical field. It is the latter which is the basic reality; its apparent discontinuity is explainable in terms of its continuous structure. This basic conviction accounts for Einstein's persistent search for the theory of the unitary field in which all the various manifestations of physical reality, including gravitation and electromagnetism, would be incorporated. It also accounts for his unshakable hope that rigorous determinism of the classical type will be eventually restored in microphysics. For if the discontinuity of quantum phenomena is only an appearance derivable from the continuous structure of the unitary field, then microphysical indeterminism is only apparent too because the main argument in its favor—the indivisibility of action—falls to the ground.

In other and more technical words, according to Einstein the whole of physical reality, including the corpuscles, should be described by the partial differential equations of the unitary field. The solutions of these equations should be free of singularities which would correspond to the *existence of the sources* within the field. Such sources would mean a disguised return to the duality corpuscles-field which Einstein resolutely rejects. The existence of the sources would mean the introduction of factors which are foreign to the structure of the field and the completeness of the description would be in danger.

The salient features of Einstein's view will stand out if we compare it with the view of Hermann Weyl. According to Einstein:

> Matter which we perceive is merely nothing but a great concentration of energy in very small regions. We may therefore regard matter as being constituted by the regions of space in which the field is extremely intense. . . . There is no place in this new kind of physics both for the field and matter *for the field is the only reality*.[38]

From this point of view, the electron represents that region of the electromagnetic field where the field intensity is incomparably higher than in its surroundings; but according to Einstein, the same field equations (though certainly more complex ones than the classical equations of Maxwell) hold *inside* the electron as well.

According to Weyl, who in this respect was probably inspired by Riemann,[39] the electrons are constituted by the *gaps* in the spatiotemporal field. For this reason it is meaningless to speak about regions *inside* the electron. Moreover, these gaps are the *sources* of the field, the sources from which the field is produced *according to statistical laws only*. Weyl thus restored the duality of matter and field. While the field represents the region of the continuous and strictly determined physical actions, matter represents a field-producing agency (*das Feld-erregende Agens*) essentially discontinuous and undetermined in its manifestations. In this way Weyl tried to do justice to the phenomena of quanta.[40]

Einstein's opposition to Weyl's view was based on a strict adherence to the idea of the all-embracing unitary field free of discontinuities as well as of singularities, all of whose changes are describable by means of partial differential equations. This is another illustration of how strict determinism and the concept of spatiotemporal continuity are practically inseparable.

The deep affinity between Einstein's theory of unitary field and various theories which try to restore determinism in microphysics by introducing "hidden parameters" is unmistakable, though it will probably appear more clearly to a future historian of scientific ideas. Even today it is not difficult to see that various "theories of hidden parameters" are species of unitary field theories, with which they share two basic assumptions: determinism and spatiotemporal continuity. In De Broglie's original interpretation of wave mechanics, particles were regarded as regions of the continuous undulatory field in which the amplitude was considerably higher than in the surrounding regions. It is significant that in 1927 Einstein encouraged De Broglie to persist in his search for a deterministic interpretation in spite of De Broglie's first unsuccessful attempt in this direction.[41]

But at that time Louis de Broglie did not follow the advice. For twenty-five years he adopted the indeterminist view of Heisenberg, Born, Bohr, and the majority of physicists. Even in recent years he still regarded his conversion to indeterminism as "final." [42] One of the reasons for his disagreement with Einstein was his correct realization that the general theory of relativity is a macroscopic theory and cannot deal adequately with the basic discontinuity of the quantum phenomena which can be disregarded on the macroscopic scale.[43] But when in 1951 David Bohm resumed De Broglie's original efforts to construct a hydrodynamical model of wave mechanics and when slightly later Jean Pierre Vigier called De Broglie's attention to the analogy between his original hypothesis in 1927 and the general theory of relativity, the situation changed radically and the time for the second conversion came.[44]

This second conversion of De Broglie was rather a reconversion to his original determinism of 1927, and philosophical motives played in it a far greater part than experimental facts. De Broglie's nostalgia for Cartesian clarity has just been mentioned. It is hardly surprising that he also returns to the idea of precise location of corpuscles and to the view that the uncertainty relations represent a *mere surface aspect* of physical reality. The corpuscles are for him, as for Einstein, *products of the field*, and their motions are in his theory guided by the undulatory field as unequivocally as the movements of "particles" (viz., local anomalies of the field) are determined in the general theory. From this point of view, the atomicity of action and the indeterminacy which it implies, instead of being an ultimate feature of reality, must be, so to speak, an *epiphenomenon* produced by hidden mechanisms whose nature does not differ basically from the mechanisms of classical physics. To believe otherwise would mean to sacrifice the two most cherished dogmas of classical thought—spatiotemporal continuity and determinism.[45]

It is loyalty to these two dogmas which makes the opposition to microphysical indeterminism so suspect. It is difficult not to see the analogy between the "theory of hidden parameters" and the *ad hoc* hypotheses which were postulated in order to explain in a classical way the negative result of Michelson's experiment; the reluctance to give up the concept of absolute space was as strong then as is now the reluctance to give up strict determinism.[46] The resistance to indeterminism of any kind is especially strengthened by the wrong assumption that *any* departure from classical determinism, no matter how slight, means the end of the possibility of *any* rational description of the world and, consequently, a suicide of reason.

We shall return to this point shortly; for the present, let us bear in mind the conclusion of our previous discussion which was only strengthened by our analysis of Einstein's and De Broglie's views: the concept of spatiotemporal continuity is *the very basis* of classical determinism and any threat to this basis is a threat to determinism itself. C. S. Peirce recognized this shortly before the beginning of the contemporary crisis of physics when he wrote:

> For the essence of the necessitarian position is that certain continuous quantities have certain exact values. Now, how can observation determine the value of such a quantity with a probable error absolutely *nil?* [47]

Today we would rephrase the last question in the following way: How can observation determine such quantities if *no sharp values* in nature exist and if the whole concept of infinite divisibility of space and time is an unwarranted extrapolation of our limited macroscopic experience?

THE INADEQUACY OF THE QUANTITATIVE VIEW OF NATURE

To speak on the microphysical level of the precise values of energy is as meaningless as to speak of well-defined positions and velocities of alleged particles. This follows from the second form of Heisenberg's principle of indeterminacy (see p. 290). But if this is so, is it still meaningful to speak of the law of the conservation of energy on the microphysical level? Evidently not. As Louis de Broglie observed as early as 1929,[48] the only way to consider energy constant is to assign it a well-defined value; the constancy of energy means always *the constancy of a certain quantity* of energy. Yet, it is precisely this quantity which, according to the second form of Heisenberg's principle, remains ill defined, i.e., retains a fuzzy edge which, though negligible on the macroscopic level, cannot be disregarded on the level of microphysical processes. The existence of this fuzzy edge prevents us from claiming absolute validity for the law of the conservation of energy, even if the traditional concept of energy is amended in the relativistic sense.

This negative consequence of the principle of indeterminacy is rarely stated explicitly. Usually *the very opposite claim* is made, viz., that the validity of the conservation laws has been verified even on the microphysical level.[49] In this respect, textbooks of atomic physics are more Spencerian in their spirit than their authors are aware of. The misunderstanding is due to our continued usage of the term "energy," even though this concept is devoid of intelligible meaning on the scale of individual subatomic events. Here is another instance of an illegitimate extrapolation of macroscopic concepts to the microcosm, or if we prefer a psychological point of view, another instance of our semantic and mental inertia which prevents us from getting rid of traditional concepts. The concept of energy is one of many traditional concepts whose realm of application does not extend below the lower limits of the zone of the middle dimensions. In Lindemann's words:

> We maintain that energy is just as much a statistical concept as temperature. It has no meaning unless averaged over a finite time, any more than temperature has a meaning unless averaged over a considerable number. In the same way that fluctuations of energy would be called fluctuations of temperature, if one attached any meaning to this term in relation to a single particle, one could refer to the fluctuations of energy in a single particle not affected by any external force. Over the time t the mean value of deviation would be h/t.[50]

In other words: to speak of a definite quantity of energy within a single subatomic occurrence is as little meaningful as to speak of the definite temperature of a single molecule of gas.

The passage quoted is interesting in one more respect; although it rejects unambiguously the adequacy of the concept of energy on the microscopic (or rather, *microchronic*) scale, at the same time it exhibits the extreme difficulty of stating this rejection in a language free of the very term which is being rejected. If we continue to use the term "energy," then the indeterminacy principle forces us to speak about "fluctuations of energy." Then we face the following dilemma: either these "fluctuations" are *caused*, and then we are back in classical determinism; or we may say, as Lindemann indicates, that microphysical energetic quantities spontaneously, that is, *causelessly*, fluctuate around certain mean values. The first alternative is favored by Einstein, De Broglie, Bohm, and Vigier; from their point of view the radioactive explosions are *determined events* whose mechanisms elude us merely because of their complexity. More specifically: alpha particles are able to leave the nucleus *only* because they possess in the moment of their ejection energy sufficient for climbing over the potential wall; and they have this sufficient energy *only* because their normally low energy is momentarily increased by intranuclear collisions. This is a corpuscular and deterministic model of the radioactive emission; there is no need to repeat the criticism of it which has already been given. But if we favor the second alternative, then we are espousing a contingentism of an almost miraculous type; for we concede the existence of *uncaused* fluctuations of energy.

There is, however, *the third way* which is apparently suggested by Lindemann himself when he shows his dissatisfaction with both alternatives considered above. The horns of our dilemma have two features in common: (1) they agree in assigning to microphysical energies *definite* values; in other words, they agree that the concept of energy remains meaningful on the subatomic level; (2) they agree that fluctuations of microphysical energies are *real changes*, real increases or decreases of energy, that is, real transitions from one definite value to another. They disagree on one important point: whether these fluctuations are caused or not. Determinists in their claim remain loyal to the spirit of Democritus, Newton, Laplace, and Spencer. Indeterminists, in opposing this claim but in retaining the definite values of microphysical quantities, implicitly or explicitly concede the possibility of absolute creation and destruction of energy. Their intellectual ancestors are few: in antiquity Lucretius with his idea of *clinamen*, in the modern era Renouvier with his assertion of the possibility of "absolute beginnings" in nature.

There are two obvious disadvantages of this kind of indeterminism: (1) it is *absolute*, that is, by accepting *creatio ex nihilo* it not only denies the Laplacian type of causality, but *any* kind of connection between the past and present events; (2) it rejects determinism while retaining its whole conceptual framework. This was especially obvious in the case of Lucretius, whose "spontaneous deviation" is clearly a foreign element incongruously grafted on his otherwise entirely mechanistic system. And the position of those who would accept causeless fluctuations of energy would not be essentially different from that of their Roman predecessor; not only would their indeterminism be equally absolute and therefore equally irrational, but it would remain equally incongruous with the remaining part of their conceptual framework. This incongruity would arise from their attempt to retain the traditional concept of energy whose constancy is at the same time being denied.

It is important to stress that such absolute indeterminism, paradoxical as it may sound, is as incompatible with the objectivist interpretation of the indeterminacy principle as is absolute determinism of the classical type. Absolute contingentism agrees with absolute determinism that there are well-defined quantities of energy on the microphysical level even if we are unable to measure them accurately. If, however, the concept of energy is inapplicable to the microcosm, there is no sense in speaking of its definite quantity whether we regard it as constant or as changing abruptly and causelessly from one definite value to another. If we admit the objective character of microphysical indeterminacy, then the idea of *creation* or *destruction* of energy *is as illegitimate as its constancy* for the simple reason that the very concept of *quantity of energy* loses its classical macroscopic meaning. We have stated the reasons which suggest that the concept of *quantity of any kind* loses its adequacy on the subatomic level.

What then would be that third way which Lindemann merely suggests without elaborating? Such a solution would be somehow intermediate between the intransigent classical determinism and the equally intransigent contingentism of absolute and miraculous kind. We concede that the words "somehow intermediate" are unsatisfactorily vague. But we can acquire a preliminary glimpse of their positive meaning by stating once more their negative significance. The fallacy common to the strict Laplacian determinism and its no less radical denial is *the quantitative view of reality*. The conviction that physical reality may be regarded as a constant substantial quantity persisting through time was the *leitmotiv* of classical thought, philosophical and scientific, as Emile Meyerson established in his works. In philosophy it inspired the search for the ultimate unchanging substratum underlying all apparent phenomenal changes; in science it found its expression in the formulation of *conservation laws:* first the law of the conservation of matter, and later the laws of the conservation of energy, of momentum, of charge. The affinity between various philosophical substances and various constant physical quantities, whether of matter or energy, is certainly undeniable; in the minds of certain naturalistically-minded philosophers like Spencer, Nietzsche, Ostwald, and Haeckel the concepts of philosophical substance and physical substance merged.

One of the philosophically most significant results of microphysics is that this view, which filled nineteenth-century minds with an almost religious emotion, is no longer valid; the correlated ideas of *quantity* and *constancy* fail at the microphysical scale. The fact that these ideas preserve their usefulness within the realm of middle dimensions does not diminish in any way the philosophical significance of their basic inadequacy. Today it is impossible to share the enthusiasm of Hippolyte Taine who, speaking of the law of conservation of energy, exclaimed: "The immutable ground of being has been attained; we have reached the permanent substance" [Le fond immuable de l'être est atteint; on a touché la substance permanente].[51] Today it would be equally impossible to make the law of the conservation of matter the cornerstone of philosophical thought, as materialists of all ages have done; nor would it be possible to repeat the magnificent attempt of Herbert Spencer to derive all more special laws of nature from that of constancy of energy. Today it would be impossible to speak hymnically about the universe as a "metallic quantity of force," as Nietzsche did in the last part of his posthumous *Wille zur Macht*.[52] The inapplicability of the concept of constant quantity to the basic elements or *rather events* of the physical world makes much of nineteenth-century thought as well as its twentieth-century prolongations obsolete. For the concept of *the quantitative constancy of nature* was always correlated with the concept of strict determinism, which in the second half of the last century radiated from physical sciences into all other areas.

There is no need to repeat what has been said about this correlation in the first part of this book.[53] Let us only briefly recall how the view of nature as a constant quantity of matter and motion had been expressed already by Lucretius, who regarded the indestructibility and uncreatability of being as the very basis of the order of nature. When twenty centuries later Herbert Spencer and Wilhelm Ostwald tried to derive the law of causality from the law of the constancy of energy, they were inspired by essentially the same idea. It is then hardly surprising that the contemporary crisis of physics equally affects the two closely correlated ideas: strict determinism and constancy of being.

We hear much less about the crisis of the conservation laws than about the crisis of determinism; but this merely shows that the prestige of the former is still great enough to make us overlook the consequences of the principle of indeterminacy. We hear even less about the crisis of the *concept of motion*, although its correlation with the two preceding concepts is beyond any doubt. We know that the classical definition of motion as a displacement of matter in space was introduced in order to preserve the principle of constancy of substance; for the only kind of change which does not threaten this principle is *change of position*. Since then the idea of the constant quantity of matter, which persists through time while only its spatial distribution is changing, has dominated—and apparently still dominates to a considerable extent—our thought. The historical connection of the concept of motion with that of determinism is as clear as their logical connection. Explicitly formulated determinism and the kinetic view of nature were born at the same time and, if we disregard the curious inconsistency of Epicurus and Lucretius, have remained conjoined since. Motions of particles were always regarded as *causally determined* motions. When modern dynamics formulated the causal laws governing the motion of particles in the form of conservation laws, it brought into focus only in another form the close connection between kinetism, determinism, and the principle of quantitative constancy of nature. Is it surprising that when doubts concerning determinism and the conservation laws appear, the kinetic model of nature is also in danger?

Chapter XV pointed out the definite, though still rather implicit and generally not yet recognized, tendency in modern physics to substitute *changes* for displacements and *events* for particles. In the context of the present chapter these trends appear in a new light. The obtrusive force with which determinism imposes itself on our mind is due to its close connection with the corpuscular-kinetic scheme, which confers its deceptive "Cartesian clarity" upon determinism itself, even in its abstract form. Is it then surprising that the rival indeterminist hypothesis can never be adequately expressed within the scheme which, historically as well as logically, has always been associated with determinism of the most uncompromising type? Any attempt to express any kind of contingentism in the terms of the kinetic model of nature is bound to fail; the corpuscular-kinetic scheme by its own nature resists any such attempt. If we persist, the result will be nothing but absurdities and incongruities, like "unsharply localized particles," "particles without definite momentum," "energy quantities with dim edges," "electrons freely choosing their future orbit" or "causeless fluctuations of energy." The new wine is poured into old bottles, with the usual results. Not only does indeterminism fail to fit the traditional conceptual framework, but when forced into it, it necessarily acquires the improbable form of *absolute* indeterminism. This is natural; indeterminism within a quantitative view of reality will always appear in the form of creation or destruction of a certain quantity of either matter or energy or momentum, that is, in the form of *creatio ex nihilo* or *reductio in nihilum*. But such absolute creations and annihilations are characteristic of *absolute*, that is, *miraculous* indeterminism which is incompatible with *any* kind of coherent universe.

If contingentism cannot be adequately expressed in the form of traditional kinetism, why not throw the old bottles away when the old wine is gone? Is there a chance of gaining a more positive insight into microphysical contingentism by trying to express it in the nonkinetic and noncorpuscular terms suggested in Chapter XV? Would it be possible thus to obtain a more rational form of contingentism which would not destroy all connection between the successive events without merging them artificially into the timeless Laplacian formula? This is the last question which we shall face in the next chapter.

NOTES

1. W. Heisenberg, *The Physical Principles of Quantum Theory*, tr. by Carl Eckart and Frank C. Hoyt (University of Chicago Press, 1930), Chaps. II, III.
2. F. A. Lindemann, *The Physical Significance of the Quantum Theory* (Oxford, Clarendon Press, 1932), p. 83.
3. *Ibid.*, p. 110.

3a. H. Margenau, *The Nature of Physical Reality* (McGraw-Hill, 1950), p. 313; E. Bauer, *L'Electromagnétisme hier et aujourd'hui* (Paris, 1949), p. 175.

4. Two notable exceptions are F. A. Lindemann and Gaston Bachelard, whose books are referred to below.
5. A. Righi, *Kometen und Elektronen*, deutsch von M. Ikle (Leipzig, 1921).
6. G. Bachelard, *Le Nouvel esprit scientifique*, pp. 140-41.
7. Lindemann, *op. cit.*, pp. 53, 68.
8. G. Bachelard, *L'Expérience de l'espace dans la physique contemporaine* (Paris, 1937), p. 74.
9. Léon Brunschwicg, "Science et la prise de conscience," *Scientia*, Vol. LV (1934), p. 334; also *La Physique du XXe siècle et la philosophie*, Actualités Scientifiques et Industrielles, No. 445 (Paris, 1936). The same view was adopted by H. Margenau (who has since abandoned it) in his early article "Causality and Modern Physics" in *The Monist*, v. 41 (1931) and by W. H. Werkmeister in his book *A Philosophy of Science* (Harper, New York, 1940), pp. 272-277.
10. Karl R. Popper, *The Logic of Scientific Discovery* (Basic Books, New York, 1959), p. 249. The characteristic passage in H. Spencer's *Prin-*

ciples of Psychology (New York, Appleton, 1897), I, p. 502, may serve as a model to our contemporary defenders of determinism: "The irregularity and apparent freedom are inevitable results of the complexity; and equally arise in the inorganic world under parallel conditions. . . . A body in space, subject to the attraction of a single body, moves in a direction that can be accurately predicted. If subject to the attractions of two bodies, its course is but approximately calculable. If subject to the attractions of three bodies, its course can be calculated with still less precision. And if on all sides of it are multitudinous bodies of various sizes at various distances, as in the middle of one of the great star clusters, its motion appears uninfluenced by any of them: it will move in one indefinable way that looks self-determined: it will seem to be *free.*"

11. Heisenberg, *op. cit.*, pp. 20 f.
12. G. Bachelard, *La Formation de l'esprit scientifique* (Paris, 1947). The subtitle is characteristic: "Contribution à une Psychanalyse de la connaissance objective."
13. Louis de Broglie, *Continu et discontinu en physique moderne* (Paris, 1941), p. 67.
14. E. Schrödinger, "Über die Unanwendbarkeit der Geometrie im Kleinen," *Naturwissenschaften*, Vol. 22 (1934), p. 519.
15. A. Einstein, "Remarks Concerning the Essays Brought Together in This Cooperative Volume" in *Albert Einstein: Philosopher-Scientist*, in particular Einstein's answer to Born, Pauli, Heitler, Bohr, and Margenau, pp. 666-670; L. de Broglie, "La physique quantique restera-t-elle indéterministe?," *Bulletin de la Société française de Philosophie*, séance du 25 avril 1953; also *Nouvelles perspectives en microphysique* (Paris, 1956), pp. 115-165; David Bohm, *Causality and Chance in Modern Physics* (London, Routledge and Kegan Paul, 1957), especially Chaps. III, IV; Jean Pierre Vigier, *Structure des micro-objets dans l'interprétation causale de la théorie des quanta* (Paris, 1956). Einstein's criticism of quantum theory and especially his identification of indeterminism with positivism and even solipsism would require a very careful and detailed analysis. A considerable part of his criticism may be accepted without agreeing with his determinism; for, contrary to his belief, contingentism is *not* equivalent to positivism or Berkeleyan idealism.
16. A. N. Whitehead, *Science and the Modern World*, pp. 172-173.
17. A. E. Eddington, *The Nature of the Physical World*, p. 225.
18. Philip Frank, *Philosophy of Science: The Link Between Science and Philosophy* (Englewood Cliffs, N.J., Prentice-Hall, 1957), p. 215.
19. P. Frank, *ibid.*, p. 230.
20. Chapter VI, above, Note 3.
21. E. T. Whittaker, *A History of the Theories of Aether and Electricity*, Vol. I, p. 21; I. Newton, *Optics*, qu. 8; B. Cohen, *Franklin and Newton*, p. 165.
22. Cf. Chapter IX, above, Note 4.
23. C. F. von Weizsäcker, *The History of Nature*, tr. by F. D. Wieck

(University of Chicago, 1949), p. 57; F. A. Lindemann, *op. cit.*, p .107.
24. Cf. Note 29 below.
25. W. Finkelburg, *Atomic Physics* (New York, McGraw-Hill, 1950), p. 236.
26. *Albert Einstein: Philosopher-Scientist*, pp. 172-73; 205.
27. P. Jordan, "Die Erfahrungsgrundlagen der Quantentheorie," *Naturwissenschaften*, Vol. XVII (1929), p. 504.
28. D. Bohm, *op. cit.*, pp. 79-80. A reserved but on the whole neutral attitude toward Bohm's view is adopted by N. R. Hanson, in *Patterns of Discovery* (Cambridge University Press, 1958), pp. 172-175.
29. R. D. Evans, *The Atomic Nucleus* (New York, McGraw-Hill, 1955), p. 61.
30. E. Schrödinger, *What is Life? and Other Scientific Essays* (Garden City, Doubleday, 1958), p. 175.
31. F. A. Lindemann, *op. cit.*, p. 88.
32. R. D. Evans, *op. cit.*, Chap. VIII, p. 276 (about the nonexistence of nuclear electrons).
33. H. Poincaré, *La Science et l'hypothèse* (Paris, 1909), p. 30: "De la célèbre formule, le continu est l'unité dans la multiplicité, la multiplicité seule subsiste, l'unité a disparu." Cf. H. Weyl, "Das Kontinuum," in *Kritische Untersuchungen über die Grundlagen der Analysis* (Leipzig, 1918): "im 'Kontinuum' der reelen Zahlen in der Tat die einzelne Elemente so isoliert stehen, wie etwa die ganze Zahlen." About the discrepancy between the intuitively given continuum and the mathematical continuum: H. Bergson, *Creative Evolution*, p. 170; E. Cassirer, *Determinism and Indeterminism in Modern Physics*, tr. by O. Th. Benfey (Yale University Press, 1956), p. 170.
34. L. de Broglie, "La physique quantique restera-t-elle indéterministe?," *Bulletin de la Société française de Philosophie*, séance du 25 avril 1953; the same author, *Nouvelles perspectives en microphysique* (Paris, 1956), p. 140.
35. Cf. Chapter VII, pp. 108-116.
36. Cf. Chapter XI, above, Note 31.
37. Cf. *Bulletin de la Société française de Philosophie*, séance du 6 avril 1922, p. 112.
38. Quoted by Louis de Broglie, *Nouvelles perspectives en microphysique*, pp. 187-88.
39. Cf. Chapter XIII, above, Note 16.
40. H. Weyl, *Was ist Materie?* p. 84; cf. also his *Philosophie der Mathematik und Naturwissenschaft* (München and Berlin, 1927), pp. 132-34.
41. L. de Broglie, *op. cit.*, p. 236.
42. L. de Broglie, "Souvenirs personnels sur les débuts de la mécanique ondulatoire," *Revue de métaphysique et de morale*, 48e année (1941), pp. 1-23; reprinted in *Physique et microphysique* (Paris, 1947), especially pp. 181-190.
43. L. de Broglie, *Matière et lumière* (Paris, 1937), pp. 234-35.
44. L. de Broglie, *Nouvelles perspectives en microphysique*, pp. 199-200.

45. L. de Broglie, *op. cit.*, pp. 220-226. D. Bohm is only consistent when he hopes that on the deeper subquantic level the alleged "atoms of energy" will be found still divisible. This is another illustration of the close connection between classical determinism and the dogma of spatiotemporal continuity. (*Op. cit.*, p. 81.)
46. Chapter X, especially pp. 146-149.
47. C. S. Peirce, "The Doctrine of Necessity Examined," *The Monist*, Vol. 2 (1892-93); reprinted in *Collected Papers of C. S. Peirce*, ed. by C. Hartshorne and P. Weiss, Vol. VI, p. 35.
48. L. de Broglie, "Déterminisme et causalité dans la physique contemporaine," *Revue de métaphysique et de morale*, année 38 (1929), p. 442: "Attribuer aux corpuscules une énergie bien determinée est la seule manière de pouvoir appliquer le principe de la conservation de l'énergie."
49. L. Goldstein, *Les théorèmes de conservation dans la théorie des chocs electroniques* (Paris, Actualités Scientifiques et Industrielles, No. 70, 1932); W. Finckelburg, *op. cit.*, pp. 279-80. Even more instructive is the case of Ernst Cassirer who while stressing the fact that the concept of energy loses its meaning on the microphysical level, still claims that the law of conservation of energy remains valid there. Cf. *Determinism and Indeterminism*, pp. 117, 191.
50. F. A. Lindemann, *op. cit.*, p. 109.
51. H. Taine, *De l'Intelligence* (Paris, 1870), I, préface. (Page 11 of the 16th edition.)
52. F. Nietzsche, *Gesammelte Werke* (München, 1926), Vol. XIX, p. 373.
53. Cf. Chapter IX, above, pp. 136-138.

CHAPTER 13

Eternal Recurrence — Once More

On at least three different occasions, Peirce insisted on the necessity to admit the cyclical, i.e. self-returning nature of time (1.274, 1.498 and 6.210). This view is altogether incompatible with one essential aspect of his philosophy, i.e. with his *tychism* which unambiguously rejected the rigorous determinism of classical science; but this, as we shall see, was one of the basic assumptions of the cyclical theory of time. The fact that Peirce did not mention it, shows that he failed to grasp the full meaning of the cyclical theory of time; and this undoubtedly explains why his own arguments in favor of this theory are so obscure and unconvincing. (I shall return to his arguments at the end of this paper.)

I analyzed the doctrine of eternal recurrence on several occasions, in particular in my article in *The Journal of Philosophy* (Vol. LVII, April 28, 1960) and in *Encyclopedia of Philosophy*, Vol. III (1967). I pointed out that the theory is based on four basic assumptions: a) that the universe is made of distinct atomic units which persist un changingly through time so that they may be identified in successive instants; b) that their number is finite; 3) that it is meaningful to speak of a definite "state" of the universe at each particular instant; d) that each particular state, embodied in a definite instantaneous configuration of self-identical units, determines unambiguously all subsequent states. In its precise modern form the theory was formulated in the nineteenth century under the impact of two closely related views – Laplacean determinism and the classical corpuscular-kinetic model of nature. Its first assumption – that of strict determinism – has *always* been present, even in the earlier forms of the theory; in truth, one reason why the Christian Fathers, such as Origen, rejected it, was because it was incompatible with human freedom. The first assumption was, however, absent in Stoicism which rejected the discontinuous structure of matter, while accepting both the cyclical theory and strict determinism.[1] The modern form of the theory described above emerged only when atomism ceased to be a mere philosophical speculation and when modern cosmology revived the interest in the theory itself. For the nebular hypothesis of Kant (1755) and Laplace (1796) implicitly raised the question of the origin of the primordial nebulae: did any of them represent a truly initial stage which came into existence by the act of divine creation or was it merely one of countless stages in unending stages of successive, but basically identical worlds? The principle of the uniformity of nature in time, proclaimed by Bruno and Spinoza, strongly favored the second answer. On the other hand, there was an embarrassing discovery of the second law of thermodynamics which apparently strongly suggested the irreversibility of the whole cosmic process. It was, however, emphasized that this law, because of its statistical character, did not exclude the reversibility of the processes on a large megacosmic scale. Such was, for instance, the view of Ludwig Boltzmann. Various hypothetical devices, some of them quite adventurous, were proposed for a "rewinding of the cosmic clock", at least on a local scale; the most popular was that of cosmic clashes by which two already cold stellar masses can be converted into another nebula which would develop into another world, in principle similar to the previous ones, even though not completely identical in all its details.

From this view there was apparently an easy step – at least psychologically – to a far more radical version of the same view: that eventually the same state of the universe, completely identical in all its details, must inevitably occur; and once it will happen, all subsequent states will follow in exactly the same order since in virtue of strict determinism no other order would be possible. The two most famous proponents of this view were such very different thinkers as Friedrich Nietzsche and Henri Poincaré. But they were not alone; about a decade before Nietzsche August Blanqui expressed this view in nearly identical words as Eudemus of Rhodos more than two thousand years before.[2] But what in Blanqui was rather a poetical vision, was for Nietzsche the conclusion based on the four premises listed above: 1) the atomistic view of the universe; b) its finiteness; c) strict classical determinism and d) the meaningfulness of the concept of instantaneous state of the world (which, prior to Einstein was universally, though tacitly, assumed). We should not be deceived by the intensely lyrical style in which Nietzsche expressed it in various places, especially in *Thus Spake Zarathustra*. Nor should we be deceived by Nietzsche's criticism of mechanism; while it is true that he rejected the notion of atom as a tiny piece of passive matter by substituting for it – under the influence of Boscovich and Faraday, – the concept of dynamic center of force, he basically retained the atomistic view of matter as consisting of discrete and persistent entities. Such entities, being finite in number, can enter only into a finite number of combinations and this necessitates an eventual reoccurrence of the same combination. This is what he says *verbatim* in *The Will to Power*:

> If the world may be thought of as a certain definite quantity of force and as a certain definite number of centres of force (*Kraftzentren*) – and every other representation remains indefinite and therefore useless – it follows that in the great dice game of existence, it must pass through a calculable number of combinations. In infinite time, every possible combination would at some time or another be realized; more: it would be realized an infinite number of times. And since between every combination and its recurrence all other possible combinations would have to take place, and each of these combinations conditions the entire sequence combinations in the same series, a circular movement of absolutely identical series is thus demonstrated: the world as a circular movement that has already repeated itself infinitely often and it plays its game *in infinitum*.[3]

The logic of this passage is obviously not basically different from the following passage of Hans Reichenbach:

> When we shuffle a deck of cards just after putting all the red cards on top of the black ones, we shall transform this ordered state into a mixture; but if we shuffle long enough, we must

> by pure chance eventually come back to the original state, because the probability of arriving at such an arrangement is larger than zero.[4]

A few years after Nietzsche Henri Poincaré proved that every mechanical system, no matter how complex it may be, provided it is made of a finite number of elements, must in a sufficiently long interval of time pass an infinite number of times through a configuration which is infinitely close to that through which it had already passed. The proof of this so-called "theorem of phases" was later simplified by Zermelo and Caratheodory. A similar consideration is applicable to the whole universe, provided, of course, that *the universe is like a pack of cards,* that is, a finite number of discrete and enduring entities, identifiable in successive moments of their existence.

This is precisely the assumption which has become very questionable in the light of the twentieth century physics. There is increasing evidence that there are no permanent dice in what Nietzsche called "the great dice game of existence"; nor are there any permanent cards. In other words, if we continue to call the microphysical entities "particles," we disregard the fact that such so-called "elementary particles" lack all essential features of the classical particles of Democritus, Gassendi and Dalton, in particular their immutability and persistence through time. To apply the term "particle" to such quasi-instantaneous entities which "endure" only for one quadrillionth or quintillionth of a second is nothing but a lazy concession to our traditional habits of thought and language. The individuality of such entities is the individuality of an event rather than that of a thing. There is no place here for what I have discussed before in several of my writings;[5] let me only say that there is overwhelming evidence that our macroscopically conditioned, object-oriented language is altogether inadequate to represent the ultimate nature of physical reality.

The concept of object is inadequate not only on the micro-physical level, but on the megacosmic scale as well. This has been pointed out very rarely, if at all, although it is a direct consequence of the negation of simultaneity of distant events. The classical view of the universe was as a set of objects of various sizes – atoms, stars and galaxies –

occupying at each particular instant different positions in what was properly called "instantaneous space." But since the arrival of relativity there is no such thing as "instantaneous space." In other words, we cannot meaningfully speak of "the state of the world at a given instant" and the very idea of megaobject as consisting of co-existing simultaneous parts loses its meaning. This notion has an approximate validity only in the realm of small spatio-temporal dimensions and small velocities, but not in the realm of fleeting galaxies, especially those whose velocities approach that of light. Thus the second basic assumption of the cyclical theory of time is without foundation.

The third essential assumption of the theory – strict classical determinism – has been seriously questioned only since 1927, if we disregard a few lonely prophets of contingentism in the second half of the last century. Again it would be otiose to repeat what has been written about the decline of classical determinism in the last fifty years; although it is always possible to search for "the hidden variables" which would restore determinism on the subquantic level, this search has been so far fruitless and the circumstantial evidence against its success is overwhelming. Thus the only assumption of the cyclical theory which remains compatible with contemporary physics, is the assumption of the finiteness of the universe; but this is hardly helpful, as the very idea of megaobject, whether finite or infinite, seems to be nothing but an illegitimate extrapolation of the concept of object formed by the pressure of our limited macroscopic experience.[6]

It is thus clear that the theory of eternal recurrence is incompatible with the present trends in physics. Furthermore, there are intrinsic difficulties within the theory itself. If recurring cosmic situations are *in all respects identical*, how can they be differentiated? The obvious answer is: by their very succession which makes them *numerically* different, even though they are otherwise completely identical. But this can mean only one thing: that the identical cosmic cycles *succeed* each other and thus they differ by successive positions in absolute time which is their container and which itself is irreversible, even though its concrete physical content is reversible. But such a distinction between container and its content underlies the Newtonian concept of time which flows irreversibly and would continue so even if the same configuration would identically occur. In other words, the all-containing time is not cyclical; changes are cyclical, not time itself.

Quite as often is the cyclical theory tied with the *relational theory of time.* According to this theory time is inseparable from its content, i.e. from changes and motions; therefore if the total cosmic situation returns, time itself returns. But if it is so, we should be consistent and we should speak only of one single cycle. For to admit the plurality or even infinity of cycles implies a non-cyclical unidirectional absolute time. Within the framework of the relational theory the concept of identical repetition is devoid of meaning – so much more an infinite number of repetitions which the doctrine of eternal recurrence requires.

For this reason Bas Van Fraassen in commenting on my early paper proposed to draw a distinction between the theory of eternal recurrence and that of cyclical time. He conceded that the former makes a covert reference to the all-embracing irreversible time in which identical cycles follow each other; while the latter does not make such a reference since it accepts only one cosmic cycle which is then adequately symbolized by a single circle. According to Van Fraassen such time is "finite, though unbounded" and quite compatible with the relational theory of time.

Now geometrical diagrams and spatial symbolisms in general are often very useful, but when applied to time, they generate endless confusions and contradictions. What is the meaning – if any – of the proposition that the world history consists of one single circle? Obviously, within the circle itself there would not be any repetition; each moment would be unique and would occur only once since its repetition would require another subsequent cycle, contrary to the original assumption. But then would it not be equally correct to represent such a finite time by a segment of a straight line instead of by a circle? Van Fraassen may protest in the following way: even though there is no repetition within a single cycle, each event within it being unique and unrepeatable, there are at least two privileged moments – the initial one and the terminal one – which do coincide. Therefore we have the right to use a circle as an adequate symbol. But how do we know that these two coinciding moments are really *two*? Can we speak meaningfully of their *twoness* without making a reference to their "different positions in time," and consequently to the irreversible time again? In other words, we are back to Newton's absolute container-like time.

What about the assertion that time may be finite and unbounded like Riemannian space? Then one might argue there should be no initial and no terminal moment since all moments would be equivalent, each of them being both initial and terminal in the same way as in an unlimited space where there are no privileged points, no boundaries at all. But even a superficial analysis will show how exceedingly misleading the analogy with Riemannian space is. If each moment in the cycle is both initial and terminal, then we face again the same alternative: either each moment is an identical repetition of the same moment in the preceding cycle – but such a cycle is excluded by the hypothesis itself which admits the existence of one cycle only; or – and this would be consistent with the relational theory of time – each moment would occur once and without repetition and then we would have only a finite series of events and it would be superfluous to symbolize it by a circle. Time can be, of course, finite (if we assume the finiteness of the total cosmic past as the philosophical finitists and some interpreters of the Big Bang theory suggest) and unlimited in the direction of the future. In truth, this is the only sense in which we can meaningfully speak of time as being both finite and unlimited; its unboundedness means that it will be forever going on and on. A time consisting of a single cycle obviously cannot go on; once the terminal moment is attained, the world history would be at its end – unless a new cycle is postulated. If one insists that every moment is both initial and terminal in the same sense as all points on a circle are, then everything would be standing still and time, change, and becoming would be completely eliminated. Zeno of Elea and modern Neo-Eleatics would be pleased – but who else?

It is perhaps not entirely fair to deal with Van Fraassen's two specific objections allegedly based on modern physics since they may not reflect his present views. He claimed that even on the basis of the theory of relativity we can still speak of the world at a given instant. In support of this view he mentioned Hans Reichenbach without restating his view very clearly. The fact is that Reichenbach replaced the term "simultaneous" by "indeterminate as to time order"; in this sense, the given event "Here-Now" is "simultaneous" with the *whole class* of events which causally do not interact with it and which constitute what Eddington called "Elsewhere" region (which equally well could be called "Elsewhen"). In Minkowski's diagram this region is wedged between two causal cones, "Absolute Past" and "Absolute Future" and any instantaneous cross-section through it remains arbitrary. More specifically, the observers in different frames of reference will divide it in different ways; each of them will have a different "instantaneous space." It is true that Reichenbach points out that they can *agree* on selecting one particular cross-section as the common "instantaneous state of the world," but this is a mere convention without any physical counterpart. We must not forget that such so-called simultaneous events are *unobservable in principle* because of the finite velocity of light. What then is the use of a conceptual entity, created by definition, i.e. by an arbitrary stipulation, and which remains intrinsically unobservable? It has a definite disadvantage of introducing into physics a term which is charged by thoroughly misleading classical associations. On this point we have an unambiguous statement of Einstein himself from the last years of his life: "There is no such thing as simultaneity of distant events."[8]

His second objection consisted in his claim that the concept of instantaneous configuration remains meaningful on the microphysical level. Even if true, this would be hardly relevant; for an identical recurrence of microscopic configuration would not guarantee a recurrence of the total megacosmic state which the cyclical theory requires. But it is *not* true; once one concedes the undiscernibility and also nonpermanence of particles (as Van Fraassen does), the very concept of configuration is devoid of meaning. Furthermore, the principle of indeterminacy in its second form $\Delta E \Delta t \geqq h$ (which perhaps I did not explicitly stress) renders a precise localization in time impossible. Thus the concept of mathematical, so to speak, "infinitely thin" instant, modeled on the concept of geometrical point, which was also suspect from a more general epistemological point of view,[9] and abandoned in psychology long ago, loses its meaning even in physics. The notion of "nature at an instant" does not have an exact physical counterpart, even though it remains a useful methodological device on the macroscopic – or rather macrochronic – scale where the duration of elementary quantum processes can be for practical purposes disregarded and replaced by "zero durations" of mathematical instants. (There is no reason to fear that the textbooks of elementary mechanics would be affected by the inapplicability of the concept of instant on both the microphysical and megacosmic scale.)

Peirce's own version of the cyclical theory of time, never systematically explained and sketched rather than formulated in a highly idiosyncratic language, is not only inconsistent with his tychism, but also riddled by intrinsic discrepancies which leave his interpreters puzzled.[10] In this sense it does not have the clarity of the classical version whose main difficulty – an oscillation between the relational and absolute theory of time – is not immediately obvious and whose empirical inadequacies appear only within the context of twentieth century physics. I can discern only one argument in favor of his theory. It is sketched in 1.274 as follows:

> The other question is whether time is infinite in duration or not. ***If it has no flaw in its continuity,*** it must, as we shall see in Chapter 4 [never written] return into itself. This may happen after a finite time, as Pythagoras is said to have supposed, or ***in infinite time***, which would be a doctrine of consistent pessimism. (Italics mine.)

This argument is merely restated in 1.498 with one complication added which shows that Peirce admitted the possibility of time being, in part at least, empty:

> What is here meant is that time has no instant from which there are more or less than two ways in which time is stretched out, whether there always be in their nature the foregoing or coming after, or not. If that be so, since every portion of time is bounded by two instants, ***there must be a connection of time ringwise.*** Events may be limited to a portion of this ring; but time itself must extend round or else there will be

> a portion of time, say future time or past time, not bounded by two instants. The justification of this view is that it extends the properties we see belong to time to the whole of time *without arbitrary exceptions not warranted by experience.* (Italics added.)

The argument is the same in the third passage where the possibility of empty intervals of time, devoid of events, is again conceded (6.210):

> You may, for example say, that all evolution began at this instant, which you may call the infinite past, and comes to a close at other instant, which you may call the infinite future. But all this is quite extrinsic to time itself. Let it be, if you please, that evolutionary time, our section of time, *is* contained between these limits. Nevertheless, it cannot be denied that time itself, unless it be discontinuous, as we have every reason to suppose it is not, stretches on beyond those limits, *infinite though they be,* returns into itself, and begins again. Your metaphysics must be shaped in accord with that. (Italics, except the first one, added.)

From the passages quoted above three facts are obvious:

1. Peirce's cyclical theory was certainly not based on the relational theory of time which does not admit the existence of empty intervals of time, completely devoid of concrete events. In its character his theory is clearly Newtonian; the only, but very important difference being that the time of Newton is linear and irreversible while the time of Peirce is symbolized by a closed curve. But as both Thomas Goudge and M. C. Murphey pointed out, Peirce was not consistent in this respect since in other places he denied the very same view upheld in the passages quoted above, insisting on the inseparability of time from concrete changes. Inconsistency is obvious; one cannot hold the absolute and relational theory at the same time.[10]

2. One peculiarity of Peirce's theory is that he was non-committal about the duration of his cycles; he does not exclude that they may

be infinitely long! Now it is difficult to see the meaningfulness of such a concept. He speaks of the beginning of the cycle "in the infinite past" and its end in the infinite future; yet, they both coincide since, as he says, "time returns into itself and begins again." Like all those who accept actual infinity, he is deceived by his own language; he apparently does not realize that the saying "the cycle began in the infinite past" is equivalent to the statement that there was no beginning; and that in saying that the cycle will end in the infinite future one can mean only one thing – that it will never end. A circle is either finite or it ceases to be a circle; already Nicolaus Cusanus knew that a circle with an infinite radius coincides with a straight line in infinity. In other words, Peirce's infinite cycle becomes indistinguishable from Newton's linear time. In this respect Peirce's cyclical theory of time is radically different from the classical version as we know it from Pythagoras to Nietzsche.

3. The only argument which Peirce adduces in favor of the cyclical theory is that it is required by – for him unquestionable – continuity of time. This is a strange argument. Certainly Newton's linear time is also continuous in the sense required by Peirce: *all* intervals of Newtonian time are also bounded by two successive instants. There is no question that mathematical continuity is not a privileged property of a circle, but of every curve, whether closed or open. How he could convince himself about the seriousness of his argument remains a real puzzle.

It is a task of Peircian scholars to determine – if possible – the exact date of Peirce's fragments about the cyclical theory of time.[11] Only then can it be decided whether the conflict between his cyclical theory of time and his cosmology with the emphasis on indetermination and spontaneity was merely a discrepancy between the beliefs which he held in different periods or whether it represented a chronic tension in his thought. (But even the second solution would not remove the discrepancies within the fragments themselves.)

In any case, this conflict does exist and it is certainly due to the transitional period in which Peirce lived. He was one of a very few thinkers such as Renouvier, Boutroux and Bergson who with a rare perspicacity, challenged the dogma of classical determinism which had been accepted nearly without exception by physicists and even by a majority of philosophers in the second half of the last century; Peirce's critical analysis of physical determinism ("Necessitarianism" as he called it) had a truly prophetic ring. Yet, this philosophical predecessor of quantum physics was a contemporary of Nietzsche and shared with him a theory which was in a sense the most extreme culmination of the rigorously deterministic and mechanistic view of the world.

NOTES

1. Samuel Sambursky, *The Physical World of Stoics*, tr. by Merton Dagut, New York, 1956, Ch. 8.

2. According to Eudemus of Rhodes: "Everything will eventually return in the self-same numerical order, and I shall converse with you staff in hand, and you will sit as you are sitting now, and so it will be in everything else, and it is reasonable to assume that time too will be the same." H. Diels and W. Kranz, *Fragmente der Vorsokratiker*, 48B34. Compare it to the words of Auguste Blanqui written in the prison: "Ce que j'écris en ce moment dans un cachot du fort Taureau je l'ai écrit et je l'écrirai pendant l'éternité sur un table, avec une plume, sous des habits, dans des circonstances toutes semblables. . .". A. Blanqui, *L. Eternité par les astres*, 1871; cité par Abel Rey, *Le retour eternel et la philosophie de la physique*, Paris, 1927, p. 7.

3. *The Will to Power, The Complete Works of Friedrich Nietzsche*, Edinburgh and London, 1913, Vol. IX, p. 430.

4. Hans Reichenbach, *The Direction of Time*, University of California Press, 1956, p. 111.

5. In particular in *The Philosophical Impact of Contemporary Physics*, Princeton, 1969, chs. XIV, XV.

6. Jean Piaget, *L'Introduction à la epistemologie génetique*, Paris 1974, pp. 204-5; M. Čapek, "The Significance of Piaget's Researches on the Psychogenesis of Atomism," *Boston Studies in the Philosophy of Science*, Vol. VIII (1971), pp. 446-455.

7. Bas C. Van Fraassen, "Čapek on Eternal Recurrence" *The Journal of Philosophy*, Vol. LIX, pp. 371-75.

8. *Albert Einstein, Philosopher-Scientist*, The Library of Living Philosophers, Vol. VIII, Evanston 1949, p. 63.

9. M. Čapek, *op. cit.*, pp. 230-243: cf. also my article "The Fiction of Instants," in *Studium generale*, Vol. 24 (171), pp. 31-43.

10. Thomas A. Goudge, *The Thought of C. S. Peirce*, Toronto, 1950, p. 244; Murray C. Murphy, *The Development of Peirce's Philosophy* (Cambridge, Mass. 1961), pp. 388-89.

11. According to the information kindly given to me by Christian J. W. Kloesel of Indiana University, the probable date of the fragments 1.498, 6.210. and 1.274 are the years 1896, 1898 and 1902 respectively. All these dates are later – the last one even a decade later – than his unambiguous rejection of classical determinism in his article "The Doctrine of Necessity Examined." Yet, "necessitarianism" is – and always has been – an essential ingredient of the cyclical theory of time. But since Peirce never returned to classical determinism, it appears that he remained unaware of this peculiar discrepancy in his own thought.

CHAPTER 14

NOTE ABOUT WHITEHEAD'S DEFINITIONS OF CO-PRESENCE*

In his *Concept of Nature* Whitehead gives the following definition of the term "co-presence":

> I call two event-particles which on some or other system of measurement are in the same instantaneous space 'co-present' event-particles. Then it is possible that *A* and *B* may be co-present, and that *A* and *C* may be co-present, but that *B* and *C* may not be co-present. For example, at some inconceivable distance from us there are events co-present with us now and also co-present with the birth of Queen Victoria. If *A* and *B* are co-present there will be some systems in which *A* precedes *B* and some in which *B* precedes *A*. Also there can be no velocity quick enough to carry a material particle from *A* to *B* or from *B* to *A*.[1]

In this paragraph Whitehead states one of the basic differences between the classical physics and the relativity theory. Classical physics accepted both absolute space and absolute time. Those events which were simultaneous in absolute sense constituted three-dimensional layers in the four-dimensional becoming; classical space was thus only a name for the class of events having the same date in the stream of the Newtonian time. The terms "space" and "the class of the events simultaneous in absolute sense" were entirely synonymous. Besides this absolute simultaneity and absolute space there were also *apparent* simultaneities and *apparent* spaces. A starry sky seen in a bright summer night was an instance of such spurious simultaneity of an apparent space; classical physics knew that in virtue of the finite velocity of light there was a difference between "seen now" and "existing now". The fixed stars which appear as luminous spots on our present sky are in truth years or even centuries older than the real stars in the depth of the absolute space. However, if the relative velocity of the stars in respect to the earth was known, it was easy to compute the real date of any celestial event which is contemporary to our present summer enjoyment; conversely, it was easy to compute the date in the history of the earth which was simultaneous with the emission of the starry light which is seen by us *now*. The possibility of adding vectorially the relative velocity of the earth and the stars with the velocity of light made it possible to find a *real* simultaneity of any couple of cosmical events no matter how far they were separated in space.

In the relativity theory the difference between "seen now" and "real now" is fully preserved with one important difference, however; the real "now" of the spatially distant events is not unambiguous as it differs for different frames of reference. This is a consequence of the principle of constant velocity of light; as

* Received July, 1956.

[1] *The Concept of Nature*, p. 177–78.

the velocity of light cannot be added vectorially to the relative velocity of the source and the observer, there is no way of establishing the objective simultaneity of distant events which would hold in all frames of reference. This means that "instantaneous spaces" are different for different observers; a couple of events lying in one instantaneous space of one observer may appear as a couple of successive events to another observer. Moreover, it may even happen that the same couple of events will appear in a reversed order of succession to certain classes of observers. This is precisely what is called the relativity of simultaneity.

But this relativity of simultaneity and succession is far from being absolute and unqualified. It can be easily shown that either from Lorentz's transformation or from Minkowski's formula for the constancy of the world-interval follows that no reversion of succession can take place for any observer if the events are causally related[2]. Moreover, the succession of causally related events can never degenerate into simultaneity in *any* frame of reference. Both these cases were not excluded by classical physics which did not forbid mechanical velocities greater than the velocity of light; for an observer moving with such velocity, the events of the place from which he started his journey would appear in a reversed order, and with a telescope powerful enough he would be able to watch even his own birth. This is precisely excluded by the relativistic principle according to which no causal action can travel with the velocity greater than the velocity of the electromagnetic vibrations, and, more specifically, that no physical body may reach this velocity.

In the light of this principle Whitehead's illustration of the relativity of simultaneity appears rather strange. According to him "at some inconceivable distance from us there are events co-present with us now and also co-present with the birth of Queen Victoria." As Queen Victoria was born in 1819 and Whitehead wrote his *Concept of Nature* in 1920, Whitehead's statement seems to be equivalent to the assertion that to some distant observers two events separated by the whole century appear simultaneous. There would be nothing fundamentally paradoxical about it if both events were causally independent; but this is precisely *not* true for the given couple of events. Queen Victoria was born in London while Whitehead wrote his book in Cambridge; their distance in space being less than their interval in time multiplied by the velocity of the fastest causal action, they are according to the relativistic definition causally related. (In a less technical language, this follows from the very fact that Whitehead *knew* about the birth of the Queen in 1819, i.e., that the information about this event reached him; this is only another way of stating that he was *influenced* by it.) But, as stated above, the succession of such events can never degenerate into simultaneity no matter what the spatial distance and the relative velocity of the observer may be.

But it is hardly credible that such a serious thinker as Whitehead would have overlooked one of the basic consequences of the special theory of relativity. It

[2] According to the relativistic definition two events are causally unrelated when their distance in space is greater than their separation in time multiplied by the velocity of light.

can be shown that his statement was misleading rather than incorrect. Its misleading character, however, depends on one incorrect assumption made unconsciously in interpreting the diagram by which the causal relatedness within the relativistic world is usually represented. The subsequent analysis intends to show how easy and psychologically almost natural it was to interpret the diagram in the way Whitehead did.

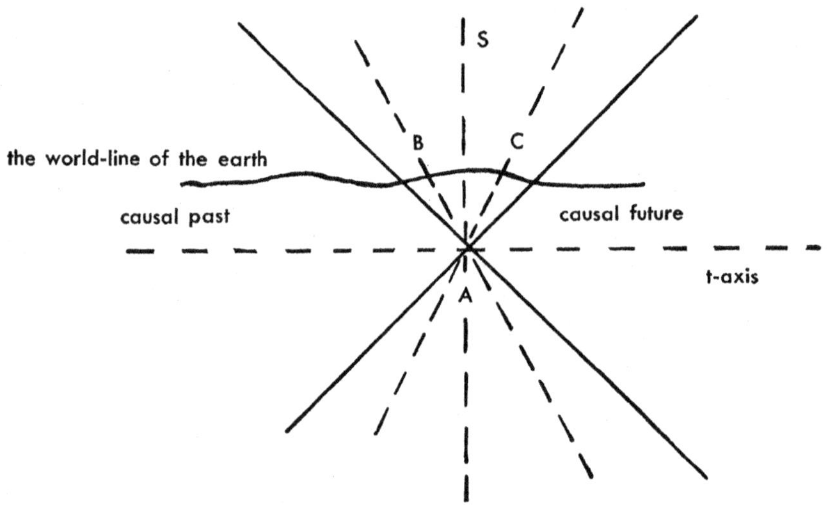

In the diagram above Whitehead's own symbolism is used. A represents "an inconceivably distant observer" (who as we shall see, may not be so 'inconceivably distant' as Whitehead claims) while B and C symbolize two successive dates in the history of the Earth or, more specifically, of England, the years 1819 and 1920. For simplicity sake the spatial distance between Cambridge and London is disregarded as it is too minute on the scale the diagram is drawn. Very approximately both successive events B and C may be regarded as being located on the same world-line. In classical physics only the perpendicular plane $\overline{AS}$ bisecting the angle between two outer surfaces of the causal cones represented the "true space" which was also the objective substrate of the truly simultaneous events. According to the special theory of relativity there are countless "instantaneous spaces" passing through the world-point A; no one has the privilege of being the "truly real" space containing the "truly simultaneous" events. The lines AB and AC in the diagram are such two instantaneous cuts across the world-becoming which are equally legitimate. But in the diagram these two instantaneous cuts intersect in the world-point (or, speaking less statically, in the *world-event*) A which apparently lies in two different instantaneous spaces; in other words, the diagram seems to suggest that the world-event A is simultaneous both with B and C, i.e., with two successive and causally related events.

But the fact is that our eye as well as our mind is deceived by the simplicity of the geometrical diagram. The true physical situation is far more complex. The deceptive simplicity of the geometrical point A hides the multiplicity of different observers whose velocities may have all possible values between zero and the velocity of light: $0 \leqq v < c$. Each of these observers will split the world-becoming along a different cleavage-plane, i.e., each of them will have a different instantaneous space containing the events simultaneous only in respect to him. It is true that according to the diagram these various observers at one particular instant have the same position in space. It is precisely the world-event A which symbolizes their spatio-temporal coincidence. But even if we grant that the frames of reference moving with different velocities coincide for an infinitesimal fraction of time (i.e., when the origins of both systems happen to be in the same point), their geometrical equivalence does *not* entail their dynamic equivalence as even a cursory inspection of Lorentz's formulae shows; their instantaneous spaces even at the instant of their geometrical coincidence do *not* coincide and the respective observers will not generally agree about the simultaneity of distant events. Using Whitehead's own example: while the present moment of one observer would be simultaneous with the year 1819, the present moment of the second observer, who moves with the fantastic velocity in respect to the first one, would be simultaneous with the year 1920. Essential is that that there are *two* distinct observers and that their corresponding present moments remain *distinct physical events*, and it is not only misleading, but plainly incorrect to designate them both by one and the same symbol in spite of their momentary geometrical coincidence. Even at the instant when their spatial distance becomes zero, we may still, using Bergson's words, speak of the "distance of their velocities" indicating thus the irreducible dynamic difference between both systems.[3]

It is to be noted also that the alleged geometrical coincidence of both observers is nothing but a convenient mathematical fiction with no concrete physical counterpart. To speak about ideal frames of reference made of three perpendicular axes intersecting in one mathematical point is possible only in abstract kinematics. In physical reality every frame of reference is made of a concrete physical body with finite dimensions. Even in the most extreme case when two frames of reference are embodied in two electrons, there cannot be a strict geometrical coincidence as their minimal distance cannot be of a smaller magnitude than 10^{-13} cm.[4] Thus not only dynamically, but even in the geometrical sense they would remain two distinct events.

It is evident that the fallacy in question is essentially of *semantic* nature; Whitehead was deceived by the apparent geometrical simplicity of the point-

[3] *Durée et Simultanéité*, p. 101.

[4] It is true that a real fusion of microphysical particles does occur in some nuclear reactions which would seem to indicate the possibility of *complete* spatial coincidence. But with the fusion of two particles into a new one their dynamical difference would disappear. We do not have to forget moreover, that the concepts of "position" and "velocity" lose their traditional meanings on the microphysical scale.

event A in the diagram used. This simplicity is only spurious as the underlying situation is more complex: there are *two* different observers with *two* different velocities and only *approximately* coinciding in space for an *excessively small* interval of time. The error is especially instructive as it was committed by a thinker belonging to the group of process-philosophers who, beginning with Bergson, were always cautious in not confusing static and abstract symbols with their concrete and dynamic designata. Everybody acquainted with the theory of relativity is aware of inestimable services which the diagram of two causal cones rendered to the better understanding of the relations between space and time in new physics. But even this diagram has its own limitations due to the fact that, like every diagram, it is *static* in its nature and refers to *one single moment* of the process; by representing the instant when two observers coincide at A, it hides their dynamical distinctness. This shortcoming may be partly eliminated by using two letters A_1 and A_2 instead of one single symbol, thus suggesting that underlying the deceptive unity of the geometrical point in the diagram is the physical duality of two momentarily coinciding frames of reference. The physical distinctness of both systems would become immediately obvious if the diagram in question were, so to speak, unfolded in time. If instead of using one single diagram representing one single phase of the process, the whole series of successive diagrams were used like in a cinematograph, the dynamical distinctness of A_1 and A_2 would become even visually obvious as both observers would be spatially far apart already in the next second and even our eye would be able to differentiate them. In our situation it was obviously a case of the correctly informed mind misguided by the eye; a warning example how even the greatest intellect may be misled by the fallacious simplicity of the visual diagram.

Two years later Whitehead significantly modified his definition of co-presence in the following way: while in *The Concept of Nature* it was defined as a dyadic relation holding between *two* events in one instantaneous space, in his *Principle of Relativity* (p. 30) it is defined as a relation between the event A *and the whole four-dimensional region separating the causal past and the causal future of the same event.* This "region of co-presence" is symbolized in our diagram by the area between two causal cones. This wedge-shaped region may be defined as *the class of the events causally unrelated to the world-event A.* This follows from the fact that no causal action can be faster than the velocity of the light-signals; or in the language of the diagram, that no world-line may be inclined to the t-axis at an angle greater than the world-lines of light by which the surfaces of both causal cones are constituted. Whitehead was aware of this when in the quotation above stated that "no velocity is quick enough to carry a material particle[5] from A to B or from B to A". The events in the wedge-shaped area can neither influence the event A, nor be influenced by it. All causal ancestors as well as successors of the event A must be looked for *within* the causal cones.

If we accept this modified definition of co-presence, then there is nothing objectionable in saying that two successive events located on one and the same

[5] He should have added: "including the 'light-particles' or photons . . ."

world-line are for some very distant observers included in the region "co-present" with them. Let us consider once more Whitehead's own example of two earthly events separated by the interval of the whole century. An observer with whom these two events are co-present does not have to be "inconceivably far" away from the earth; the distance of Polaris, which is fifty light-years away from us would be sufficient. For the position of Polaris the temporal thickness of the wedge-shaped area corresponding to the world-line of the earth is equal to one hundred years. The point A in our diagram would then correspond roughly to the position of Polaris. But again we have to be careful in not using the misleading language used by Whitehead in his *Concept of Nature*. Even in his *Principle of Relativity* when he speaks about the region of co-presence, instead of defining it by the relation of causal independence in respect to the world-event A, he calls attention to the fact that the region is constituted by the class of all possible "moments", i.e., instantaneous spaces, passing through the world-event A. This is basically correct, but unless we are on guard, we may slip into the same fallacy as before. The diagram seems to suggest that all "moments" intersect at the event A, which would mean that one and the same event lies in an infinite number of different instantaneous spaces and is thus simultaneous with an infinite number of successive and causally related events. It would be needless to repeat the criticism already given. The true meaning of co-presence has already been stated: using Whitehead's example, when we say that one event on Polaris is co-present with the whole century of the earthly history, it means that it is causally unrelated to the whole series of the earthly happenings which fill up the interval between the year 1819 and 1920. According to classical physics there is always one privileged moment of this earthly history which is objectively simultaneous with one particular event on Polaris. According to new physics it is not so; various observers at Polaris moving with different velocities would each of them claim a different earthly date as being simultaneous in respect to them; the interval of one century will measure the widest possible divergence in their computations. (This, of course, does *not* mean that *one and the same observer* may claim to be simultaneous with *two* successive events on the earth.) Classical simultaneity of spatially separated events is gone and is replaced by their causal independence.

If it is so, is it not misleading to call the wedge-shaped area "the region of co-presence"? It is easy for us to think about the co-presence of instants or co-presence of temporal intervals; but it is contrary to our habits to think or even to speak about an event being co-present with the whole interval of time or *vice versa.* If we say that the Declaration of Independence was co-present with the year 1776, we would be accused of using a fuzzy language and we would be told that this event, though occurring *within* the year 1776, is truly co-present with one single date only: July 4th 1776. It is obvious that the word *co-present* suggests almost automatically the classical notion of simultaneity and we have a natural tendency to apply this notion even to the realm where it is not legitimate any longer, that is to the realm of cosmic distances and large velocities. This is hardly surprising; our language is so thoroughly moulded by our limited physical environment within which the pre-relativistic physics holds, that it remains utterly inadequate for describing the structure of the widened experience on which the new physics is based. Eddington's embarrassment, as reflected in the following passage, is precisely due to this linguistic problem:

> It suggests itself that the neutral wedge might be called the Absolute Present; but I do not think that it is a good nomenclature. It is much better described as Absolute Elsewhere. We have abolished the Now-lines, and in the absolute world the present (Now) is restricted to Here-Now. (*The Nature of the Physical World*, p. 49)

Why the term "Absolute Present" would be unsatisfactory? Because it would apply to the events which are successive for *every* conceivable observer like the segment *BC* of the world-line of the earth. But if the term "Present" is not satisfactory, the composite words which contain it, like for example the word "co-presence", would be hardly more appropriate. It is significant that Whitehead in his latter writings preferred the term *contemporary* to *co-present*. At the same time he stressed the corresponding causal unrelatedness more explicitly.[6] It is always dangerous to pour new wine into old vessels; the old words, in spite of new definitions associated with them, never lose entirely their old connotations or they retain at least a tinge of their traditional meaning which seriously interferes with understanding of the significance we intend to convey by them. The term "independent contemporary" is relatively the best expression we can find in our limited language for characterizing in an abbreviated fashion the relation, or rather a *lack of relation*, between the world-lines *not yet interacting*. The word "contemporary", unlike the word *co-present*, suggests less the outdated idea of classical simultaneity or *co-instantaneity*. The word itself is, of course, powerless to convey the adequate meaning of the relativity theory without a thorough understanding of Lorentz's and Minkowski's formulae.

On the other hand it would be a mistake to believe that a complete mathematical mastery of the relativity theory means necessarily a full understanding of its physical and philosophical meaning. Numerous examples can be given showing how within one and the same mind the true grasp of the mathematical side of the theory may co-exist with some serious misapprehensions of its physical and especially philosophical meaning. Let us only remember all popular misinterpretations concerning the "fourth dimension" and its identification with time. This is psychologically understandable for two reasons: 1) the mental habits characterizing our Newtonian thinking, whose roots lie deep in the phylogenetic heritage of man, are too obstinate to be modified except by a conscious and sustained effort; 2) this effort is not to be found in a bare mastery of the mathematical formalism which modifies little, if at all, our intellectual automatisms to which we referred. Ernest Mach once wrote that in performing mathematical operations he sometimes had an uncomfortable feeling that his pencil was thinking instead of his mind. He thus merely referred to the well-known fact that even the highest and most complex mathematical operations may be performed almost mechanically and effortlessly once their basic rules

[6] *Process and Reality*, p. 486–88; *Adventures of Ideas*, p. 251.

are known and automatized by repetition and habit. From a purely mathematical point of view Lorentz's equations are hardly anything more than an elementary algebra; all mathematical consequences can be derived from them without effort by a simple mechanical application of the familiar algebraic rules. The real difficulty begins when we try to grasp the *physical* meaning of these mathematical consequences and to express it in our unreformed Newtonian and Euclidian language. Then we shall find out that our own pencil will *not* perform this task for us as the required radical revision of the traditional forms of thought cannot be realized without sustained and vigorous effort. For this reason the choice of more adequate words and, if necessary, the creation of new terms, is of such fundamental importance. The previous discussion of Whitehead's word "co-presence" showed it, I hope, sufficiently.

CHAPTER 15

BERGSON AND LOUIS DE BROGLIE

It is true that all previous considerations presuppose that microphysical indeterminacy has an objective status and therefore is not a mere result of temporary technological limitations of our present measurements. I discussed this particular problem at length in a chapter entitled 'The End of the Laplacean Illusion'[1] in my previous book in which I listed all the facts supporting the objective status of indeterminacy: not only the general bankruptcy of *all* the ideas constituting the classical deterministic model of the physical reality, but also the peculiar character of radioactive explosions, whose statistical character and indeterminacy cannot by their very nature depend on the intervention of the observer. I also pointed out that this character is not confined to radioactive processes only. The emissions of photons have essentially the same 'radioactive' character, and this is true of spontaneous disintegrations of all recently discovered "particles" as well. I also pointed out that resistance to the concept of the objective contingency of microphysical events is due mainly to the tenacity of certain classical beliefs, including the belief in the Laplacean-Spinozist concept of causality, which is still wrongly regarded as the *only* type of rational order.

In this context I shall confine myself to a brief discussion of the philosophical thought of Louis de Broglie, whose discovery of the undulatory nature of matter had as revolutionary a significance for the development of physics as did the early discovery by Planck and Einstein of the corpuscular character of light. I am choosing de Broglie not only because he shows for a physicist a rather uncommon knowledge of Bergson's work, but also because his successive hesitations between a subjectivist and objectivist interpretation of quantum indeterminacy exemplify in an abbreviated form the present controversy about it.

The philosophy or, more particularly, the philosophy of science of Louis de Broglie can be gathered from a number of essays contained in five of his books: *La physique nouvelle et les quanta* (1937), *Matière et lumière* (1937), *Continu et discontinu en physique moderne* (1941), *Physique et microphysique* (1947) and *Nouvelles perspectives en microphysique* (1956). With the exception of the last book, he consistently maintained the indeterministic interpretation of microphysics, not hesitating also to use philosophical and epistemological arguments against the plausibility of any return to classical deterministic models. In the first four books mentioned above he consistently maintained that the principle of indeterminacy of Heisenberg is an inevitable consequence of the indivisibility of the atom of action h, and that the very existence of this constant makes it impossible to speak of the precise localization of physical entities in space and time.[2] This is what makes both the concepts of the classical particle as well as of the classical wave inapplicable on the microphysical level. Thus the extension of the classical principle of determinism to microphysical phenomena is nothing but an illicit extrapolation of macroscopic experience beyond its legitimate limits.

He also pointed out that his own interpretation, though incompatible with rigorous Laplacean causality, nevertheless remains compatible with a *widened* or generalized notion of causality which would still preserve some continuity or connection between successive events. In this type of connection the antecedent "cause" would be a *necessary, though not a sufficient, condition of the* "*effect*". Such causal continuity is describable, *not* by the classical dictum "*causa aequat effectum*", but by its emended version, *sublata causa tollitur effectus.*[3]

His view, then, was fairly close to Reichenbach's view of "ambiguous causal nexuses" and to the prevailing tendency among physicists to regard each event as merely *probably* implied by its antecedents instead of being deducible from them in strict Laplacean fashion. This makes it understandable that in this period he regarded his original commitment to determinism as a result of mere habit rather than of conviction. He firmly rejected the hypothesis of "hidden variables" by means of which determinism could be restored on the sub-quantum level. He concluded that "the laws of mechanics, with their apparent rigour, are nothing but a macroscopic illusion due to the complexity of the objects on which our direct experience bears and not to a lack of precision of our measurements".[4]

These views of Louis de Broglie showed themselves in his attitude toward the three twentieth century French thinkers who were concerned about the problem of physical determinism, even though they approached it from different angles and arrived at different conclusions: Émile Meyerson, Léon Brunschvicg and Henri Bergson. Louis de Broglie dealt with each of them in separate essays and expressed, with his habitual tact and gentleness, his critical reservations with respect to each of them. Yet, when we re-read these essays now, we see clearly that his maximum disagreement was with Brunschvicg, who in various places stated his firm belief that microphysical indeterminacy is due exclusively to the disturbing intervention of the observer in the phenomenon observed.[5] As far as Meyerson was concerned, he merely observed that, had he had more time to study the revolutionary implications of modern physics, he might have modified some of his views. (He probably meant Meyerson's main thesis about the immutability of human reason in the sense that the search for explanation in science will always be synonymous with the search for identification in space and time.)[6] As far as Bergson's view of physical indeterminacy was concerned, we already mentioned de Broglie's approval and admiration of the passage quoted above, according to which macroscopic determinism is due to our *macrochronic* perspective, that is, to the fact that our mnemic span is incomparably wider than the temporal span of the elementary physical events. After quoting it, Louis de Broglie concludes his essay with the following remark:

> A curious suggestion according to which living beings would necessarily have a 'macroscopic' perception, for in the macroscopic only there prevails the apparent determinism which renders possible their action on things. In reading this isolated text, how much one regrets that the great philosopher was unable to survey with his piercing glance the unforeseen horizons of the new physics![7]

Several pages before this conclusion de Broglie observed that:

> If Bergson had been able to study the quantum theories in detail, he would undoubtedly have joyfully stated that in the picture they offer us of the evolution of the physical world, they show at each instant nature hesitating between several possibilities, and without doubt he would have repeated, as in his last work, that "time is this hesitation itself or it is nothing."[8]

In the concluding footnote de Broglie calls attention to the notion of "weakened causality" (we would prefer to call it 'generalized causality'), formulated by Bergson in his very first book. According to this notion, "the effect will no longer be given in the cause. It will reside there only in a state of possibility...." After quoting this Louis de Broglie concludes:

"The analogy with the probabilistic concept of causal relation of the present quantum physics is obvious."[9] The agreement between Bergson and Louis de Broglie on this point could hardly have been more complete.

Another interesting rapprochement which Louis de Broglie makes in this essay is that between Bergson's analysis of Zeno's paradoxes and the Heisenberg principle. Already in his book *Matière et lumière* in the essay "Les idées nouvelles introduites par la mecanique quantique" de Broglie pointed out that the impossibility of correlating a precise spatio-temporal location together with a definite dynamic state of a moving body has a certain affinity with Zeno's famous paradoxes of motion. The essence of Zeno's argument is that if a 'flying' arrow really *is* at each particular instant at a certain place, that is, if it really *coincides* with its position, then it cannot be moving, it cannot fly at all. De Broglie pointed out that the traditional, and still current, answer to this paradox is that based on the concept of spatio-temporal continuity. Consideration of the successive positions infinitely close to each other makes it possible to define the instantaneous velocity at each point in the same way as we determine the derivation of the usual continuous functions at each of its points. But the basic assumption of this rebuttal of Zeno, that of the spatio-temporal continuity, is now threatened by the discovery of the constant h – the atom of action – which introduces an element of discontinuity absent in the classical theories:

Without wishing, then, to make Zeno a precursor of Heisenberg, and without forgetting the part played in this problem at the present day by the finite value of Planck's constant, we may still say that the impossibility, which recent theories reveal, of assigning simultaneously to a moving body an exact spatio-temporal localization and also a completely defined dynamic state, appears to have some kinship with a philosophical difficulty which has long been familiar. To use Bohr's terminology, the exact location in Space and Time is one "idealization," and the concept of a completely defined state of motion is another, so that the two "complementary idealizations," while almost quite compatible on the macroscopic scale, are not strictly so on the microscopic scale.[10]

It was nearly a decade later when Louis de Broglie brought forth this reflection on Zeno and Heisenberg's principle in relation to Bergson's criticism of Zeno. Both Bergson and Zeno believed that motion cannot be built of immobilities. While Zeno concluded from this that motion does not exist, Bergson concluded that motion – and, more generally, any change – is real and therefore does *not* consist of immobilities.

A moving body *is* as little at any particular point of its trajectory as, according to James, the stream of thought *is* at any particular idea or as, according to Koffka, a melody *is* at any particular tone. Louis de Broglie then underlines the fact that forty years before the formulation of the indeterminacy principle Bergson insisted on the impossibility of motion *being* at any static point of its trajectory. In truth, borrowing the language of the quantum theories, Bergson could have said: "If one seeks to localize the moving object in a point of space, one will obtain nothing but a position, and the state of motion will completely escape." De Broglie then added that, if Bergson erred, it was by excess of caution.[11] He still spoke of a continuous trajectory, although such a concept loses its definiteness on the microphysical level. (In fairness to Bergson, however, it must be pointed out that not only were the instances in which he referred to trajectories macroscopic in nature, but also that he regarded spatial displacement as a mere outward manifestation of qualitative becoming. See Chapter 9 of this part.)

We know today that Louis de Broglie does not uphold the objectivity of microphysical indeterminacy any longer. In the session of the French Philosophical Society on April 25, 1953, he announced his return to *basically* the same philosophical position which he upheld prior to 1928.[12] For twenty-five years he defended the indeterministic position of Heisenberg, Born, Bohr and the majority of physicists. In truth, in his article, 'Souvenirs personnels sur les débuts de la mécanique ondulatoire', the concluding paragraph describes his "final conversion" to the probabilistic interpretation of microphysical phenomena.[13] It was not as 'final' as he believed when he wrote it in 1947; for his reconversion to determinism came only five years later.

Our main concern in the present context is the role of a *philosophical* motive in Louis de Broglie's reconversion to determinism. That philosophical motives played the main and almost exclusive role in it, was conceded by de Broglie himself. Certainly no new discovery in physics induced him to abandon the view which he held for a quarter of a century. His two main reasons were both of a philosophic nature: first, distrust of the phenomenalistic and even idealistic tendencies present in the orthodox, probabilistic interpretation of quantum physics; second, nostalgia for "*clarté cartésienne*" of the classical, deterministic models.[14]

It is true that de Broglie's reconsideration of his own position was triggered by two external factors: the articles of David Bohm in *The Physical Review* in 1952 and the theoretical work of Jean-Pierre Vigier. But it would hardly be appropriate to call these factors 'external'. David Bohm resumed de Broglie's original efforts in 1927 to construct a quasi-hydrodynamical model of wave-mechanical phenomena while J. P. Vigier called de Broglie's attention to the affinity between his original hypothesis of 1927 and the general theory of relativity.[15] It is precisely this affinity which sheds light on both Einstein's and de Broglie's negative attitude toward physical indeterminism as well as on the Cartesian character of their thought.[16]

Needless to say that this character is 'Cartesian' only in a very broad and sophisticated sense which can be conspicuous only to a historian of ideas acquainted with both the history of physics and philosophy. What is common to both Einstein's unitary field theory and de Broglie's "hypothesis of hidden parameters" is, besides the firm belief in rigorous determinism, also the idea that the existence of corpuscles and, more generally, the grain-like structure of the fields, should be explained by the structure of the corresponding field itself.[17]

In this respect, this view is akin to the plenist or fluid theories of matter which from Descartes and Hobbes to William Thomson regarded the particles as mere structural complications of the all-pervading aether. It is true that the concrete idea of the aetherial medium was gradually replaced by a more abstract idea of the field which in the general theory of relativity was identified with a still less intuitive concept of the non-Euclidian continuum – with locally variable curvature. Correspondingly, the concrete aetherial "vortex-atoms" of William Thomson were finally replaced by the singularities of the field or the local "humps" of the non-Euclidian continuum. The motion of the "particles" in de Broglie's theory was guided by their dynamic surroundings (i.e. the so-called "pilot waves") as fully as the motion of the Cliffordian "humps" is by their non-Euclidian medium or the motion of the planets by the rotation of the aether in the original, naive Cartesian model.

It is interesting that de Broglie, who prior to 1953 stressed the *macroscopic* character of the general theory of relativity (and of the field theories in general) which cannot do justice to the quantum phenomena, reversed his position: instead of regarding the continuity of fields as a macroscopic appearance resulting from basic discontinuities, he now regards the spatio-temporal continuity of fields as *basic* and the discontinuity of particles, including photons, as mere local irregularities in the underlying continuum. In this respect both Einstein's unitary field theory and de Broglie's hypothesis of "hidden parameters" are Cartesian in a broad sense.

In fairness to Louis de Broglie it must be stressed that the term 'reconversion to determinism' is perhaps too strong. While not hiding his preference for a deterministic re-interpretation of the quantum phenomena, he is aware of the great difficulties of its concrete and convincing elaboration. All that he asks is an open-minded attitude toward exploring this possibility.[18] This cautious attitude is even more evident in the thought of David Bohm. Although he considers the possibility of re-establishing physical determinism on the sub-quantum level, he explicitly rejects any kind of Laplacean determinism which would regard the postulated, causal, sub-microscopic levels as "the ultimate bottom of nature". This aspect of Bohm's thought, which he himself calls "the principle of qualitative infinity of nature", is not always stressed in cursory references to his criticism of the "orthodox" indeterministic interpretation.[19]

It is interesting and even amusing to contrast the cautious attitude of de Broglie and Bohm with the enthusiastic welcome which they received from classically oriented rationalistic philosophers. Nothing shows their *a priori* commitment to determinism more than the jubilation with which they hailed de Broglie's "reconversion" to determinism at the meeting of *Société française de philosophie* referred to above. Thus M. Salzi praised Louis de Broglie for accepting the Cartesian view, according to which nature, instead of containing a surd irrational element, as Plato erroneously believed, is completely transparent to rationality. That he means by 'rationality' the narrow, Spinozist-Laplacean type of rationality is clear from his illustrations borrowed from classical physics: the kinetic theory of gases shows that statistical laws are always reducible to hidden causal processes. His concluding words have a ring both naive and bombastic:

> Such, Mr. Louis de Broglie, are some of the reasons why philosophers, no less than scientists, rejoice to learn that you are closing your ears to the song of the Sirens of the probabilistic contingentism. Your researches reacquired the fruitfulness by which you made the country of Descartes glorious.[20]

It is the same deterministic attitude as that of Léon Brunschvicg in 1940, mentioned above. Understandably it was the attitude of one of Bergson's severe critics, René Berthelot in 1934.[21] More surprisingly, it was also the attitude of Edouard Le Roy in 1935 in spite of his Bergsonism.[22] It was also the attitude of Bertrand Russell, at least in his book *Scientific Outlook*, when he wrote: "The principle of Indeterminacy has to do with measurement, not with causation... There is nothing whatever in the Principle of Indeterminacy to show that any physical event is uncaused." It is then only natural that Brand Blanshard, whose commitment to rigorous determinism approaches a religious fervor, approvingly quotes this passage from Russell.[23] There would hardly be any quarrel with this quotation, provided that the term 'causation' were not understood in a narrow, Laplacean sense. This is what the traditional rationalists are constitutionally unable to do. Neither Boutroux nor Peirce nor Whitehead nor Bergson ever claimed that any event is "uncaused" as Russell insinuates. All they claim is that it is *not* equivalent with, and reducible to, its antecedents.

It is also unfortunately true that the current, indeterministic interpretation is most frequently associated with, and justified by phenomenalistic and even idealistic tendencies. On this point Louis de Broglie's reminder that a physicist, as Émile Meyerson stressed, is always a realist is fully justified.[24]

As I tried to show in my previous book, it is the usage of phenomenalistic language which, in the official interpretation, makes the denial of determinism so confusingly ambiguous. If indeterminacy is an *objective* feature of nature, it must be understood and stressed as such, that is, without exaggerated emphasis on the mind of the observer and the 'disturbing' effect of the act of observation. Very few philosophies offer an appropriate framework for such an objectivist interpretation: the contingentism of Boutroux, tychism of Peirce, Whitehead's ideas of the "self-creativity of actual occasions" are among them. Needless to say that Bergson's emphasis on the elementary indetermination of physical processes belongs with this group.

Finally, one interesting concluding footnote. We said that Einstein also – like Lorentz, Planck, Schrödinger and de Broglie – retained the hope that microphysical indeterminism is not the last word of physics. It is interesting, however, that in his preface to the English translation of *Physique et Microphysique* he wrote:

What impressed me most, however, is the sincere presentation of the struggle for a logical concept of the basis of physics which finally led de Broglie to the firm conviction that all elementary processes are of statistical nature. I found the consideration of Bergson's and Zeno's philosophy from the point of view of the newly acquired concepts highly fascinating.[25]

This was written a few months before Einstein's death. It is significant that in spite of the brevity of the preface, he did not fail to include the passage quoted above. This passage constitutes nearly a half of the whole preface! It hardly seems to be a mere conventional politeness. Could it be that the great physicist in encountering Bergson's criticism of Zeno in the light of Louis de Broglie's comment began to have some second thoughts about his life-long commitment to classical determinism?

NOTES

1 *The Philosophical Impact of Contemporary Physics*, Ch. XVI.

2 L. de Broglie, *Continu et discontinu en physique moderne* (Paris, 1940), p. 61.

3 *Ibid.*, p. 64.

4 L. de Broglie, *Physics and Microphysics*, pp. 199–200.

5 L. de Broglie, 'Léon Brunschwicg et l'évolution des sciences', *Revue de métaphysique et de morale* **50** (1945) 72–76. Here the author gently, but firmly rejected the deterministic interpretation of the Heisenberg's uncertainty relations in Brunschwicg's book *La physique nouvelle et la philosophie*, Hermann, Paris, 1936.

6 Cf. 'To the Memory of Émile Meyerson' in de Broglie's book *Matter and Light. New Physics* (transl. by W. H. Johnston), Dover, New York, 1939, p. 286.

7 *Physics and Microphysics*, p. 193.

8 *Ibid.*, pp. 191–192.

9 The quotation from Bergson is from *T.F.W.*, p. 212. The footnote of de Broglie is not translated in Davidson's translation: its full text is in the French original version *Physique et microphysique*, Paris, 1947, p. 211.

10 *Matter and Light*, p. 255.

11 *Physics and Microphysics*, pp. 189–200.

12 *Bulletin de la Société française de philosophie* (séance du 25 avril 1953). Cf. also *Nouvelles perspective en microphysique*, Paris, 1956, pp. 199–201.

13 *Physique et microphysique*, pp. 181–190. (In the English translation the term 'conversion finale' is translated rather obscurely as 'final transformation', p. 156).

14 *Nouvelles perspectives en microphysique*, p. 140.

15 *Ibid.*, p. 116.

16 Cf. *La physique quantique restera-t-elle indeterministe*?, Gauthiers-Villars, Paris, 1953, esp. pp. 65–111 with several contributions of J.-P. Vigier. On Einstein's satisfaction with which he reacted to the attempts ro revive the deterministic interpretation of quantum physics cf. *Nouvelles perspectives en microphysique*, pp. 200–201.

17 Cf. Note 6 of Chapter 11 of this part; also *The Philosophical Impact of Contemporary Physics*, pp. 316–322.

18 *Nouvelles perspectives en microphysique*, pp. 142–143.

[19] D. Bohm, *Causality and Chance in Modern Physics*, Harper Torchbook, 1961, pp. 158–160.
[20] Cf. Salzi's intervention in the discussion following de Broglie's lecture in *Bulletin de la Société française de philosophie*. See Note 12 of this chapter.
[21] *Bulletin de la Société française de philosophie* **34** No. 5 (octobre-december 1934), 172–183.
[22] Edouard le Roy in his article 'Ce que la microphysique apporte et suggère a la philosophie', *Revue de métaphysique et de morale* **42** (1935), 345–347 rejects – like Brunschwicg – any "reification" of quantum indetermination.
[23] Brand Blanshard, *The Nature of Thought*, George Allen & Unwin, London, 1948, II, p. 493. Russell's attitude to this problem was far from consistent; while in the passage quoted by Blanshard he claimed the principle of indeterminacy has to do with measurement not with causation (*Scientific Outlook*, 109–110), he was far more cautious in *Analysis of Matter* where he conceded that "there may be the limits to physical determinism" and that this possibility 'interposes a veto on materialistic dogmatism" (p. 393). In *Human Knowledge*, Simon and Schuster, New York, 1962, he explicitly concedes the objective character of microphysical uncertainty, but belittles its significance since it, according to him, does not affect the macrophysical determinism. (Cf. pp. 23–24.)
[24] L. de Broglie, *op. cit.* in Note 18, p. 141, J. Ullmo, *La crise de la physique quantique*, Paris, 1955, accuses Heisenberg of "pure subjectivism" (p. 36) and the language of certain passages of Heisenberg's *Physics and Philosophy*, New York, Harper & Brothers 1958, easily yields to this interpretation (see in particular pp. 133, 144.)
[25] L. de Broglie, *Physics and Microphysics*, p. 7.

CHAPTER 16

WHAT IS LIVING AND WHAT IS DEAD IN THE BERGSONIAN CRITIQUE OF RELATIVITY

If we do not count *The Mind Energy* (L'Energie spirituelle) which was a mere collection of lectures and articles previously published, *Duration and Simultaneity*, published in 1922, is the fifth book of Bergson, – the first one published after *Creative Evolution*. The second edition in 1923, with a new preface and three appendices, was incorporated in the texts published by André Robinet in 1968 under the title *Mélanges*. Henri Gouhier in his introduction to this book stated the reasons why *Duration and Simultaneity* was included in it because Bergson's attitude toward the relativity theory remained unchanged; in other words, that the book remains an authentic part of his thought. This becomes quite clear when we re-read a long footnote on this subject in his last book *La pensée et le mouvant.*[1] Bergson at the end of his life, in a conversation with Edouard Le Roy, expressed some doubts about a new re-printing of his *D.S.* since he – according to the same witness – viewed it as incomplete rather than erroneous because in it the general theory of relativity had been hardly dealt with.[2]

Thus, whatever one may think about the value of his book, it must be viewed as a real part of his whole work. But, as I am going to show, it contains some views which are incompatible not only with the spirit, but even with the letter of his previous writings, in particular, with *Matter and Memory* and *Creative Evolution*. What is even more significant, it is that those views represent what is precisely dead in his criticism of relativity. On the other hand, what is truly living in it is implied in Bergson's basic philosophical theses. In other words, his views on relativity should be neither wholly accepted nor rejected *en bloc*. The extreme attitudes are always risky and never fair; the truth is always more complex and too *nuancé* to be characterized by the sweeping declarations and facile condemnations. But until recently what we have heard were judgements more or less severe.[3] Thus Louis de Broglie while appreciating some Bergson's anticipations of modern physics, did not hesitate to call *Duration and Simultaneity* 'less good' (*moins bon*). Olivier Costa de

Beauregard, in spite of his pronounced sympathies for Bergson's thought, was even more severe when he claimed that Bergson's arguments in the book were 'absolutely erroneous.'[4] Ilya Prigogine in his account of the discussion between Bergson and André Metz (which was recently translated in English) treated Bergson's arguments with a hardly disguised condescendence.[5] No less severe and often quite unfair were the criticisms of André Metz and A. d'Abro, published a few years after the publication of *Duration and Simultaneity*. The present prevailing tendency to accept – almost without any discussion – the view that Bergson's critical comments about the relativity theory are 'absolutely false' is certainly due to the influence of these two authors. This is why it is especially important to analyze carefully their own objections to Bergson's own critical comments.

But before analyzing them it is important to have a clear idea what Bergson himself said and, more importantly, what he did *not* want to say. Now on this point the pertinent texts of *Duration and Simultaneity* are unambiguous. Taken, for instance, the following passage in the final pages of the book:

> In short, there is nothing to change in the mathematical expression of the theory of relativity. But physics would render a service to philosophy by giving up certain ways of speaking which lead the philosopher into error, and which risk fooling the physicist himself regarding the metaphysical implications of his views.[6]

It was impossible to be more explicit. Unlike some other critics of the relativity theory such as, for instance, Jacques Maritain, Hans Driesch, René Berthelot, Bergson on the other hand accepted the fact of constant velocity of light independent of the motion of the observer and its mathematical expression – the Lorentz transformation. He accepts without reservations Einstein's negation of the absolute and privileged frame of reference which implies an elimination of the motionless aether: "But there is no longer any aether, no longer absolute stability anywhere."[7] Briefly, it is impossible to see Bergson as an 'adversary' of Einstein. He sincerely believed that he fully accepted the relativity theory while rejecting merely its false philosophical interpretations. Some of his errors were no more serious than those of his critics; in any case, they do not diminish the value of his other subtle and judicious analyses.

Let us first analyze the first chapter of his book called 'Half-relativity.' No dispute is possible about its content and so far as I know, nobody contradicted it. In this chapter the author discussed the problem of simultaneity as it emerged *after* the Michelson-Morley experiments, but *before* it's formulation by the special theory. The problem of simultaneity in general had hardly existed before the discovery of the finite velocity of light by Olaf Römer in 1675; only then was it realized that there is a difference between 'seen now' and 'Now', between the simultaneity *perceived* and the *real* simultaneity. But the belief in the existence of the *real*, i.e. *objective, absolute* simultaneity of distant events persisted, even though it was difficult to find it, because of a complex superposition of the velocity of the earth, of the velocity of the stars and that of the light. The Michelson experiment seemed to be a unique opportunity to determine definitely the absolute velocity of the Earth with respect to the absolute space; this would also permit us to determine definitely the absolute simultaneity of distant events. The negative result of the experiments gave to the problem of simultaneity an entirely new aspect and Bergson describes very accurately the attempts at reconciling its result with the classical concepts of space, time and absolute simultaneity. These attempts are associated with the names of G.F. Fitzgerald and H.A. Lorentz. Their principal aim was to retain the absolute frame of reference as well as to explain the impossibility to confirm empirically its existence. But this was not enough. It was necessary to add another hypothesis in order to explain why an observer moving with respect to the motionless aether – whose existence was still then accepted – finds the same velocity of light as an observer who is at absolute rest or, as Bergson said, 'one who is motionless with respect to aether.' This was the reason why 'the dilatation of time' was introduced, i.e. the retardation of all clocks in all frames of reference *moving* with respect to aether.

Such a second hypothesis was logically independent of the first one as Bergson, following in this respect C.D. Broad, correctly stressed.[8] It is important to underline the fact that both hypotheses still accepted that the classical theorem of addition of velocities applied also to the velocity of light and that there was one single privileged frame of reference – that of the absolute space which was, so to speak, embodied in the motionless aether in which neither the contractions of length nor dilatations of time could occur. In other words, both hypotheses retained the concept of absolute time whose each moment contained the objectively simultaneous events; thus 'dislocation of simultaneity' is merely *apparent* since it is due to a retardation of the clocks in *all* the systems moving with respect to absolute space. The same fact can be expressed also by saying that the flow of the universal time is correctly measured only by the observers who remain *stationary* with respect to absolute space.

I insisted on the truly Newtonian spirit which inspired the first assumptions of Fitzgerald and Lorentz in order to indicate the distance which separates them from the special theory of relativity. It is true that the latter also speaks of 'longitudinal contractions' and 'time dilatations,' but their meaning is altogether different. The principal aim of the second chapter of Bergson's book was precisely to insist on this fundamental difference. Its title is 'Complete Relativity.' Allow me to sum up briefly its content; we shall see that it is difficult to imagine a serious physicist accepting Einstein's theory who would not agree with Bergson's exposition of it.

According to the Lorentz-Fitzgerald hypothesis the longitudinal contractions and the retardations of the clocks were *real physical changes* which supposedly occur – in different proportions – in all systems moving with respect to absolute space. On the other hand, according to the same hypothesis, the constancy of the velocity of light is, in a sense, a mere appearance. For we must not forget that the equations of Lorentz were obtained on the assumption that the classical law of composition of velocities applies also to the propagation of the electromagnetic waves. Thus it was believed that the velocity of light was *really modified* in the systems which were in motion with respect to the aether, but that the contractions of the length and the slowing down of clocks prevent us from observing these modifications. (In truth, the Lorentz-Fitzgerald hypothesis was *invented* in order to save the classical theorem of the composition of velocities.)

The situation is entirely different in the Einstein theory. The constancy of the velocity of light is viewed as a *basic reality* which does not need to be explained by more or less artificial assumptions. In other words, we must give up not only the classical law of the composition of velocities, but also to supersede the whole Newtonian mechanics by a new one which would agree with the primordial fact of the constant velocity of light. The classical principle of relativity – the so-called Galileo principle – is broadened in such a way that all inertial principles be equivalent not only *dynamically*, but also *optically* so that all electromagnetic vibrations would spread in all of them with the same velocity. Thus it is impossible to avoid a consequence on which Bergson so much insisted: a complete reciprocity of appearances in all inertial systems as well as the character *apparent* (referential) of longitudinal contractions and the time-dilatations. Let us see why.

Let us imagine two frames of reference moving with a constant velocity with respect to each other. An observer in the system S observes that in the system S′ the dimensions parallel to the direction of its motion are shortened and that all its clocks are slowed down. But because of complete equivalence of both systems, the observer in the system S′ observers the *same* modifications of lengths and of the rhythm of the clock in the system S. There is thus a *complete equivalence of appearances* which looks absurd only as long as the longitudinal contractions and the retardation of the clocks are regarded as *real* physical changes since it is evident that the clocks in S cannot be *both* slower and faster than the clocks at S′. The situation is comparable to the *effects of perspective* which in this particular case may be called *the perspective of velocity*. This is precisely the point on which Bergson so strongly insisted. He chose a particularly appropriate analogy to illustrate symmetry of appearances in the inertial systems. Suppose that there are two persons, John and James, separated by a certain distance; each of them will see the size of the other reduced and this will be seen in the sketches which one will make of the other; each of them will reduce the size of the other. But it is clear that both of them preserve their normal size and that neither John nor James shrank to the size of a dwarf. The only difference between the perspective of distance and that of velocity is that in the former the observers are separated by the distance while in the latter they are separated, as Bergson says, by 'the distance of their velocities.'[9] It is precisely the distance of their velocities which is the cause of the apparent symmetrical modifications of the spatial and temporal intervals.

On this point there is no doubt: all outstanding physicists agree on it without having read Bergson. Thus Jean Becquerel says explicitly:

> Clearly, the contraction is reciprocal; if two stems are identical, each of them at rest in his own system, i.e. one in the system S and the other in the system S′, each observer finds that the stem in the other system is shorter than one in his own. The fact that a moving object is contracted in the direction of its own motion, does not mean that it is really modified by its own motion.[10]

Similarly Max Born:

> Thus the contraction is merely a consequence of the way of our regarding the things; it is not a real physical change.

This is why, according to Born, the objections of those who claim that such modifications, arising apparently without any physical cause, contradict the law of causality, are baseless because there is no question of physical changes.[11] One finds the same reasoning in other physicists and even – what is especially important – in those who criticize Bergson. Thus A. d'Abro writes:

The phenomenon of the Einsteinian contraction is thus similar to that which affects the image of all bodies, whether hard or not, placed in front of a concave mirror. The same reasoning applies in an analogous way to the dilatation of duration.[12]

Obviously, the metaphor used by d'Abro is not so much different from that used by Bergson: there is no question of any real change, but of an appearance comparable to an optical illusion – caused by the effect of distance, according to Bergson, by the reflexion of the light by a concave mirror, according to d'Abro.

The reciprocity of appearances is also conceded by André Metz, at least in some of his writings, but his thought is, as we shall see, rather wavering and uncertain. (I shall return to this point.) Nevertheless, he concedes 'the principle of proper durations' in saying that 'the velocity of the frames of reference does not modify the proper time of the phenomena.'[13] But this is just another way of affirming the apparent nature of 'the dilatations of time' on which Bergson insisted so much. All effects resulting from 'the perspective of the velocity', including 'the dilatation of time,' vanish for an observer inside of his own system. A. d'Abro is also on this point in agreement with Bergson and other physicists, but says, somehow peevishly, that this is obvious and that Bergson, in insisting on the equality of local times, 'is merely breaking the door opened long ago.'[14]

Far more disputable is Bergson's view according to which the dilatation of time and dislocation of simultaneity of the distant events are merely *ideal* since they result from an imagined duplication of the real observer. On this point d'Abro is unquestionably right; the confusion of 'apparent' and 'unobservable' is a serious error which Bergson could have avoided. Even his illustration of 'the perspective of velocity' by that of distance does not agree with his notion of 'phantasmatic observer',[15] is it not true that in his own illustration James *sees* really the size of John reduced by perspective and *vice versa*? The observations of lengthened durations, of longitudinal contractions and of the dislocations of simultaneity are as real as those of the sizes diminished by the perspective of distance. Nothing shows better that 'appearance' and 'unobservable' are two very different things. D'Abro was right to say that the experiment of Fizeau on the propagation of light in the moving media confirmed the dilatation of time, at least indirectly;[16] today the slowing down of the disintegration of mesons observed by B. Rossi and D.B. Hall is regarded as a direct confirmation.[17] If Bergson were right, an increase of the mass of the electron would be impossible or unobservable, but it is known that such a phenomenon was observed at the beginning of this century and that one cannot view the physicists such as Kauffmann, Bucherer and Guye as 'phantasmatic observers'! The fact that an increase of the mass cannot be observed within the system of the particle does not mean that it is unobservable by an observer *outside* of that system. The situation is entirely similar to that of the dilatation of time and the contraction of the length.

But it would go too far to place the proper times of the events on the same level as the times lengthened by the perspective of velocity. This is what d'Abro does and there is some ambiguity even in Born when he does not want to apply the adjective 'apparent' to the dilatations of time.[18] Yet, did he not himself concede the *apparent* character of these effects? Evidently, he, no less than Bergson, failed to differentiate 'apparent' from 'unobservable.' Why then do we *prefer* the proper duration of the atomic vibration even though we concede that it appears dilated to an observer who moves with a great velocity with respect to the same atom? Why do we believe that the mass of the electron measured by an observer who at rest with respect to it represents something more essential than the same mass enlarged by the perspective of velocity? Why do we call it 'proper mass'? Even though the mass increase as well as the dilatation of time are observable and, in this sense *real* for an observer in motion, they rest, so to speak, outside the intrinsic nature of the phenomenon.

It is true that the hyperrelativist, such as d"Abro, would object that such distinctions are altogether contrary to the spirit of relativity and that in calling the proper time of the phenomenon 'more real,' we would view a system in which the vibrating atom is at rest as the privileged frame of reference; we would be thus returning unconsciously to the absolutism of Newton. But such a conclusion is altogether unjustified. Even the relativists do not doubt that each observer is *at rest with respect to himself* without identifying such rest – relative, though real – with the absolute immobility of the fictitious Newtonian space. This unique relation 'of being motionless with respect to himself' which is nothing but a physical or, rather, kinematic expression of the law of contradiction, was correctly recognized by Bergson as an *absolutist element of relativism*; from it the privileged character of the proper time, proper length and proper mass follow, inevitably.

According to d'Abro, "to ask which of these durations are the real one is equivalent to the question which is the true color of a piece of opal. It can be yellow if we look at it from a certain angle; red when we move toward the left, green or blue if we move ourselves toward the right."[19] This is quite true, but a more attentive analysis shows that this analogy is quite superficial and even false. From the physicist's standpoint the color is nothing but a light reflected by the material structure of the illuminated object; the reflected light depends also on the incident light, i.e. on its wave length, on the angle of incidence, on the milieu in which the object itself is placed, but all these additional factors are *external* to the object. One less misleading analogy would stress the close relation of local, proper time to the vibratory structure of atom; in the same way as the latter is, so to speak, the *causal core* of the all reflected rays, the proper time is also the *causal core* of all times lengthened in other inertial systems. Furthermore, the intrinsic character of the proper duration is suggested by another illustration used by d'Abro when he compares the slowing down of time to a distortion of its image in a concave mirror; in the same way as the identity of the object is not affected by its deformations in a concave mirror, the identity of the proper time persists despite its apparent dilatations in the perspectives of other systems – precisely because of its intrinsic character. D'Abro's hesitations indicate that Bergson in insisting on the identity of proper times was doing something more than 'breaking the door already opened.'

As I mentioned above, the only error Bergson committed was that he believed that the terms 'apparent' and 'unobservable' are synonymous. He took the expression 'inside of his own system' too literally. Without realizing it, he assimilated every frame of reference to a 'monad without windows,' impervious to the influence of any external events taking place in other systems; a rather strange view of the thinker who insisted so much on the universal interaction and who anticipated Whitehead's criticism of the fallacy of simple location! On this point Bergson's error was incompatible not only with Einstein's theory, but also with the spirit – and even with the letter – of his own philosophy. Let me underline this point once more: the fact that an observer inside of his own system cannot observe a dilatation of his own duration, does not prevent him from perceiving apparent dilatation of time in other systems.

The second error of Bergson was to believe that the time retardation in the general theory of relativity has the same referential character as in the special theory and that there is thus the same reciprocity of appearances. This assumption was the basis of his criticism of Paul Langevin's *Gedankenexperiment*. – his famous 'journey in the projectile' (*le voyage en boulet*) called also 'the paradox of the twins' or simply 'the clock paradox.' Langevin was the first who expounded it in 1911 and it was what attracted Bergson's attention to it.[20] Let me indicate its main features: let us assume that there are two brothers of the same age one of whom remains on the earth while the second departs with the velocity approaching the velocity of light and after turning around at the farthermost distance he returns to the earth with the same velocity. While his roundtrip lasted two years, he found out after his return that two centuries have elapsed on the earth and his twin brother died long ago.

Bergson's objections, which in this respect were accepted by A.O. Lovejoy, Herbert Dingle and, more recently, by Mendel Sachs,[21] were based on the view that the alleged difference between a slowed down aging of the interplanetary traveller and 'normal' aging of his brother on the earth is incompatible with the reciprocity of appearances and a complete equivalence of all frames of reference which is the very essence of the relativity theory. Since there is no privileged system, the moving projectile (today we would say: the rocket-ship) itself could be chosen as being at rest; then it would be the earth which would be moving with an equal velocity in opposite direction. Then by applying the same reasoning used by Langevin, we would conclude that two centuries would elapse in the spaceship while the earth itself would grow only two years older! But this is absurd since both clocks – that of the space-traveller and that of the earthly observer – cannot be simultaneously slower and faster than each other. The only way to avoid such contradiction is, according to Bergson, to regard the retardations of the clock as merely referential, apparent, mere 'mirages' produced by the perspectives of velocity. They cannot be *real* and *effective* dilatations or retardations of time; Langevin's reasoning – Bergson concludes – is just another example of 'half-relativity' which is nothing but a residual unconscious Newtonianism or even implicit geocentrism.

Unquestionably, Bergson's argument would have been correct, if the situation described by Langevin could be adequately treated within the scope of the special theory. But this is precisely not so. For in the situation envisioned by Langevin enormous accelerations take place at the point of the farthest distance when the projectile begins its return trip; not only the original enormous speed should be reduced to zero, but it should be then increased to a near velocity of light in the opposite direction. This is why the projectile (or spaceship) cannot be regarded as an inertial frame of reference; in other words, an analysis of its motion is beyond the scope of the special theory. The true solution of the clock paradox can be found only on the terrain of the general theory.

This was recognized by Einstein himself, then by Hans Thirring, Jean Becquerel, A.N. Whitehead, Hans Reichenbach and more recently by David Bohm. According to the principle of equivalence the effects of acceleration are equivalent to the effects of the gravitation field which 'slow down effectively the course of time.'[22] Thus it is a *dilatation of the proper time which is real* and thus altogether different from a mere referential dilatation in the special theory. This is a fundamental difference between the special and the general theory and ignored by Bergson and it has not been always stressed enough – not even by his critics in a rather confused discussion which took place after the publication of *Duration and Simultaneity*. It is true that d'Abro insists correctly on the fact that a 'slowed-down time' is lived by an accelerated observer'[23] and that his own proper duration and that of an stationary observer are not interchangeable. He is also right when he points out that Bergson is consistent when he accepts the curvature of space while denying an effective retardation of time without realizing that the latter is implied by the former.[24] On the other hand, his discussion of 'the traveller in the projectile' (today we would say: 'of the astronaut') is confused since it is exclusively confined to analysis of the displacement of two *inertial* systems; thus he pretends to obtain the asymmetry of two proper times within the framework of the special theory! He is more careful in his second book, even though he persisted in treating the clock paradox almost exclusively within the framework of the special theory. His hostile attitude toward Bergson remained the same.[25]

The case of André Metz and his critique is, if not worse, certainly not better. He expounded his views in two books: *La relativité: Exposé élementaire des théories d'Einstein et réfutation des erreurs contenus dans les ouvrages les plus notoires* (1923) and *Temps, espace, relativité* (1928) as well as in a number of articles. In the first book he insisted, without realizing that he agrees with Bergson, on the identity of proper times: "the velocity of any system of reference has no influence on the proper time of phenomena." (p. 18). Like d'Abro, he stressed correctly, that the reciprocity of appearances exists only in the inertial systems and thus it is absent in the round-trip of Langevin's astronaut; nevertheless, also like d'Abro, he applies the formulae of Lorentz to it. (p. 69). What is worse is that on the following page he asserts that a *real dilatation of time* occurs during the astronaut's motion away from the earth *before* his point of return, i.e. before any acceleration takes place! But on the next page (p. 71) he returns to a more correct reasoning in concluding that there is a reciprocity between the system of Peter (the stationary observer) and that of John (who continues his journey without return). This is doubtless true since in this situation we have two *inertial systems*; obviously, he overlooks that the same situation exists in the case of Langevin's traveller *before* his point of return. At the same time he claims that Peter finds *in his own system* a *retardation* of his own clock with respect to the clock of John which defiles in front of him. This would be a flagrant contradiction: the clocks of John are falling behind the clock of Peter and the clocks of Peter fall behind the clock of John ...was it perhaps a misprint?[26]

The same confusions are even more visible in his polemic with Bergson one year later.[27] One finds in it entirely correct views alongside with some almost incredible confusions. Again he emphasizes correctly the fact that the proper dimensions and the proper time are 'intrinsic properties' of the system and that the longitudinal contractions are only apparent, though they are observable (p. 70–71); also that such reciprocity is absent when the *whole* journey of Langevin's traveller is considered (i.e. his trip away from the earth and back) as his own motion ceases to be inertial. But his own illustration is rather unfortunate: he places his astronaut on a moving sidewalk parallel to the road along which there are the synchronized clocks of the stationary observer. He thus forgets that such sidewalk – as well as the road – would be *solid bodies*, each being a physical realization of an *instantaneous signal*, i.e. *absolute simultaneity* which was definitely eliminated by Einstein. For there are no instantaneous connections or interactions in the physical world; this is why there are no perfect solid bodies – the fact emphasized first by Langevin, then by Reichenbach and, later, even by André Metz who, nevertheless, never gave up his unfortunate illustration of a sidewalk moving along the stationary road stretching along a distance of million of kilometers.[28] Misled by his own illustration, he was getting entangled in contradictions; first he concluded that the clock of the traveller appears to fall behind the clocks on the road (proposition 1); then that the clock of the stationary observer 'progressively retards with respect to the clocks on the moving sidewalk as it meets them one after another' (prop. 3)! This is clearly a contradiction unless he would view the retardations of the clocks as merely *apparent*; but this he refuses to accept as it would bring him in agreement with Bergson. Thus he has no other choice than to accept a flagrant impossibility: that the clocks on the stationary road and those on the moving sidewalk both *really* fall behind each other.[29]

One can easily comprehend that Bergson could not be convinced by such clearly incoherent arguments of his critics whose uncertainty betrayed their oscillation between the special and the general theory; thus he in his answers to A. Metz continued to reaffirm all theses of his book concerning the difference between 'the real and conscious observers' and 'phantasmagoric observers' as well as the difference between 'the real and lived time' and one 'merely attributed.'[30] He persisted in considering the accelerated systems exclusively from the kinematic point of view in closing his eyes to the basic difference between the physical behavior of the inertial systems and that of the accelerated systems. It is true that a not very clear argument of his critics did not help him to understand the importance of this distinction. This was true not only of the writings of Metz and d'Abro, but also, to some extent at least, of Jean Becquerel who, in applying the Lorentz transformation to Langevin's astronaut, remained exclusively within the confines of the special theory; it is then hardly surprising that d'Abro as well as Metz followed Becquerel on this point. What is more curious is that Becquerel in his book which appeared before that of Bergson, insisted correctly on the difference between the merely *apparent* and *referential* retardation of time in the special theory and its *real* and *effective* retardation in the gravitational fields and, consequently, in the accelerated systems. The fact that Becquerel failed to mention this distinction in his polemic with Bergson hardly helped to clarify this problem.[31] What excuses Becquerel – as well as Bergson and his critics – is the fact that even Paul Langevin in his classic exposition of 'the journey in the projectile' used the same procedure.[32] But let us not forget that this was in 1911, before the formulation of the *general* theory.

What is less excusable for Bergson is his unsatisfactory answer when Metz pointed out that Fizeau's experiment on the propagation of light in moving liquids verified Einstein's new theorem about the composition of velocities and thus confirmed indirectly the dilatation of time. It seems that Bergson failed to grasp the significance of this experiment and that of the confirmation of the increase of the mass of the fast moving electrons. He is undoubtedly correct when he insists that an observer attached to a fast moving electron does not perceive any dilatation of his proper time.

But this is hardly relevant since he does not perceive an increase of his proper mass either; but it does not prevent an observer *outside of his system* from observing it as the experiments of Kaufmann, Bucherer and de Guye established. Bergson failed to understand that there was a complete analogy between the observability of the mass increase and that of the dilatation (retardation) of time.[33]

We can now see clearly that the most persistent cause of misunderstandings and confusions which occurred not only in the discussion of Bergson with his critics, but also in nearly all discussions of the 'clock paradox' was the absence of the distinction between 'apparent' and 'intrinsic'. As Marie-Antoinette Tonnelat observed: "One can ask why do so many exegeses of relativity seem to avoid systematically the adjective *'apparent'?*"[34] She found three reasons for it. First, the word 'apparent' suggests the notion of something illusory or unreal. But there is nothing illusory in relativistic effects; we have seen that 'apparent' and 'unobservable' are *not* synonymous. Yet, this does not eliminate the distinction between the proper mass, proper time, proper length, measured within the frame of reference, and their, so to speak, 'projections' – apparent, yet still observable – in other system. Another reason was the influence of positivism which placed all observations and measures on the same level; thus the distinction between 'reality' and 'appearance' – which had been always suspect to positivists – supposedly does not exist. Finally, the same distinction may appear doubtful in the light of equivalence of all inertial systems; it seems to be a disguised return to the hypothesis of the privileged system. This was precisely d'Abro's argument – as we have seen, not consistently upheld – against Bergson's insistence on the equality of proper times. But he ignored the basic difference between the privileged character of Newton's absolute space and the 'privileged' character of *all* observers with respect to their own systems. In the words of M.-A. Tonnelat, "If an observer who belongs to the same system as the object itself can be called *privileged*, it is evidently so in relation to the proper time of the object and to the corresponding associated space."[35] It is probable that the word 'intrinsic' instead of 'real', and 'referential' instead of 'apparent', would have caused much less of misunderstanding. In any case, a precise enough definition of 'apparent modification of quantity' should suffice to dispel all unnecessary confusions. Such a precise definition was proposed by the author quoted above, M.-A. Tonnelat, in the following words: apparent modification of quantity is equal to the difference between its quantity measured in any inertial system and one measured within its proper system.[36]

This does not mean that the awareness of this distinction has been always absent. It certainly was known to Paul Langevin who while stressing the difference between the proper time of his astronaut and that of his terrestrial brother, insists that, for instance, *during the first half of the astronaut's journey* 'both brothers would see each other live two hundred times slower than usually.'[37] In other words, there is a symmetry of appearances *as long as the motion remains inertial* in contrast to the asymmetry of the proper times when the *whole* round trip is taken into account. This means that an *effective* dilatation of the astronaut's proper time takes place at the point of his return, i.e. at the maximum distance from the earth where drastic accelerations (first negative, then positive) of his system occur; on this point the thinkers as different as Hans Reichenbach, A.N. Whitehead and David Bohm agreed with Langevin – and with Becquerel who, strangely enough, did not use it as an argument in his polemic with Bergson.[38]

Langevin also pointed out that the presence of a 'conscious and living passenger' on which Bergson so much insisted, is not essential; we can replace two living observers by two specimens of radium to discover that the travelling piece of radium after being subject to accelerations has aged less than the terrestrial specimen; the proper times of two specimens will be different.[39]

Allow me to add a few remarks on Bergson's criticism of one illustration by which Einstein tried to explain relativity of simultaneity in his popularizing book *Relativity, The Special and General Theory* which is reproduced *verbatim* in *Duration and Simultaneity* (pp. 92–97). It is a well known example of the train of the length A′B′, moving with constant velocity. An observer placed in the middle of the train at a point M′; just when he passes a *stationary* observer, placed at the point M, two luminous signals are sent from two points of the railroad A and B, equidistant from M at the precise moment when the point A and B coincide with the extremities of the train A′ and B′ respectively. The observer at M, being equidistant from the points of A and B of the railroad, will perceive the light signals at the same time and will conclude that they are simultaneous with respect to his system, i.e. to the system of the railroad. Einstein then continues:

Are two events (e.g. the two strokes of lightning A and B) which are simultaneous *with reference to the railway embankment* also simultaneous *relatively to the train*? We shall show directly that the answer must be negative... If an observer sitting in the position M′ in the train did not possess this velocity, then he would remain permanently at M and the light rays emitted by the flashes of lightning A and B would reach him simultaneously, i.e. they would meet just here he is situated. Now in reality (considered with reference to the railway embankment) he is hastening toward the beam of light coming from B, whilst he is riding on ahead of the light coming from A. Hence the observer will see the beam of light emitted from B earlier than he will see that emitted from A... We thus arrive at the most important result: Events which are simultaneous with reference to the embankment are not simultaneous with respect to the train and *vice versa*. (Relativity of simultaneity).[40]

Bergson used the same diagram as Einstein; he only added the supplementary arrows indicating the relative movement of the railroad embankment *with respect to the train*. The essence of his criticism is contained in the following passage:

Let us now emit two flashes of lightning. The points from which they set out no more belong to the ground than to the train: *the waves advance independently of the motion of their source*.[41]) It then becomes evident at once that the two systems are interchangeable, and that exactly the same thing will occur at M′ as at the corresponding point M. If M is the middle of AB, and if it is at M that we perceive a simultaneity on the track, it is at M′, the middle of B′A′, that we shall perceive this same simultaneity in the train.

Let me say at once that the illustration used by Einstein is rather unfortunate. Since it is pervaded by the macroscopic imagery of classical physics, it cannot replace the precise mathematical apparatus of the Lorentz transformation and thus it can easily mislead the reader. (It was probably Einstein's notion of 'extremely long train' that suggested to A. Metz an absurd idea of a moving sidewalk, stretching for a distance of millions of kilometers in space, thus obscuring to him the impossibility of solid bodies in the special theory.) Contrary to Einstein's intentions, his illustration can easily be misinterpreted by an uninformed reader in the sense that the velocity of light is *affected* by the relative motion of the observer with respect to the source of light. The words of Einstein that 'in reality' the observer on the train is 'hastening' toward the beam of light coming from B 'whilst riding on ahead of the light coming from A' can be easily misunderstood in the sense that the classical theorem of the composition of velocities is being applied to the velocity of light as well. This was pointed out by Maritain who accused Einstein of using an *antirelativistic* argument for establishing the relativity of simultaneity.[42] This is implicitly Bergson's view too; in reading the passage from *Duration and Simultaneity* quoted above, one has the impression that Bergson seems to be more faithful to the principle of constant velocity of light than Einstein himself! We have seen that – unlike Jacques Maritain or Hans Driesch and other antirelativists – he accepted that principle without any reservations. Also Bergson was right in insisting on a complete equivalence of two inertial systems and on the fact that whatever is said of the so-called stationary system applies equally to the system of the moving train. In other words, that there is a complete reciprocity of appearances in both systems.

On the other hand, Bergson never responded satisfactorily to one very serious objection raised by both d'Abro and Metz who tried to show that Bergson in his polemic with Einstein ignored the *absolute character of spatio-temporal coincidences*.[43] If two rays of light meet each other at a certain point – the point M in the system of the railroad embankment – no change of the frame of reference can abolish such encounter; this particular spatio-temporal coincidence remains intact in all frames. But is it certain that this is what Bergson tried to deny? Is it necessarily true that Bergson claims that *the same two rays* (or the same couple of photons) intersect at the points M and M′? I do not see any evidence for it in the quoted passage. The caricatures of d'Abro are nothing but cheap deformations of Bergson's thought.[44] If the rays meeting each other at M are different from those which meet at M′, there would be two *different* spatio-temporal coincidences whose absolute character would not be violated.

We must not forget that one faces insurmountable difficulties in trying to provide 'popular' explanations of the relativity theory and, in particular, by trying to 'clarify' its paradoxical features by the visual diagrams or models. One easily forgets such difficulties as his claiming that the space of the special theory is Euclidian. But this is a dangerous half-truth. It is true that the curved non-Euclidean space was introduced by the general theory, but on the other hand, the kinematics of the special theory is *not* Euclidian and the paradoxical character of its formulae – such, for instance, $c+v = c$ – cannot be made intuitive by any visual diagram.[45] This is only natural; while our reason does accept the mathematical apparatus of the special theory, our geometrical subconscious which remains obstinately Euclidian is opposed to it. Whenever we try to illustrate the non-Euclidean kinematics by diagrams which remain Euclidian, we are getting inevitably entangled in contradictions. We have seen such confusions in Bergson's critics; but the very same habits of the classical imagination prevented Bergson from accepting the negation of absolute simultaneity:

We claim that a single time and an extension independent of duration continue to exist in Einstein's theory considered in its pure state; they remain what they have always been for common sense.[46]

This passage contains two theses: the first one which asserts that the unity of time can be reconciled with the Einstein theory is correct only if we do *not* mean that unity in the metrical sense, but only in a broader, topological sense. But Bergson does not draw this distinction, or at least does not formulate it clearly and explicitly, even though it is compatible with his philosophy. The second thesis which posits a sharp separation between space and time in the sense in which the common sense understood it, is entirely false from the standpoint of relativity and, furthermore, incompatible not only with the spirit, but also with the letter of Bergson's own thought. Let us consider first the second thesis.

The sharp separation of space from time in the Newtonian physics was a consequence of their absolute character. It is known that these two concepts were closely connected; indeed, their conjunction constituted what may be called 'the Newtonian space-time' whose character was naturally profoundly different from space-time of Minkowski. The meaning of the former was accurately expressed by Lagrange when he characterized mechanics as a 'geometry of four dimensions.' In other words, the history of the universe was regarded as a *continuous succession of instantaneous states each of which constituted an instantaneous space*, in other words, an instantaneous state of Euclid and Newton was nothing but an instantaneous cut accross the four-dimensional becoming of the world. All events occurring at a certain instant t are located in an instantaneous Euclidean space called by Eddington 'a world-wide instant' and only such events are simultaneous in an absolute sense of the word.[47] One sees at once the very close connection between *absolute space* and *absolute simultaneity*, the former being the *substratum* of the latter. In other words, the instantaneous space of Newton is nothing but the totality of points – or, rather, of *point-events* – objectively simultaneous, i.e. occurring in the same instant of the universal time. This is the meaning of the profound expression of Newton, anticipated already by Gassendi, on the ubiquity of one and the same moment in the whole space: *Unumcumque durationis indivisibile momentum ubique* (Newton); *quodlibet temporis momentum idem est in omnibus locis* (Gassendi).[48] It is rather unfortunate that Bergson was not aware of the close correlation between the Newtonian space and absolute simultaneity; while he denied the former, he was trying to preserve the latter. Both concepts were preserved in classical physics and both of them were negated by the physics of relativity; one cannot have one without the other.

It is even more curious that the retention of absolute simultaneity is incompatible not only with the relativistic physics, but also with the implications of Bergson's own philosophy. The impossibility of absolute simultaneity of distant events results from the impossibility of instantaneous cuts in the four-dimensional becoming, i.e. the cuts by which one obtains at each instant 'the extension independent of duration,' constituting the locus of simultaneous events. The relativity theory showed that such cross-sections are merely fictions, based on the hypothesis of instantaneous connections; but such connections do not exist since there are no infinitely fast signals; if one tries to introduce them, one obtains different instantaneous spaces for different observers which have nothing in common with the definite and unique character of the Newtonian simultaneity. But is it not true that Bergson said the same thing in holding the view that all instantaneous cuts across the universal becoming are mere fictions resulting from 'the cinematographic mechanism of thought'? What is even more curious is that even in *Duration and Simultaneity* Bergson came close *to an explicit negation of absolute simultaneity* – evidently without drawing all the consequences from it. Here is the text in which he stated that in nature there are only *the simultaneities of flows*, but never the *simultaneities of instants*:

> Now from the simultaneity of two flows, we would never pass to that of two instants, if we remained within pure duration, for every duration is thick; real time has no instants.[49]

This is clearly not a language of a relativistic physicist; yet its content is not different from the relativistic affirmation that the only real simultaneity is *the simultaneity of intervals*; that of instants is physically meaningless precisely because of the impossibility of 'instantaneous cross-sections' in space-time. Let me explain it more since this will help us to understand in what sense and to what extent the first Bergsonian thesis on the universal character of becoming is true.

The expression 'simultaneity of time intervals' (or: 'simultaneity of flows') cannot be found in the textbooks of relativity, but nothing is easier than to give a precise definition of this concept in the language of the same theory. Two spatio-temporal intervals are simultaneous* when, while being different, they have only two events in common: the initial event and the final one. These two events – for instance, the event of the departure of one twin from the earth and the event of his return to the earth – are evidently successive and their succession remains such in all frames of reference. This is what Paul Langevin called the irreversibility of 'the world-lines,' i.e. of causal links.[50] On the other hand, the proper times of both contemporary world-lines between their two common events (the initial and the terminal) are *metrically different*; when their difference is very pronounced, we have Langevin's 'paradox of twins' or 'clock paradox.' It is thus important to differentiate two kinds of invariance of spatio-temporal intervals: the *metrical* and *topological*; in contrast to the physics of Newton, the relativity theory retains only the latter. On the other hand, there is no such invariance of the relations of co-existence of juxtaposition for one simple reason that there are no such relations in the world of Einstein-Minkowski. In other words, there are no barely spatial distances; this is a simple consequence of the negation of absolute simultaneity. From what has been just said one very important consequence follows: *there are some successions* (i.e. *the successions of causality related events*) *which remain successions in all frames of reference while this is not true of the relations of coexistence (juxtaposition)*. It is thus clear – contrary to the wide-spread opinion, especially among philosophers, – that the union of space and time, proposed by Minkowski, works in favor of time rather than that of space, in other words, that it can be characterized as a *dynamization of space* rather than a *spatialization of time*. This is why the term 'time-space' is probably far more appropriate than 'space-time.'

Once the dynamical character of time-space is recognized, it can be better understood in what sense the universal character of time – or rather of becoming – is preserved. Obviously, it is not the time of Newton; this is at once clear when we compare the structure of Newtonian space-time with that of the time-space of Einstein. In the latter the past and the future are not separated by the infinitely thin present, i.e. by an instantaneous Now. Yet, their separation remains unambiguous; in truth, it is *even more effective* in the light of Minkowski's diagram, which shows that the boundary between the absolute past and the absolute future is *not* constituted by the infinitely thin line of the absolute present, but by the four-dimensional region of *causal independence*, called 'Elsewhere' by Eddington. No physical signal, no causal action from this region can reach the present event (Here-Now). A more attentive analysis shows that with respect to every present event its future is in a physical sense *empty* since all events 'located' (*sit venia verbo*) in its own future, remain intrinsically unobservable – not only by an observer in 'Here-Now,' but by any other conceivable observer located anywhere in his 'Elsewhere' region. Such events, being intrinsically unobservable, cannot possess any character of physical reality. Thus the *physical emptiness of the future* in the universe of Minkowski is a rather unexpected justification of the Bergsonian notion of 'the open future.'[54]

Unexpected because the universe of Minkowski was frequently interpreted as a sort of static hyperspace in which matter is, in the terms of O. Costa de Beauregard, 'statistically unfolded in space-time.'[55] Such an interpretation is merely the most recent form of the spatialization of time which Bergson denounced in all his writings. It was only natural that he attacked it also in *Duration and Simultaneity*, especially in Chapter VI, without being aware of the possibility of a dynamic interpretation of the relativistic time-space. It is true that he agreed with Whitehead's dynamic interpretation of the relativity theory when he underscored the kinship of Whitehead's 'Creative advance of nature' with his 'universal duration.'[56] But he failed to realize that Whitehead's concept of 'advance of nature into novelty' is *not*, properly speaking, serial and measurable because it underlies the metrically different series,[57]; in the twin paradox, for instance, there is the same 'advance of nature,' the same 'universal duration', underlying both the proper time of the astronaut and the proper time of the Earth despite the fact that the two temporal series are metrically different. Bergson reached *almost* the same conclusion when he wrote:

> Without this unique lived duration, without this real time common to all the mathematical times, what would it mean to say that they are contemporaneous, that they abide within the same interval? What meaning could we really find in such a statement?[58]

True, but Bergson continued to regard the unity of his cosmic duration as the *metrical* unity in the Newtonian sense and persisted obstinately to view 'the mathematical times' as merely 'phantasmagoric' or imagined. Nevertheless he always accepted the absolute simultaneity of distant events and the existence of 'the extension independent of duration' since those two concepts imply each other. We do not find in him the Whiteheadian distinction between the topological unity of the underlying duration and the plurality of different temporal series, the latter being merely different metrical complementary manifestations of the former.[59]

And yet, not only the implications of his thought, but also numerous explicit passages of his own writings indicate that Bergson was really giving up the Newtonian concept of the metrical time. Had he not *always* emphasized that the real time by its own nature is *not* measurable, i.e. not metrical? Even more important is the fact that the idea of different temporal series, called by him 'different rhythms of duration,' is the central theme of *Matter and Memory* as well as of *Creative Evolution*. Without this central theme it is altogether impossible to comprehend either of these books and all its implications. Thus it is even stranger to see the same author in 1922, while admitting that 'different durations, differently rhythmed, might coexist,' still is denying 'any reason for extending this theory to the physical universe.' Evidently, what we see here is, if not a contradiction, at least a certain distraction within Bergson's thought as Z. Zawirski showed in his unjustly forgotten book long ago.[60]

The fact that the relativistic union of space with time was frequently interpreted – especially by philosophers – in a static way, as a 'spatialization of time,' prevented Bergson from becoming aware of a real affinity of his 'extensive becoming' with the time-space of Minkowski. The latter, as we have seen, does not allow any instantaneous three-dimensional cross-section; the notion of instantaneous space, separated from time, is a mere abstraction and nothing but a residue of the conceptual apparatus of classical physics. In the language of Bergson, "the instant is what would terminate a duration if the latter came to halt. But it does not halt."[61] In the language of relativity, such a 'suspension of time' would be realized – *per impossibile* – by the instantaneous physical actions, propagated with an infinite velocity; their 'world-lines' in the diagram of Minkowski would be 'perpendicular to the time axis' in any system of reference; they would constitute space at each particular instant. Such an impossible situation in relativity was explicitly rejected by d'Abro who was aware that the instantaneous character of space could remain in the Newtonian classical physics, but *not* in the physics of Minkowski and Einstein.[62] He, on the other hand, remained unaware how close was his view to that of Bergson denying the durationless character of physical space in which *no* extreme limit of 'distension of time' could exist. For the physical extension, unlike geometrical spaces, has still a *duration*, even though a 'diluted' one. Hence, Bergson's distinction between *space* and *extension*: while the former retains in geometry and classical mechanics its instantaneous character, the latter does not since it contains *durational* character; its intervals or micro-intervals, can never be identified with durationless pseudo-entities. In Bergson's words, "matter extends itself in space without being absolutely extended"; or "although matter stretches itself out in the direction of space, it does not completely attain it."[63]

Such expressions appeared rather mysterious in 1896 or even in 1907; but they cease to be obscure when they are translated into the language of relativity. It is known today that there are no pure durationless spatial distances, for only *the spatio-temporal distances* are real. As Whitehead observed as early as in 1919 that while 'the relations which constitute (classical) space are instantaneous,' since the advent of relativity they 'must now stretch across time.' He recognized that the belief in the existence of 'a definite present instant at which all matter is simultaneously real' is a very tenacious hereditary illusion which was the basis of the classical scientific materialism; in the physics of relativity 'there is no such unique present instant.' Whitehead was aware that such denials of instants appeared to our visual imagination, but it loses its incomprehensibility when we turn to auditory models. Nature 'like a note of music, is nothing at any instant.'[64] It is true that the auditory models cannot be substituted for the physical theories; they nevertheless can neutralize the influences of our visual habits which are adapted only to 'the middle dimensions' of the universe where the physics of Newton remains practically valid. Outside of the zone of the middle dimensions – in the microcosmos or the megacosmos – our visual models cease to be applied; thence the coming of relativity as well as that of quantum theories.

CONCLUSION

We can now see what is living and what is dead in Bergson's critical comments on relativity. But some similar difficulties are present in his opponents, such as A. d'Abro and, even more serious, in André Metz as I pointed out.

Bergson was definitely right when he insisted on a complete equivalence of all inertial systems and on the *apparent* and *referential* character of longitudinal contractions and the time dilatations; in other words, on a complete reciprocity of apparent modifications in the special relativity theory. But he was incorrect when he identified 'apparent' with 'unobservable,' although he admitted the osbervability of perspective appearance in his own example!

He is right in insisting on the equality of proper (local) times in the special theory, but he erred when he insisted on the equality of proper times even in the general theory. Thus he excluded the possibility of different *proper* times of two successive events which are joined by two different world-lines; such difference of the local times is due to the presence of gravitation or an accelerational field. This error marred seriously his criticism of Paul Langevin and Jean Becquerel; he claimed the proper times are metrically identical, although their difference is merely 'apparent'. Bergson, followed later by Herbert Dingle, claimed that there is a complete metrical symmetry of two time intervals – not only in inertial systems, but also in the accelerated and gravitational systems; hence he dismissed Paul Langevin's discussions of the paradoxes of twin travellers or twin clocks as mere illusions.

His critic, A. d'Abro, correctly pointed out Bergson's one occasional inconsistency in his approach to the general relativity: while Bergson recognized the *curvature of space*, he failed to see what it implied when he denied the influence of gravitation on the spatial-temporal measurements. On the other hand, d'Abro seriously misinterpreted Bergson's notion of 'absolute mobility' as a defense of the concept of absolute motion in the Newtonian sense. D'Abro apparently ignored Bergson's explicit rejection of the classical concept of the motionless space; such a rejection implied a negation of absolute motion. By 'absolute mobility' Bergson meant 'real change', not merely in the classical kinematic sense.[65]

Thus Bergson was approaching modern physics without reaching all its implications. He unfortunately retained two classical concepts closely related – absolute simultaneity and the separation of absolute (instantaneous) space from absolute time. He overlooked the fact that both such classical concepts are incompatible with the principle of constant velocity of light which he explicitly accepted. But he was not the only one of all such thinkers who accepted the relativity theory without being aware of some of its implications. We have seen André Metz who persistently charged Bergson with his alleged opposition to relativity. Yet, Metz maintained the notion of an enormously large *solid* body (by which he magnified Einstein's railroad) while not being aware of the incompatibility of a megacosmic *solid* body with the relativity theory.

Bergson was entirely correct in writing in his book on *Matter and Memory* that the contemporary metrically discordant series presupposed the unity of time or duration which underlies them all; unfortunately, he in *Duration and Simultaneity* – unlike in *Matter and Memory* – interpreted such underlying unity of time in a *metrical* and not in *topological* sense. He also in the same book claimed no reason to accept the physical reality of such multiple temporal series which the general theory requires and which he himself in his *Matter and Memory* admitted! This earlier book is philosophically more insightful than *Duration and Simultaneity* since it foreshadowed some essential features of the wave mechanics and of indeterminism in microphysics thirty years before; its significance was fully recognized by Louis de Broglie, at least before his return to classical determinism later.

Nevertheless, the most important significance of *Duration and Simultaneity* remains Bergson's correct rejection of the static interpretation of Einstein-Minkowski's time-space. Bergson's view pervades all his thought from *Time and Free Will* in 1889 to *The Creative Mind* in 1934. But in 1923 he should have perceived more clearly the affinity of a dynamic interpretation of modern time-space with the structure of his 'extensive becoming' (*devenir extensif*) in his *Creative Evolution.*[66]

NOTES

1. *La pensée et le mouvant*, Paris, Alcan, 1934 (2nd ed.), pp. 46–49. *The Creative Mind*, tr. by Mabell L. Andison (Philosophical Library, 1966), pp. 301–303.
2. Letter-preface of E. LeRoy of September 29, 1953 in *Ecrits et paroles*, Paris, P.U. Press, 1957, pp. VII–VIII. Even André Metz admitted that it was probably an error not to include *Duration and Simultaneity* into Bergson's Compl. Works. (*Bull. de la Soc. fran. de Philosophie*, 6e année 1957, p. 52.
3. The book of M.-A. Tonnelat, *Histoire du principe de relativité*, in which she proposed 'to re-open a brief (dossier) which seemed to be definitively closed' is an exception.
4. Louis de Broglie, *Physique et microphysique*, Paris, A. Michel (1947), 197; O. Costa de Beauregard, 'Le Principe de relativite et spatialisation du temps,' *Revue des questions scientifiques* (1947), p. 63.
5. Cf. Prigogine's review of P.A. Gunter's book *Bergson and the Evolution of Physics in Nature* (Nov. 19, 1971): 'Bergson's struggle with the Lorentz transformation in *Duration and simultaneity* is as pathetic as it completely misses the point.
6. *Durée et simultaneité*, 3e ed. (Paris, Alcan, 1926). Appendix III, p. 278. Reprinted in *Mélanges* (P.U.F. Paris, 1972), p. 237. Translated in English by Léon Jacobson *Duration and Simultaneity* (The Library of Liberal Arts, 1965) with an Introduction by Herbert Dingle p. 185. Hence referred to as *D.S.*
7. *D.S.*, p. 33; *Mélanges*, pp. 86–7.
8. *D.S.*, p. 14; C.D. Broad, 'Euclid, Newton and Einstein,' *Hilbert Journal* (April, 1920).
9. D.S., pp. 73–74; Mélanges, pp. 126–7.

10. J. Becquerel, *Le principe de la relativité et la théorie de la gravitation*, (Paris, 1922), p. 45.
11. Max Born, *Die Relativitätstheorie Einsteins und ihre physikalische Grundlagen* (Berlin, Springer, 1921), p. 189. The same view in A.E. Eddington, *Space, Time, and Gravitation* (Cambridge Univ. Press, 1920), p. 23–25; René Dugas, *History of Mechanics*, tr. by J.R. Maddox (Central Book Co., New York, 1955), p. 495; J.L. Synge, *Relativity: The Special Theory* (North Holland Publ. Co., Amsterdam, 1956), pp. 119–20.
12. A. d'Abro, *Bergson ou Einstein* (H. Goulon, Paris, 1927), p. 75.
13. A. Metz, *La relativité* (Chiron, Paris, 1923), p. 18.
14. A. d'Abro, *Bergson ou Einstein*, p. 138.
15. *D.S.* pp. 79–82; 109–113; *Mélanges*, p. 130, 135, 163–7.
16. A. d'Abro, *op. cit.*, pp. 117–18.
17. If we regard the motion of the mesons as uniform, the dilatation of their proper duration is only referential; if not, even their proper time must be slowed. The experiments mentioned above do not allow us to decide whether the observed phenomenon should be interpreted by the special or the general theory. (I am indebted to Professor Costa de Beauregard for this information.)
18. M. Born, *op. cit.*, p. 189–90.
19. A. d'Abro, *op. cit.*, p. 214.
20. *D.S.*, pp. 74–78; also Appendix I; Mélanges, p. 127–131, 216–225. Paul Langevin expounded his paradox at The Philosophical Congress at Bologna in 1911, then in the session of *Soc. française de philosophie* (October 19, 1911) under the title 'Le temps, l'espace et la causalité,' and finally in the article 'L'évolution de l'espace et du temps' in *Revue de mét. et de moral*, **XIX** (1911), p. 455–66.
21. A.O. Lovejoy, 'The Paradox of the time-retarding journey,' *The Phil. Review* XL (1931), pp. 48–68, 152–167; H. Dingle in his Introduction to the English translation of *D.S.* (1965); Mendel Sachs, 'A Resolution of the Clock Paradox,' *Physics Today 24* (September 1971), pp. 23–29.
22. J. Becquerel, *op. cit.*, p. 240; A. Einstein, 'Dialog über die Einwände gegen die Relativitätstheorie,' *Naturwissenschaften*, VI (1918), pp. 697–702; Hans Thirring, *Naturwissenschaften*, IX (1921), p. 209; A.N. Whitehead, 'The Problem of Simultaneity' in *Aristotelian Soc. Suppl.*, Vol. III (1923), pp. 34–41; Hans Reichenbach, *Philosophie der Raum-Zeit-Lehre* (Berlin & Leipzig, 1928), pp. 222–225; David Bohm, *The Special Theory of Relativity* (New York, W.A. Benjamin, 1965), Ch. 3.
23. A. d'Abro, *op. cit.*, p. 196.
24. *Ibid.*, p. 280.
25. Ibid., pp. 126–136; cf. also his book *The Evolution of Scientific Thought from Newton to Einstein* (New York, Dover Publ., 1950), Chap. XXI, pp. 212–227.
26. There was one such misprint in his article published in 1968 (cf. Note 2). Fortunately, in a copy which the author kindly sent to me, the misprint was corrected by the author's pen.
27. 'Le temps d'Einstein et la philosophie. A propos de la nouvelle édition de l'ouvrage de M. Bergson *Durée et simultanéité*. *Revue de philosophie*, vol.

XXXI (1924), pp. 56–87.

28. The absolutely solid bodies would be able to transmit the physical actions instantaneously; their impossibility from the standpoint of relativity was recognized by P. Langevin (cf. Note 20), then by H. Reichenbach (*op. cit.*., p. 127). A. Metz, who conceded the impossibility of instantaneous signals (cf. his article 'De la durée intuitive au temps scientifique', *Dialogue* V, 1966–7, p. 203), nevertheless retained his moving sidewalk and solid road in his most recent article 'Interpretation philosophique de la théorie de la relativité' (*Bull. de la Soc. fr. de Philosophie*, avril-juin 1967, p. 68.)
29. Cf. his art. quoted in Note 27, p. 74–5; also two incompatible theses – the inequality of proper times and the reciprocity of appearances – in his last article pp. 43–4.68 (Cf. Note 27).
30. H. Bergson, 'Le temps fictif et le temps réel' in *Revue de philosophie*, XXXI (1924), pp. 241–260.
31. Cf. *D.S.*, Appendix I, pp. 163–67: which contains a letter of Jean Becquerel to Bergson; *Mélanges*, pp. 218–20.
32. This fact is emphasized by M.-A. Tonnelat, *op. cit.*, p. 285.
33. Cf. Note 16 and A. Metz, *Revue de philosophie*, XXXI, pp. 70, 438–39; Bergson's answer on pp. 256–260.
34. M.-A. Tonnelat, op. cit., pp. 253–4.
35. *Ibid.*, p. 254.
36. *Ibid.*, p. 251.
37. Paul Langevin, *La physique depuis vingt ans* (Paris, 1923), p. 294 f. This book contains also two of the author's articles, referred to above.
38. The analysis of the twin paradox by Reichenbach, Whitehead and Bohm are reprinted in *The Concepts of Space and Time. Their Structure and Their Development* (D. Reidel, Dordrecht, 1976), pp. 441–453.
39. *Op. cit.* (cf. Note 36 above), p. 292.
40. Albert Einstein, *Relativity. The Special and General Theory.* (Crown Publ., New York, 1961), pp. 25–6.
41. My italics.
42. Jacques Maritain, 'Les nouveaux débats einsteiniennes, *Revue universelle* (avril 1923), pp. 65–67.
43. It was apparently Einstein himself who accused Bergson of ignoring the absolute character of spatio-temporal coincidences in a letter to André Metz, July 2, 1924, quoted in *Revue de philosophie*, v. XXXI, p. 440
44. A. d'Abro, *op. cit.*, pp. 119–128, 'Le principe de reciprocité fausse de M. Bergson.' One example of d'Abro's caricatures: 'If Paul holds a sword in his hand and if we immobilize Peter in our thought so that it is Paul who moves to Peter, it will be Peter who will be transpierced. If we now immobilize Paul, it will be Paul who, in virtue of the principle of reciprocity, will be transpierced.'
45. G.N. Lewis established long ago the identity of the relativistic kinematics with the geometry of asymptotic rotation. (*Proceedings of American Academy of Arts and Sciences* (1912); *The Anatomy of Science* (Yale Univ. Press, 1926), Ch. II and III. In the special theory it is only the three dimensional space which remains Euclidian; but this is only an instantaneous cut in the time-space of Minkowski.

46. *D.S.*, p. 30; *Mélanges*, p. 85.
47. A.S. Eddington, *The Nature of the Physical World* (Cambridge Univ. Press, 1933), pp. 40–47. Cf. Paul Langevin's definition of space as 'the class of simultaneous events' (*op. cit.*, p. 275).
48. I. Newton, *Opera*, ed. by Horsley, II, p. 172, P. Gassendi, *Syntagma philosophicum* (1658), p. 224.
49. *D.S.*, p. 52, *Mélanges*, pp. 105–6.
50. P. Langevin, *op. cit.*, pp. 287–88, 338–342.
51. On this point cf. E. Meyerson, *The Relativistic Deduction*, tr. by David A and M.A. Sipfle (D. Reidel, Dordrect, 1985), Ch. VII; H. Reichenbach, *op. cit.*, p. 134; A.S. Eddington, *op. cit.* (cf. Note 45), pp. 47–52. A more complete list of those who rejected 'the spatialization of time' is in my book *The Philosophical Impact of Contemporary Physics* (Van Nostrand, Princeton, 1969), Ch. XI 'The Fusion of space with time and its misinterpretation.'
52. According to the information given to me by later Professor Giorgio Santillana the term 'chronotopo' is used in Italian.
53. A.S. Eddington, *op. cit.*, p. 47, 57–8.
54. Cf. my arrticles 'The Myth of Frozen Passage' in *Boston Studies in the Philosophy of Science*, 1964, pp. 441–463; 'Relativity and the Status of Becoming,' *Foundations of Physics*, V (1975), pp. 606–16.
55. O. Costa de Beauregard, *Le second principe de la science du temps* (Paris, Ed. du Seuil, 1959), p. 132.
56. *D.S.*, p. 62; *Mélanges*, p. 115. Bergson there refers to *The Concept of Nature* (Cambridge Univ. Press, 1920) where Whitehead expressed his agreement with him.
57. A.N. Whitehead, *ibid.*, p. 178.
58. *D.S.*, p. 118, *Mélanges*, p. 171.
59. A.N. Whitehead, *op.cit.*, p. 178: 'The various time-series each measure some aspect of creative advance, and the whole bundle of them express all the properties of this advance which are measurable.'
60. *D.S.*, p. 46; *Mélanges*, pp. 99–100; Z. Zawirski, *L'Evolution de la notion du temps*, (Cracovie, 1934), pp. 305–5.
61. *D.S.*, p. 53; *Mél.*, p. 106. Cf. the negation of instantaneous space in *Matter and Memory*.
62. A. d'Abro, *op. cit.*, pp. 305–6.
63. *Creative Evolution*, tr. by Arthur Mitchell (The Modern Library, New York, 1944), pp. 223, 227; in a new reprinting and introduction by P.A. Gunter (Univ. Press of America, 1983), p. 203, 207. On this point of my articles 'La genèse idéale de la matière chez Bergson' in *Revue de metaphysique et de morale* (Vol. 57, 1952) esp. p. 313, in 'Bergson et l'esprit de la physique contemporaine,' *Actes du Congrès Bergson* (Paris, 1959, pp. 53–6.)
64. A.N. Whitehead, *The Principles of Natural Knowledge* (Cambridge University Press, 1925), p. 6; *Science and the Modern World* (Macmillan, 1925), p. 172, 54. See Bergson's agreements with Whitehead: *D.S.*, p. 62, *The Creative Mind*, p. 84. (where he accepts Whitehead's 'melodious continuity' of a material particle). On the significance of the auditory models cf. my book *Bergson and Modern*

Physics, part III, Ch. XV, pp. 313–334.

65. A. d'Abro, *op. cit.*, pp. 43–52. D'Abro obviously misunderstood the principal thesis of Bergson, inspired by Leibniz, according to which motion in space is a surface manifestation (*phaenomenum bene fundatum*) of the underlying qualitative change.
66. The English translation 'the extensity of becoming' is hardly an adequate rendering of the French term 'le devenir extensif'. Cf. p. 317 of *L'Evolution créatrice* (62 ed., 1946) and compare it with p. 345 of the English edition. Concerning the affinity of Bergson's 'extensive becoming' with Einstein's time-space cf. my articles and the books quoted above. Jean Piaget was right when he wrote: "It is significant that Bergson, instead of applauding the fact that Einstein introduced in physics a model of time far closer to the time of psychology (we might say to Bergsonian time) than to the time of Newton; on the contrary he tried to challenge relativity in a rather questionable way as if relative time should be reserved only for psychological and biological time!" Cf. *Le developement de la notion de temps chez l'enfant* (Paris, P.U.P., 1946), p. 298, my own translation.

* 'Contemporary' would be a more proper word.

CHAPTER 17

TIME-SPACE RATHER THAN SPACE-TIME

The Classical Space-Time

Hardly any other problem has been discussed more than that of the status of time in modern physics. This is only natural since there are not many other more important problems in philosophy of science and in philosophy in general. There are also few other areas where controversies as well as confusion were more frequent. This is true not only of popular and semi-popular expositions of the Minkowski concept of space-time but also of a number of its philosophical interpretations. Generally we do not find anything of this kind in the writings of physicists, at least as long as they confine themselves to strictly mathematical and physical expositions; but when they sometimes venture beyond a strictly mathematical approach, they often do not escape certain unconscious or semi-conscious prejudices which are contrary not only to the spirit but sometimes even to the letter of relativity. The true significance of the relativistic fusion of space and time can be understood only when we contrast it with its classical counterpart, i.e., with what may be called the Newtonian space-time. Only on such a contrasting background will the revolutionary meaning of the new concept clearly stand out.

The term "space-time," coined by Minkowski in 1908, probably had never been used before that date; but although the word was missing, the concept itself was present, even though, as we shall see, its meaning was altogether different from that of Minkowski. Thus Descartes called time a "dimension," d'Alembert even "the fourth dimension;" while Lagrange called mechanics "*une géométrie à quatre dimensions.*"[1] There is no question that classical science as well as classical philosophy had a definite notion about the way space and time were related; in this sense one can speak about the classical i.e., Newton-Euclidian, space-time or of the "four-dimensional space-time continuum."

Because of our psychological inability to visualize the fourth dimension, the four-dimensional continuum, whether classical or relativistic, can be symbolically represented only by three-dimensional models; only on such models can the structure of both concepts as well as their most significant differences be conveniently studied. It is true that there are certain implicit dangers in using such graphical representations, because in every geometrical diagram time appears to be misleadingly spatialized. On the other hand, such diagrams, provided we do not forget their symbolic nature, have a definite advantage of disclosing more clearly the relation of space and time in their respective synthesis, whether classical or modern. A similar procedure was successfully used when the properties of Riemannian space, which by their own nature remain non-intuitive, were illustrated by the properties of a two-dimensional spherical surface.

Now in the three-dimensional model of space-time, its spatial component is represented by a Euclidian plane, either horizontal or vertical, while "the fourth dimension of time" is symbolized by a straight line, perpendicular to this plane. All successive instantaneous spaces are represented by parallel planes, all orthogonal to the time-axis, each of which contains a state of the world history at that particular instant. In other words, each instantaneous space contains all events which are

[1] See the references in Émile Meyerson, *La déduction relativiste*, Paris, 1925. pp. 107-8.

simultaneous in the absolute sense; in truth, there is a mere terminological difference between the expression "the class of simultaneous events" and "instantaneous space." Let us recall the lucid words of Hermann Weyl: "All simultaneous world-points form a three-dimensional *stratum*, all world-points of equal location, a one-dimensional *fiber*."[2] Thus the classical space-time can be defined as a continuous succession of instantaneous spaces.

It is hardly necessary to stress how fundamental this model of space-time had been in classical physics. It underlay the concept of absolute simultaneity which was absolutely essential to the classical models of matter. In classical corpuscular-kinetic models every state of the world at each particular instant was defined as a huge instantaneous configuration of an enormous number of particles, each of which was moving according to the laws of classical mechanics while preserving its physical identity through time. World history was thus viewed as a continuous succession of such instantaneous configurations. It is obvious that without the concept of absolute simultaneity that of "instantaneous configuration" would be devoid of meaning; in truth, both concepts are nearly synonymous. I say "nearly" since the state of the world at an instant could also be defined in the sense of the field theories as a set of instantaneous field intensities; what is important is that in both models the assumption of absolute simultaneity was the same. This assumption of "Everywhere Now" seemed so natural and obvious that it was rarely stated explicitly; it was one of those tacit assumptions underlying the conceptual structure of classical theories. What was more natural than to believe that simultaneous with my present moment on the Earth there is a definite moment on Mars, Neptune, Alpha-Centauri, Andromeda nebula and even on any remote galaxy, no matter how enormous its distance may be?[3]

[2] Hermann Weyl, *Philosophy of Mathematics and Natural Science*, Princeton University Press, 1949, p. 95.

[3] Cf. Pierre Gassendi, *Syntagma philosophicum* (ed. Lugduni 1658), I, p. 224: "*quodlibet temporis momentum idem est in omnibus locis*;" Isaac Newton. *Opera*, ed. by Horsley, III, p. 72: "*unumcumque temporis indivisibile momentum ubique*." On this point cf. my article "Was Gassendi a Predecessor of Newton?" in *Proceedings of the X International Congress of History of Science*, Paris, 1964, pp. 705-9.

The concept of absolute simultaneity was also one of the most essential parts of the classical deterministic scheme. Let us recall its famous formulation by Laplace: a complete knowledge of *the present state of the world*—which in prevailing mechanistic models meant a complete knowledge of the position and velocities of *all* the particles in the universe—would make possible in principle the knowledge of all future states of the world. Technical impossibility of such prediction was irrelevant; what was important was the conviction that the present state of the universe—in truth *any* particular state—entails all its subsequent states, even their most insignificant details. The Laplacean "omniscient mind" was merely a metaphor illustrating the universal impersonal order of nature in which everything is rigorously predetermined from eternity and without any shade of ambiguity. This was stated by Democritus at the very dawn of Western thought: "By necessity are foreordained all things that were and are and are to come." Classical physics, twenty-two centuries later, expressed the same view in a more precise form by substituting the laws of Newtonian dynamics for the general term "necessity." There were difficulties in this view, as I shall try to point out; but what is important in the present context is that without the concept of objective, i.e., absolute simultaneity, the very concept of a state of the world at an instant—what Eddington later called "world-wide instant"[4]—loses its meaning. Yet, it is this concept on which the Newton-Laplacean determinism was based.

The Structure of Minkowski's World

One of the most fundamental differences between Newton-Laplacean space-time and its modern Einsteinian counterpart is that the latter cannot be defined as a succession of instantaneous spaces. In other words, in the latter it is impossible to make instantaneous cuts perpendicular to "the axis of time;" if we try to do it, we obtain different instantaneous spaces in dif-

[4] A. S. Eddington, *The Nature of the Physical World*, New York, Macmillan, 1933, p. 47.

ferent inertial systems. Using a similar three-dimensional model used before, we see at once a significant difference between the old and the new model: in the latter instead of one single orthogonal plane, representing one single instantaneous space at that particular instant, we obtain a multiplicity of such planes, inclined at different angles with respect to the time-axis. What is more important is that *none of them has any privileged status over any other.* In other words, there is no unique cosmic "Everywhere-Now," no "world-wide instant;" this is the meaning of the famous *relativization of simultaneity.* This is the term commonly used; but as such it does not convey the truly radical meaning which underlies it. A far more appropriate term is "denial of simultaneity." In Einstein's own final words: "There is no such thing as simultaneity of distant events."[5]

These words of Einstein's are apparently little known and, so far as I know, hardly ever stressed; yet they refer to the fact which becomes evident by a mere inspection of Minkowski's diagram. While in the classical diagram a single instantaneous space, symbolized by a plane surface perpendicular to the time dimension, separated at each particular moment the objectively past events from the objectively future events, the situation is quite different and far more complex in Minkowski's scheme. In it there is also an objective difference between the past and the future at each particular Here-Now event. But, unlike the classical scheme, the past is separated from the future not by a single instantaneous space but by *the whole four-dimensional region* which Eddington called "Elsewhere" and which in Minkowski's diagram is represented by a three-dimensional region lying between the conic regions of "Absolute Past" and "Absolute Future." By "Absolute Past" are designated those events which causally influence the particular Here-Now event; by "Absolute Future" those events which will be or may be influenced by the present Here-Now. This is why such events are also called "causal past" or "causal future" respectively. The conical shape of these regions is due to the limit character of the velocity of light; unlike in classical physics, no physical action can move with a velocity larger than that of electro-

[5] A. Einstein, "Autobiographical Notes," in: *Albert Einstein, Philosopher-Scientist*, ed. by Paul Schilpp, Evanston, Ill., 1949, p. 61.

magnetic waves. Consequently, the world-lines of photons lie on the surface of the cone which has its vertex in the Here-Now event and which separates the causal past from the Elsewhere region. The same is true of the forward causal cone, separating the Absolute Future from the Elsewhere region, except that we have to be on guard against taking our spatial diagram too literally; in other words, we must keep in mind that for each Here-Now future events are only *potentially real*; otherwise we would slip unwittingly into the fallacy of spatialization. (Such potentiality of the world lines can be indicated in our diagram by drawing *the dotted lines* in contrast to the fully drawn lines in the rearward cone of the past events.)

The following points should be stressed when we want to interpret the physical and—let us not be afraid of the word—*philosophical* meaning of Minkowski's time-space.

1. *Simultaneity of distant events is not only made relative but is simply denied.*

Let us recall again Einstein's *ipsissima verba.* This will be hotly denied by all those who claim that the term "simultaneity" becomes now a three-term relation: instead of speaking of simultaneity of two events as classical physics did, we must specify the inertial system in which such simultaneity occurs—and then such simultaneity can be unambiguously defined. In other words, each observer can legitimately bisect his own Elsewhere region by an instantaneous cross-section—his own instantaneous space—on which the events simultaneous with his own Here-Now are located. On this point two following remarks should be made. First, all such so-called simultaneous events are *ex definitione* unobservable since they are located in the observer's Elsewhere region from which not even the fastest signal can reach his particular Here-Now. In other words, the existence of such events is merely *stipulated*, never perceived or registered. Second, the physical entities cannot be created by a mere stipulation. Which physical meaning can be assigned to the entities, intrinsically unobservable and which, furthermore, are different in different inertial systems?[6] The attribute of "being present"

[6] Cf. M. Capek, *The Philosophical Impact of Contemporary Physics*, enlarged ed., Princeton, Van Nostrand, 1969, pp. 189-90.

is confined to each particular "Here-Now" and can never be extended beyond its limits to become "Everywhere Now." The word "present" should now be taken in its original etymological sense of *prae-esse* in both its spatial and temporal sense. Each "Here-Now" is simultaneous with itself (which, as we shall see, is not as trivial as it sounds). As A. A. Robb, an unjustly forgotten "Euclid of relativity," expressed it more than three-quarters of century ago, "the present instant properly speaking does not extend beyond itself." In other words, still his, "we cannot strictly identify the same instant in two distinct points of space."[7] Eddington expressed the same view when he dismissed the existence of the "world-wide instants;" and Whitehead when he stated that "there is no such unique present instant" at which all matter is simultaneously real.[8]

This is one of the most paradoxical results of relativity against which our Newton-Euclidian subconscious vigorously protests. As we shall see, the notion of absolute simultaneity persists in the imagination of not a few physicists and cosmologists despite their explicit verbal denials; "the notion of absolute simultaneity is so deeply ingrained in the way most people think about space-time that it even takes a great deal of effort to be consciously aware of when and how one is using this assumption."[9]

2. *Elimination of simultaneity does not mean an elimination of the successive character of the physical world.*

Unfortunately, the very opposite view is widely spread and can be found not only in popular and semi-popular expositions of relativity but not infrequently among physicists and even more among philosophers. Emile Meyerson in his book *La déduction relativiste*, which Einstein praised as one of the best philosophical interpretations of relativity, gave a long list of those who interpreted Minkowski's space-time in a static sense, as a sort

[7] A. A. Robb, *The Absolute Relations of Space and Time*, Cambridge University Press, 1921, pp. 7, 12-13.

[8] Eddington, *loc. cit.*; A. N. Whitehead, *Science and the Modern World*, New York, MacMillan, 1926, p. 172.

[9] Robert M. Wald, *Space, Time and Gravity*, Univ. of Chicago Press, 1977, p. 30.

of four-dimensional hyperspace whose fourth, so-called temporal dimension was not essentially different from the three spatial ones.[10] Even such an outstanding man as Ludwig Silberstein claimed that the theory of relativity was anticipated by H. G. Wells in his famous novel *Time Machine,* in which the fictitious inventor makes a machine on which he can ride in either direction of time, either into the past or into the future.[11] For such a traveller, time would obviously cease to exist as its "successive" phase would co-exist simultaneously, i.e., would not be successive at all. What we call "future" would really be a hidden present, an unknown territory not yet discovered but already existing prior to our discovery. Fortunately, such fantasies have not the slightest basis in the physics of relativity. Yet, the very persistence of such misinterpretations must have some deep, underlying cause which a historian of ideas can easily identify: the perennial tradition of both Western and Eastern thought which regards time as merely apparent and not genuinely real. The tendency to spatialize time is only a more concrete form of the same traditional trend. But in addition to this tradition the very fact of relativity of simultaneity is often used as an argument for a static interpretation of Minkowski's continuum as, for instance, by Kurt Gödel:

> The existence of an objective lapse of time, however, means (or, at least is equivalent to the fact) that reality consists of an infinity of layers of "now" which come into existence successively. But, if simultaneity is something relative in the sense just explained, reality cannot be split up into such layers in an objectively determined way. Each observer has his own set of "nows," and none of these various systems of layers can claim the prerogative of representing the objective lapse of time.[12]

[10] E. Meyerson, *La déduction relativiste*, pp. 97-108. Einstein's comment on Meyerson's book was published in *Revue philosophique de la France et de l'étranger* v. 105, 1908, pp. 161-66. Its English translation by Mary-Alice and David A. Sipfle was published in my anthology *The Concepts of Space and Time. Their Structure and Their Development*, Dordrecht, D. Reidel, 1976, pp. 361-367.

[11] L. Silberstein, *Theory of Relativity*, London, 1914, p. 134. Quoted by H. Bergson, *Durée et simultanéité*, Paris, 1923, p. 223.

[12] Kurt Gödel, "A Remark About the Relationship Between Relativity and Idealistic Philosophy," in *Albert Einstein: Philosopher-Scientist*, ed. by Paul Schilpp, Evanston, Ill., 1949, p. 558.

Gödel's argument sounds at first very plausible; for if there is no universal "Eyerywhere-Now," there would be no objective boundary separating the past from the future and thus the very distinction between successive phases of the universe would apparently disappear. What Gödel overlooked was the fact that in the world of Minkowski the future is separated from the past even *more effectively* than in the classical space-time as even a superficial inspection of the relativistic space-time diagram shows when we compare it to the classical diagram. While in the latter the boundary separating the future from the past is an "infinitely thin," durationless layer (i.e an instantaneous space at each particular moment), in the former it is the whole four-dimensional region of "Elsewhere" which separates them. The very existence of the Elsewhere region is a direct consequence of the limit velocity of light which may be properly called the velocity of causal propagation; as Eddington observed long ago, "the limit to the velocity of signals is our bulwark against the topsy-turvydom of past and future, of which Einstein's theory is sometimes wrongly accused."[13]

But it is precisely such "topsy-turvydom of past and future" which Gödel explicitly advocates. He is only consistent when he seriously considers the possibility of a Wellsian trip to the past —and to the future—and back to the present. (He even computes the weight of a fuel which a rocket ship would need for such a round trip!) He is equally consistent when he is aware of his intellectual kinship with Parmenides, McTaggart and the tradition of timeless idealism.[14] But he errs when he confuses the elimination of Newtonian time with an elimination of time in general. What he does not realize is that Newtonian time is only a special case of time in general in a similar sense as classical Euclidian space is merely a special instance of space or spatiality in general. To deny temporality in general because its specific Newtonian form proved to be unsatisfactory is as little justified as the claim of some Kantians that a denial of Euclidian space destroys the possibility of *any* geometry; or

[13] Eddington, *op. cit.*, pp. 57-58.
[14] K. Gödel, *loc. cit.*, pp. 558-561.

that a rejection of classical determinism excludes the possibility of *any* causation.[15]

3. *The causally related events which are successive in one frame of reference remain so in all other inertial systems.*

In other words, while the juxtaposition of events (which is just another term for their simultaneity) is fully relativized—I would prefer to say, with Einstein, denied—the succession of the events mentioned above remains *absolute*, independent of the observer; it is *topologically, though not metrically* invariant. This is one of a few absolutes preserved by relativity; in truth, it is more correct to say that this particular absolute was *discovered* by relativity, for in classical physics the situation was different. Since there was no upper limit to mechanical velocities, for an observer moving with the velocity of light world history would be standing still while an observer moving faster than light would perceive—with a sufficiently powerful telescope—the earth's history in a reversed order; to him "Waterloo would precede Austerlitz" as he would be gradually overtaking the earlier and earlier wave-fronts of light.[16] It is true that in classical physics such inversion of causal relation would be merely *apparent* because for the privileged observer, at rest with respect to absolute space, the events would appear in their true and objective order. But such a situation would be far more serious in the theory of relativity which eliminated the privileged frame of reference—*if* there were no limit to the velocity of light. But fortunately this is not so; thus because of the unattainability of the velocity of light *not even an apparent* inversion of cause and effect can ever occur.

All this follows inescapably from Minkowski's formula for the constancy of the world interval as was pointed out long ago by Paul Langevin. It would be otiose to restate it again in a specific mathematical form.[17] It may be summed up in the

[15] On this problem cf. my articles "The Doctrine of Necessity Re-examined," *The Review of Metaphysics* V, 1951, pp. 11-44; "Toward a Widening of the Notion of Causality", *Diogenes* No. 28, Winter, 1959, pp. 63-90.

[16] This possibility was envisaged, for instance, by Flammarion as recalled by H. Poincaré in his *La science et la méthode*, Paris, 1909, Ch. 4.

[17] Paul Langevin, "Le temps, l'espace et la causalité dans la physique moderne," *Bulletin de la Société française de la philosophie*, Séance du 19 octobre 1911;

following way: the sequence of the events whose spatial separation is smaller than their separation in time multiplied by the velocity of light—in other words, the temporal order of causally related events—can never degenerate into simultaneity by any choice of the frame of reference; *a fortiori* it can never be inverted. In the usual language of relativity, their time separation is *absolute.* In this respect the temporal order of such events is basically different from that of causally unrelated events which Hans Reichenbach appropriately called "unreal temporal sequences" (*die irreellen Zeitfolgen*)[18] whose inversion can be obtained by a convenient change to a different referential system. Reichenbach's term is especially well chosen since it indicates their unreal fictitious character. Their status is purely conceptual, comparable to the status of simultaneity of distant events; no world lines, no concrete physical connections correspond either to "unreal sequences" or to "simultaneity lines." This is obvious from Minkowski's space-time diagram: both the simultaneity lines and unreal sequences lie in the Elsewhere region, *causally not interacting* with the Here-Now.

It is thus clear that the relativistic space-time—whose more appropriate name should be time-space—consists of a network of the causal line ("world-lines") *whose successive, irreversible character is absolute*, i.e independent of any choice of the frame of reference. Such time-space is obviously *toto coelo* different from the static, becomingless hyperspace which exists more in the imagination of some philosophers than in the thought of physicists.

The Physical Unreality of the Future

The reality of succession and the unreality (or "virtuality") of the future are logically correlated terms; one cannot have one without the other. Conversely, a denial of succession always went hand in hand with the view that the future is *somehow real*, even though still hidden to our consciousness which remains

"L'evolution de l'espace et du temps," *Revue de métaphysique et de morale*, vol. XIX, 1911, pp. 455-466; also in *Scientia*, vol. X, 1911, pp. 31-54.

[18] Hans Reichenbach, *Die Philosophie der Raum-Zeit Lehre*, Berlin, 1928, p. 175.

blindfolded by the illusion of time which prevents it from perceiving what timelessly exists and what only human ignorance calls "future." Historical examples abound and to mention all of them would mean to give a survey of the whole history of Western thought from Parmenides to Bradley; the contemporary doctrine of "the mind dependence of becoming" is its last version. Thus the temptation to interpret Minkowski's diagram in the terms of the time-honored Eleatic tradition was naturally very strong. But it can be convincingly shown by attentively analyzing Minkowski's formula and his space-time diagram that no verifiable physical reality corresponds to future events; in other words, that the future is *physically empty*.[19]

One important distinction is not infrequently overlooked and rarely stressed explicitly, yet without it confusions cannot be avoided. It is the distinction between *nominally "future" events* and the genuine future. To the first category belong the events whose relations to my "Here-Now" are temporally indeterminate since they belong neither to my absolute past nor to my absolute future. Such events are *declared* to be future if they lie on the forward side of my "now line" by which I arbitrarily divide my own region of Elsewhere (which could also be called "Elsewhen"). Another observer, who shares with me my "Here Now," but belongs to an inertial system different from mine, will draw a different "now line;" to him the events which I regard as future, will appear as either simultaneous or in his non-causal past. (We should not really use the term "appear" since *all* events in Elsewhere are not only never perceived but are *unperceivable in principle.*) On the other hand there is the *authentic* future—my own *causal* and *absolute* future, symbolized by the forward cone radiating from my Here-Now. Its absolute character follows immediately from Minkowski's formula for the constancy of the world interval which entails the irreversibility of the temporal order of causally connected events.

For consider any event in my causal future; it is (more accurately: it *will* be) a causal successor of the event Here-Now and in this sense it will occur *after* my own experienced present.

[19] I dealt with this problem most recently in the article "Relativity and the Status of Becoming," *Foundations of Physics*, vol. IV, December, 1975, in particular in its last part "The Physical Emptiness of the Future." (pp. 610-17).

(My own perception and, more generally, my own psychological present is irrelevant since even a "merely physical" Here-Now event will exhibit the same relationship to its own causal future). This means that not only can I not perceive any event on my own future world-line, *but neither can I perceive any event on that segment of any other world-line different from mine which is included in my own causal future.* For the necessary condition for the observability of any event is its inclusion in the causal past of the observer in question. No event in my causal future can affect my Here-Now for the simple reason that my own causal past and causal future do not overlap. This conclusion is so truistic that it may sound silly—until we remember the persistence of "time-tunnel" fantasies about the round trips to the past, the "messages" from the future, etc.

But could my future events perhaps be perceived by the observers located on some world lines which are far enough from my present Here-Now? Such hypothetical observers can be divided into three classes: a) those included in my own causal past; b) those in my own Elsewhere region and, finally; c) those who belong to my causal future. As we are going to show, my own absolute future remains unobservable in principle *in all these three cases.* The group a) is automatically excluded by the fact that no signal can be sent to the past. For in the light of Minkowski's diagram my Here-Now—and *a fortiori* my absolute future—are included in the absolute future of all my causal predecessors; no signal can reach them from my own Here-Now—*a fortiori* none from any event in my absolute future. "We cannot send wire messages into the past," as Einstein observed long ago.[20] For common sense this is obvious; but it is known that that is not always a reliable judge, as the whole development of modern physics shows. But in this particular case the conclusion of common sense coincides with that of Einstein.

One will reach the same conclusion in considering a hypothetical observer in the Elsewhere region. By the very definition of this region, no causal influence, no signal coming from my Here-Now, can reach him. If, *per impossibile*, he should receive a message from my causal future *before* the signal coming from

[20] Quoted by Meyerson, *op. cit.*, p. 104.

my present Here-Now, he would perceive the causally related events in a *reversed temporal order* in contradiction to Minkowski's formula. The impossibility of such a situation follows immediately from Minkowski's diagram: every signal coming either from my present Here-Now or from my absolute future could reach an observer in Elsewhere only by the velocity greater than that of light in violation of the basic principle of relativity.

There remains the third class of hypothetical observers—those who belong to my absolute future. Common sense and the majority of scientists will dismiss the very idea of "future observers" as a self-contradictory fiction: are not "future observers" unreal by their very definition? But, as mentioned above, one must be on guard against deceptive intuitive certainties of traditional common sense; this is why those thoroughly acquainted with relativity will prefer to stress an *intrinsically unobservable*, i.e. *counter-empirical* character of the notion of "future frame of reference." *All* events in my causal future are intrinsically unobservable since they are not included in the causal past of my Here-Now; and this is obviously true of everything happening to hypothetical future observers. Thus the notion of observability by the observers who themselves are intrinsically unobservable remains meaningless physically as well as philosophically.

This conclusion is hotly challenged—it is true more by some philosophers than by physicists—by the counter-arguments which have a plausible relativistic ring. They do not deny that every particular Here-Now event divides unambiguously its absolute (causal) past from its absolute (causal) future. Neither would they deny (though they hardly ever stress it) that the events in absolute future are intrinsically unobservable for that particular Here-Now present. But they point out that this is true of *every* present and since the basic idea of relativity is the equivalence of *all* frames of reference, the very concept of Here-Now is relativized, which also means a relativization of the dividing line between the past and the future. To single out any particular present as absolute is contrary to the principle of relativity which denies the existence of any privileged frame of reference. There is an infinite number of different "Here-Nows" and, consequently, an infinite number of different ways to separate the past from the future; my own particular "Here-Now" is as relative as any other. Furthermore, it is continually shifting and its very movement makes its choice arbitrary. All individual Here-Nows are equivalent and in this sense equally real.[21] Another argument for a static interpretation of Minkowski's world. We would be back to Parmenides!

Let me omit the historical reasons, already mentioned, which make this view thoroughly suspect. More important are the following reasons: a) If by "relativity of the present" is meant the fact that it is non-stationary, there is no real disagreement; for the present moment is by its very nature transitory, "perishing," as Whitehead said, and to say that something is passing is another way of saying that something *become*s something else, that becoming is real. It is another argument for the dynamic nature of space-time as the supersession of the present by its causal successor is the very essence of becoming. b) If by "relativization of Here-Now" is meant the fact that there will be a future Here-Now which will include in its causal past my own present event, then again there is no disagreement; it is obvious that the events in 1984 will be influenced by my present Here-Now. It is equally obvious that the events in 1984 will be witnessed by the observers at that time; but then the whole thesis is reduced to a harmless truism that "future events will be observed in future frames of reference." But the disagreement begins as soon as we replace the future tense "will be observed" by the tenseless "*is* observed." If the latter is understood in the timeless Eleatic sense, then all frames of reference, including those now in the future would be, indeed, on equal footing, being all "equally real." But they cannot be.

For the word "arbitrary" applied to the present "Here-Now"

21 This argument was put forth by Hugo Bergman, *Der Kampf um das Kausalgesetz in der jüngsten Physik*, Braunschweig, 1929, pp. 25-28. This argument was adopted by A. Grünbaum, *Philosophical Problems of Space and Time*, New York, 1963 and *Modern Science and Zeno's Paradoxes*, Wesleyan Univ. Press, 1967, Ch. I. My answer to Grünbaum is in "The Myth of Frozen Passage" in *Boston Studies in the Philosophy of Science*, II, 1965, pp. 441-453 and "Relativity and the Status of Becoming," *Foundations of Physics* V, 1975, pp. 607-616. Some other defenders of the static interpretation of Minkowski's world ignore the basic difference between Newtonian and Einsteinian spacetime; for instance Donald Williams "The Myth of Frozen Passage" *Journal of Philosophy*, v. 40, 1951, p. 457 and W. Quine in his *Word and Object*, Cambridge, Mass., 1967, p. 160.

is certainly out of place. For my present living "now" is unescapable and in this sense *absolute*; as Hans Reichenbach pointed out, in reformulating, perhaps unwittingly, the Cartesian *Cogito* in a dynamic, temporalistic sense, the very act by which we deny it reasserts it:

> Un acte de pensée est un événement et définit donc une position dans le temps. Si mes expériences se produisent toujours dans le cadre d'un "maintenant" cela veut dire que chaque acte de pensée définit un point de référence. Nous ne pouvons pas échapper au "maintenant" parce que la tentative d'y échapper signifie un acte de pensée et donc définit un "maintenant." Une pensée sans un point de référence n'existe pas, parce que la pensée elle-même le définit.[22]

It is certainly significant that Reichenbach, who warned against the spatialization of time in his earlier writings,[23] reasserted so definitely the reality of becoming in one of his last articles. In truth, what he asserted in the passage quoted above is hardly denied by the opponents who nevertheless insist that while "Now" is *psychologically* real, it neverthless does not have any *physical* status.[24] Yet, we are certainly living in the twentieth century, more specifically in the year 1983; is it possible to claim with any degree of seriousness that such statements are devoid of any physical meaning? We are certainly not living in the Cretacious period nor in the time of the Norman invasion, nor are we living in the year 2000. My present living is certainly not merely psychological since it is roughly co-extensive with the present physical state of the Earth.

The doctrine of "mind dependence of becoming" which denies any objective status to Here-Now in excluding becoming

[22] H. Reichenbach, "Les fondements logiques de la mécanique des quanta," *Annales de l'Institut Henri Poincaré*, v. XIII, 1952, p. 157.

[23] H. Reichenbach, "Die Kausalstruktur der Welt und der Unterschied Vergangenheit und Zukunft," *Sitzungsberichte der math.-naturw. Abteilung der bayerischen Akademie der Wissenschaften*, Munich, 1924, pp. 133-175. Also *The Philosophy of Space and Time*, Dover Publ., 1958, esp. § 16, "The Difference between Space and Time" and § 43, "The Singular Nature of Time."

[24] H. Bergmann, *op. cit.*, p. 25: "Darum hat dieser rein subjektive Begriff der Jetz, der Gegenwart, in der Physik keine Stelle." Against Bergmann's view cf. G. J. Whitrow, *The Natural Philosophy of Time*, 2nd ed., Oxford, 1980, pp. 348-350.

from the physical world and confining it in a subjective, mental realm, advocates, probably unwittingly, a far more radical kind of dualism than the traditional Cartesian dualism. Its intrinsic difficulties are, consequently, far more serious. In Descartes' "bifurcation of nature" the physical and mental realms, despite their heterogeneity, shared at least one important feature: *the temporal character.* The physical as well as the mental events occur in time. In the new dualism they do not share even this single feature: on one side there is the physical four-dimensional world devoid of change as its so-called temporal dimension is, in virtue of its static character, a thinly disguised spatial distance; on the other side, there is the mental world into which change is confined. No intelligible relation or interaction between such radically heterogeneous regions is conceivable. At the same time, there is a strange ambiguity inherent in this doctrine. By its insistence on the static, timeless character of the objective world, it has an affinity with idealism which Kurt Gödel, for instance, admits quite openly; on the other hand, by claiming that the objective world is physical, it tends toward materialism which, for instance, J. J. Smart explicitly accepts. This ambiguity was already inherent in the first historical version of this theory—in the thought of Parmenides—and thus it is hardly surprising that there are similar hesitancies in the revived Eleatism of the twentieth century. Among those who are clearly aware of the dualistic character of this doctrine is M. Olivier Costa de Beauregard when he writes:

> Whitehead et M. Čapek parlent à ce sujet d'une avance créatrice de la nature. Le lecteur voit en quel sense nous n'adhérons pas à cette vue: *L'avance de Whitehead est écrite dans l'espace-temps, et en tant qu'avance* (pour nous), *et en tant que créatrice.* De la *matière seule*, qui est statiquement déployée dans l'espace-temps, on ne peut pas dire qu'elle avance; si donc on prononce le mot d'avance, et même d'*avance créatrice, c'est que la Nature dont on parle ne se réduit pas à la matière.* (Author's italics.)[25]

In other words, the author of this passage regards the allegedly static character of space-time as an *argument for dualism*; the

[25] Olivier Costa de Beauregard, *Le second principe de la science du temps*, Paris, Editions du Seuil, 1963, p. 132.

reality of becoming within the mental realm shows that not everything is reducible to matter. On the other hand, other thinkers leaning toward materialism such as, for instance, Mario Bunge, dismiss the static interpretation of space-time (without naming any of its representatives) contemptuously.[26] Such diverse reactions toward the Neo-Eleatic doctrine are due to its metaphysical ambiguity: it can be interpreted idealistically or physicalistically or dualistically. But all these interpretations have one postulate in common: that "the true reality exists unchangingly," or, as Bergson put it lucidly, "the totality of the real is postulated complete in eternity." In such a view "the apparent duration of things expresses merely the infirmity of a mind that cannot know everything at once."[27] But if succession is a mere "infirmity" or "illusion" of mind, the existence of "mind" or "mental realm" is tacitly assumed—and this makes the position of materialists or, as they prefer to be called today, physicalists, especially difficult. For idealists such as Bradley, or McTaggart, or Kurt Gödel, the alleged illusion of succession has at least a certain *locus* since it exists in the subjective realm. The physicalistic Neo-Eleatics, however, deny such realm or—what is the same—reduce it to the brain, i.e., to a part of the physical, in their view, becomingless, world; consequently, in their view, even the very *illusion* or *appearance* of becoming is impossible! They literally cut a branch on which they are sitting. The Neo-Eleatism of idealists or of Costa de Beauregard is strange because it leads to an irrational and needless bifurcation

[26] Mario Bunge, *Foundations of Physics*, New York, Springer Verlag, 1967, p. 206: "It is often claimed that SR [Special Relativity] has wiped out the difference between space and time and even between what has been and what may be: that it has spatialized time and that it pictures the world as a block given once and for all, so that nothing ever happens: everything would exist already in some region of the Minkowski space, which would be thoroughly homogeneous and isotropic. This is preposterous. SR cannot be even stated without the notion e.m. signal, and even e.m. signal is a process (sequence of events), not a static being."

[27] H. Bergson, *Creative Evolution*, tr. by A. Mitchell, New York, 1911, p. 45. The usual, uncritically accepted claim that Bergson "completely misunderstood relativity," has been recently challenged by Marie-Antoinette Tonnélat, *Histoire du principe de relativité*, Paris, 1971, p. 280-93; M. Capek, *Bergson and Modern Physics*, Dordrecht, 1971, esp. pp. 237-256; the same author "Ce qui est vivant et ce qui est mort dans la critique bergsonienne de la relativité," *Revue de Synthèse*, 1980, pp. 313-344.

of nature; the Neo-Eleatism of materialists is not only strange but also self-contradictory.

To sum up: all that was said before leads inescapably to one definite conclusion: the physics of relativity *does not eliminate* becoming. The allegedly existing—or pre-existing—future events are nothing but gratuitous and artificial constructions, intrinsically unobservable, inspired by an unconscious—and very ancient—metaphysics; they are as useless as other discarded and unobservable entities such as phlogiston, caloric, aether etc. This is why the term "time-space" is far more appropriate than "space-time" or "four-space." The last one is especially misleading.

The present trends in cosmology, in particular the theory of the expanding universe, is an additional indication of the fact that it is space which is incorporated into becoming rather than *vice versa.* But this would require another extensive analysis which would much increase the dimensions of this article.

The Dynamic Structure of Time-Space

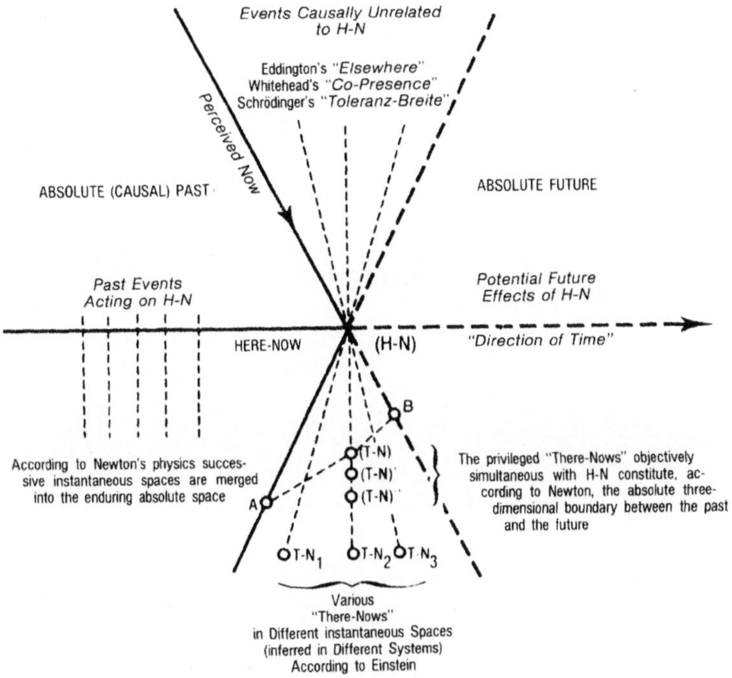

TIME-SPACE DIAGRAMS IN CLASSICAL AND MODERN PHYSICS

The main difference between classical and relativistic space-time is that in the former it was possible to have instantaneous three-dimensional cuts (instantaneous spaces) identical in all frames of reference while in the latter it is not. In relativistic space-time, absolute simultaneity was replaced by the wedge-shaped zone of the events causally independent of (H-N). But the elimination of absolute simultaneity of distant events does not destroy the reality of temporal relations. Although it is impossible to claim that (H-N) is objectively simultaneous with (T-N), (T-N)' and (T-N)'', the event (H-N) is *preceded* by the event A and *followed* by the event B in *all* frames of reference. Moreover, the worldline A—(H-N)—B is in *all* systems of reference *contemporary* with the worldline A—(T-N)—B. Thus the simultaneity of instants (co-instantaneity) has been replaced by the "contemporaneity of fluxes" or "contemporary intervals."
From the diagram it is clear that all "events" in the absolute future are intrinsically unobservable not only at H-N, but also by any conceivable observer in "Elsewhere"; their alleged existence (or rather pre-existence) is devoid of any physical meaning.

BIBLIOGRAPHY OF MILIČ ČAPEK

BOOKS

A Key to Czechoslovakia. The Territory of Kladsko (Glatz). A Study of a Frontier Problem in Middle Europe. R. Vogel, New York, 1946. Pp. 170, 9 maps.

The Philosophical Impact of Contemporary Physics. Van Nostrand, Princeton, 1961; the second printing 1963; enlarged paperback edition 1969. Pp. XVII + 419.

El Impacto Filosofico de la Fisica Contemporanea. Coleccion 'Estructura y Function', Madrid, Editorial Tecnos, 1965, pp. 411.

Bergson and Modern Physics. [*Boston Studies in the Philosophy of Science.* vol. VII.] Dordrecht, Holland, 1971. Pp. VII + 414.

The Concepts of Space and Time. Edited by Milič Čapek. [*Boston Studies in the Philosophy of Science*, vol. XXII] Dordrecht, Holland, 1976. Pp. LVII + 570.

T.G. Masaryk in Perspective. Comments and Criticisms. Edited by Milič Čapek and Karel Hruby. New York: SVU Press (Czechoslovak Society for Arts and Sciences), 1981. Pp. X + 282.

Note: Czech books not listed.

ARTICLES

1. 'Stream of Consciousness and '*durée réelle*',' *Philosophy and Phenomenological Research 10 (1950), pp. 331–353.*
2. *'The Doctrine of Necessity Re-examined,' The Review of Metaphysics 5* (1951), pp. 11–54.
3. 'La genèse idéale de la matière chez Bergson,' *Revue de métaphysique et de morale 57* (1952), pp. 325–348.
4. 'La théorie bergsonienne de la matière et la physique moderne.' *Revue philosophique 163* (1953), pp. 28–59.
5. 'The Reappearance of the Self in the Last Philosophy of William James.' *The Philosophical Review 62* (1953), pp. 526–544.
6. 'James' Early Criticism of the Automaton Theory,' *Journal of the History of Ideas 15* (1954), pp. 260–270.
7. 'Relativity and the Status of Space,' *The Review of Metaphysics* IX (1955), pp. 169–99.
8. 'Note about Whitehead's Definition of Co-presence,' *Philosophy of Science 24* (1957), pp. 79–85.
9. 'The Development of Reichenbach's Epistemology,' *The Review of Metaphysics*

11 (1957), 42–67.
10. 'Reichenbach's Early Kantianism' *Philosophy and Phenomenological Research 9* (1957), pp. 86–94.
11. 'La theorie biologique de la connaissance chez Bergson et sa signification actuelle' *Revue de métaphysique et de morale.* 44e année (avril-juin 1959), pp. 194–211.
12. 'Vers l'elargissement de la notion de causalité' *Diogène*, No. 28 (1959), pp. 75–108.
13. 'Toward a Widening of the Notion of Causality' *Diogènes*, No. 28 (winter 1959), pp. 63–90.
14. 'Bergson et l'esprit de la physique contemporaine' *Bulletin de la Société française de Philosophie*, 53 année (1959), pp. 53–56. Actes du Xe congres des sociétés de philosophie de langue française; 54 année (1959), pp. 64–85.
15. 'Process and Personality in Bergson's Thought,' *The Philosophical Forum 17* (1959–60), 25–42.
16. 'The Theory of Eternal Recurrence in Modern Philosophy of Science, with Special Reference to C.S. Peirce' *The Journal of Philosophy 57* (1960), pp. 289–296.
17. 'The Elusive Nature of the Past' In: *Experience, Existence and the Good. Essays in Honor of Paul Weiss*, (Southern Illinois University Press, 1961), pp. 126–142.
18. 'Silent Revolution' *The Carleton Miscellany*, 3 (1962), pp. 50–65.
19. 'La signification actuelle de la philosophie de James' *Revue de métaphysique et de morale*, 67e année (1962), pp. 291–321.
20. 'Simple Location and Fragmentation of Reality,' *The Monist 48* (1964), 195–218.
21. 'Memini ergo fui?' *Memorias del XIII Congresso Internacional de Filosofia* (Mexico, 1963), 5, pp. 415–426.
22. 'Was Gassendi a Predecessor of Newton?' *Proceedings of the Xth International Congress of History of Science at Ithaca 1962* (Paris 1964), pp. 705–709.
23. 'The Myth of Frozen Passage: The Status of Becoming in the Physical World.' *Boston Studies in the Philosophy of Science, 2* (New York, 1965), pp. 441–463.
24. 'Time in Relativity Theory: Arguments for a Philosophy of Becoming.' In: *The Voices of Time* (New York, Braziller 1966), pp. 434–454.
25. 'Leibniz's Thought Prior to the Year 1670: From Atomism to a Geometrical Kinetism.' *Revue internationale de philosophie*, No. 76–77, Fascicule 2–3 (1966), pp. 249–256.
26. 'Time and Eternity in Royce and Bergson.' *Revue internationale de philosophie*, No. 79–80 fasc, 1–2 (1966), pp. 22–45.
27. 'André Marie Ampère', *The Encyclopedia of Philosophy*, ed. by Paul Edwards, (New York, MacMillan and Free Press, 1967), vol. I, pp. 93–94.
28. 'Change', *The Encyclopedia of Philosophy*, vol. II, pp. 75–79.
29. 'Czechoslovak Philosophy,' *The Encyclopedia of Philosophy*, vol. II, pp. 287–88.
30. 'Dynamism,' *The Encyclopedia of Philosophy*, vol. II, pp. 444–447.
31. 'Eternal Return,' *The Encyclopedia of Philosophy*, vol. III, pp. 61–63.
32. 'Wilhelm Ostwald', *The Encyclopedia of Philosophy*, vol. VI, pp. 6–7.

33. 'Ribot, Théodule Armand', *The Encyclopedia of Philosophy*, vol. VII, pp. 191–192.
34. 'Taine, Hyppolite Adolph', *The Encyclopedia of Philosophy*, vol. VIII, pp. 76–77.
35. 'Limits of Methodology' in: *Abstracts of the Papers of the Third International Congress for Logic, Methodology and Philosophy of Science* (Amsterdam, 1968), pp. 86–87.
36. 'Einstein and Meyerson on the Status of Becoming in the Physical World', *Actes du XIeme Congres de l'histoire des sciences* (Warsaw, 1968), I, pp. 129–139.
37. 'The Present Significance of Zawirski's History of the Concept of Time', *Actes du XIe Congres International d'histoire des sciences* (Warsaw, 1968), pp. 323–27.
38. 'La deuxième revolution scientifique', *Diogenes*, No. 63 (juillet-septembre 1968), pp. 122–141.
39. 'The Second Scientific Revolution,' *Diogenes*, Fall, 1963, pp. 114–133.
40. 'Two Critics of Newton Prior to Mach', *Resumés et Communications du XII Congrès d'histoire des sciences*, Paris, 23–28 aout 1968, pp. 35–36.
41. 'Ernst Mach's Biological Theory of Knowledge,' *Synthese*, 18 (1968), pp. 171–191.
42. 'La deuxième revolution scientifique', *Diogène*, No. 63, 1968, pp. 122–141.
43. 'Masaryk in the Present Perspective' in: *Fifty Years of Czechoslovakia* (Toronto, 1968), pp. 17–22 (In Czech).
44. 'The Main Difficulties of the Identity Theory' in *Scientia*, Annus 63, vol. CIV, Luglio-Agosto 1969, pp. 388–403.
45. 'Bergson's Theory of Matter and Modern Physics' in: *Bergson and the Evolution of Physics*, ed., by Peter A. Gunter, The University of Tennessee Press, Knoxville, 1969, pp. 297–330.
46. 'Freedom and Causality,' *Religious Humanism*, vol. 14, no. 2. (Spring, 1970), pp. 70–72. Published also in the symposium *Freedom of Choice*, ed. by E.H. Wilson (Humanist Center, Yellow Springs, 1970), pp. 15–17.
47. 'Professor Blanshard on Kierkegaard,' *The Modern Schoolman 48*, No. 1 (November 1970), pp. 44–52.
48. 'The Fiction of Instants,' *Studium Generale 24* (1971), pp. 31–43.
49. 'The Significance of Piaget's Researches on the Psychogenesis of Atomism,' *Boston Studies in the Philosophy of Science 8*. (1971), pp. 446–455.
50. 'Two Critics of Newton Prior to Mach: Boscovich and Stallo,' *Actes du XII Congrès International de l'histoire des sciences 4*, (1971), pp. 35–37.
51. 'Natural Sciences and the Future of Metaphysics,' *Main Currents in Modern Thought 28*, No. 5 (May-June 1972), pp. 167–71.
52. 'Das Raum-Problem im westlichen Denken,' *Sowjetsystem und Demokratische Gesellschaft*, Freiburg, 1972, pp. 463–69.
53. 'Der Zeitbegriff in der abendländischen Tradition,' *Sowjetsystem und Demokratische Gesellschaft*, Freiburg, 1972, pp. 469–77.
54. 'Bergson and the Evolution of Physics,' (A review article of the book of the same name by Peter A. Gunter), *Process Studies 2* (summer 1972), pp. 149–59.
55. 'Leibniz on Matter and Memory' in: *The Philosophy of Leibniz and the Modern*

World, ed. by Ivor Leclerc (Vanderbilt University Press 1973), pp. 78–113.
56. 'Comenius and the Moral Significance of Exile,' in: *Comenius*, ed. by Vratislav Busek (Czechoslovak Society of Arts and Sciences, New York, 1972), pp. 33–47.
57. 'Time,' In: *Dictionary of the History of Ideas*, ed. by Philip P. Wiener (Charles Scribner, New York, 1973), *4* pp. 389–398.
58. 'The Problem of Space in Western Thought,' in: *Marxismus, Communism and Western Society. A Comparative Encyclopedia*, ed. by C.D. Kernig (Herder & Herder, New York, 1973), *8*, pp. 59–62.
59. 'The Concept of Time in Western Tradition,' in: *Marxismus, Communism and Western Society* (the same as above), *8*, pp. 62–66.
60. 'Two Types of Continuity,' *Boston Studies in the Philosophy of Science 13*. (Reidel, Dordrecht, 1974), 361–374.
61. 'Sur quelques résistances philosophiques à la physique du vingtième siecle,' *Dialectica 28* (1974), no. 3–4, pp. 211–222.
62. 'Johann Bernard Stallo,' in: *Dictionary of Scientific Biography*, ed. by Charles C. Gillispie (Ch. Scribner's Sons, New York, 1975), *12*, pp. 606–610.
63. 'Relativity and the Status of Becoming,' *Foundations of Physics 5*, (December 1975), pp. 607–616.
64. Introduction to: *Concepts of Space and Time. Their Structure and Their Development*. Ed. by Milič Čapek. *Boston Studies in the Philosophy of Science 23* (Reidel, Dordrecht, 1976).
65. 'Temporal Order and Spatial Order. Their Differences and Their Relations,' in: *Mind in Nature. Essays on the Interface of Science and Philosophy*. Ed. by John B. Cobb Jr. (University Press of America, Washington, 1977), pp. 51–59.
66. 'Immediate and Mediate Memory,' *Process Studies 7* (1977), pp. 90–96.
67. 'La pensée de Bergson en Amérique,' *Revue internationale de philosophie 31* (1977), pp. 329–350.

BOSTON STUDIES IN THE PHILOSOPHY OF SCIENCE

1. Marx W. Wartofsky (ed.), *Proceedings of the Boston Colloquium for the Philosophy of Science 1961–1962*. 1963.
2. Robert S. Cohen and Marx W. Wartofsky (eds.), *In Honor of Philipp Frank*. 1965.
3. Robert S. Cohen and Marx W. Wartofsky (eds.), *Proceedings of the Boston Colloquium for the Philosophy of Science 1964–1966. In Memory of Norwood Russell Hanson*. 1967.
4. Robert S. Cohen and Marx W. Wartofsky (eds.), *Proceedings of the Boston Colloquium for the Philosophy of Science 1966–1968*. 1969.
5. Robert S. Cohen and Marx W. Wartofsky (eds.), *Proceedings of the Boston Colloquium for the Philosophy of Science 1966–1968*. 1969.
6. Robert S. Cohen and Raymond J. Seeger (eds.), *Ernst Mach: Physicist and Philosopher*. 1970.
7. Milic Čapek, *Bergson and Modern Physics*. 1971.
8. Roger C. Buck and Robert S. Cohen (eds.), *PSA 1970. In Memory of Rudolf Carnap*. 1971.
9. A. A. Zinov'ev, *Foundations of the Logical Theory of Scientific Knowledge (Complex Logic)*. (Revised and enlarged English edition with an appendix by G. A. Smirnov, E. A. Sidorenka, A. M. Fedina, and L. A. Bobrova). 1973.
10. Ladislav Tondl, *Scientific Procedures*. 1973.
11. R. J. Seeger and Robert S. Cohen (eds.), *Philosophical Foundations of Science*. 1974.
12. Adolf Grünbaum, *Philosophical Problems of Space and Time*. (Second, enlarged edition). 1973.
13. Robert S. Cohen and Marx W. Wartofsky (eds.), *Logical and Epistemological Studies in Contemporary Physics*. 1973.
14. Robert S. Cohen and Marx W. Wartofsky (eds.), *Methodological and Historical Essays in the Natural and Social Sciences. Proceedings of the Boston Colloquium for the Philosophy of Science 1969–1972*. 1974.
15. Robert S. Cohen, J. J. Stachel, and Marx W. Wartofsky (eds.), *For Dirk Struik. Scientific, Historical and Political Essays in Honor of Dirk Struik*. 1974.
16. Norman Geschwind, *Selected Papers on Language and the Brain*. 1974
17. B. G. Kuznetsov, *Reason and Being: Studies in Classical Rationalism and Non-Classical Science*. 1987
18. Peter Mittelstaedt, *Philosophical Problems of Modern Physics*. 1976
19. Henry Mehlberg, *Time, Causality, and the Quantum Theory* (2 vols.). 1980.

20. Kenneth F. Schaffner and Robert S. Cohen (eds.), *Proceedings of the 1972 Biennial Meeting, Philosophy of Science Association*. 1974
21. R. S. Cohen and J. J. Stachel (eds.), *Selected Papers of Léon Rosenfeld*. 1978.
22. Milic Čapek (ed.), *The Concepts of Space and Time. Their Structure and Their Development*. 1976.
23. Marjorie Grene, *The Understanding of Nature, Essays in the Philosophy of Biology*. 1974.
24. Don Ihde, *Technics and Praxis. A Philosophy of Technology*. 1978.
25. Jaakko Hintikka and Unto Remes, *The Method of Analysis. Its Geometrical Origin and Its General Significance*. 1974.
26. John Emery Murdoch and Edith Dudley Sylla, *The Cultural Context of Medieval Learning*. 1975.
27. Marjorie Grene and Everett Mendelsohn (eds.), *Topics in the Philosophy of Biology*. 1976.
28. Joseph Agassi, *Science in Flux*. 1975.
29. Jerzy J. Wiatr (ed.), *Polish Essays in the Methodology of the Social Sciences*. 1979.
30. Peter Janich, *Protophysics of Time*. 1985.
31. Robert S. Cohen and Marx W. Wartofsky (eds.), *Language, Logic and Method*. 1983.
32. R. S. Cohen, C. A. Hooker, A. C. Michalos, and J. W. van Evra (eds.), *PSA 1974: Proceedings of the 1974 Biennial Meeting of the Philosophy of Science Association*. 1976.
33. Gerald Holton and William Blanpied (eds.), *Science and Its Public: The Changing Relationship*. 1976.
34. Mirko D. Grmek (ed.), *On Scientific Discovery*. 1980.
35. Stefan Amsterdamski, *Between Experience and Metaphysics. Philosophical Problems of the Evolution of Science*. 1975.
36. Mihailo Marković and Gajo Petrović (eds.), *Praxis, Yugoslav Essays in the Philosophy and Methodology of the Social Sciences*. 1979.
37. Hermann von Helmholtz, *Epistemological Writings. The Paul Hertz/Moritz Schlick Centenary Edition of 1921 with Notes and Commentary by the Editors*. (Newly translated by Malcolm F. Lowe. Edited, with an Introduction and Bibliography, by Robert S. Cohen and Yehuda Elkana). 1977.
38. R. M. Martin, *Pragmatics, Truth, and Language*. 1979.
39. R. S. Cohen, P. K. Feyerabend, and M. W. Wartofsky (eds.), *Essays in Memory of Imre Lakatos*. 1976.
40. B. M. Kedrov and V. Sadovsky. *Current Soviet Studies in the Philosophy of Science*. Forthcoming.
41. M. Raphael, *Theorie des Geistigen Schaffens auf Marxistischer Grundlage*. Forthcoming.
42. Humberto R. Maturana and Francisco J. Varela, *Autopoiesis and Cognition. The Realization of the Living*. 1980.
43. A. Kasher (ed.), *Language in Focus: Foundations, Methods and Systems. Essays Dedicated to Yehoshua Bar-Hillel*. 1976.
44. Trân Duc Thao, *Investigations into the Origin of Language and Consciousness*. (Translated by Daniel J. Herman and Robert L. Armstrong; edited by Carolyn

R. Fawcett and Robert S. Cohen). 1984.

45. A. Ishimoto (ed.), *Japanese Studies in the History and Philosophy of Science.*
46. Peter L. Kapitza, *Experiment, Theory, Practice*. 1980.
47. Maria L. Dalla Chiara (ed.), *Italian Studies in the Philosophy of Science*. 1980.
48. Marx W. Wartofsky, *Models: Representation and the Scientific Understanding*. 1979.
49. Trân Duc Thao, *Phenomenology and Dialectical Materialism.* 1985.
50. Yehuda Fried and Joseph Agassi, *Paranoia: A Study in Diagnosis*. 1976.
51. Kurt H. Wolff, *Surrender and Catch: Experience and Inquiry Today*. 1976.
52. Karel Kosik, *Dialectics of the Concrete*. 1976.
53. Nelson Goodman, *The Structure of Appearance*. (Third edition). 1977.
54. Herbert A. Simon, *Models of Discovery and Other Topics in the Methods of Science*. 1977.
55. Morris Lazerowitz, *The Language of Philosophy. Freud and Wittgenstein.* 1977.
56. Thomas Nickles (ed.), *Scientific Discovery, Logic, and Rationality*. 1980.
57. Joseph Margolis, *Persons and Minds. The Prospects of Nonreductive Materialism.* 1977.
58. G. Radnitzky and G. Andersson (eds.), *Progress and Rationality in Science*, 1978.
59. Gerard Radnitzky and Gunnar Andersson (eds.), *The Structure and Development of Science*. 1979.
60. Thomas Nickles (ed.), *Scientific Discovery: Case Studies*. 1980.
61. Maurice A. Finocchiaro, *Galileo and the Art of Reasoning*. 1980.
62. William A. Wallace, *Prelude to Galileo*. 1981.
63. Friedrich Rapp, *Analytical Philosophy of Technology*. 1981.
64. Robert S. Cohen and Marx W. Wartofsky (eds.), *Hegel and the Sciences*. 1984.
65. Joseph Agassi, *Science and Society*. 1981.
66. Ladislav Tondl, *Problems of Semantics*. 1981.
67. Joseph Agassi and Robert S. Cohen (eds.), *Scientific Philosophy Today*. 1982.
68. Władysław Krajewski (ed.), *Polish Essays in the Philosophy of the Natural Sciences*. 1982.
69. James H. Fetzer, *Scientific Knowledge*. 1981.
70. Stephen Grossberg, *Studies of Mind and Brain*. 1982.
71. Robert S. Cohen and Marx W. Wartofsky (eds.), *Epistemology, Methodology, and the Social Sciences*. 1983.
72. Karel Berka, *Measurement*. 1983.
73. G. L. Pandit, *The Structure and Growth of Scientific Knowledge*. 1983.
74. A. A. Zinov'ev, *Logical Physics*. 1983.
75. Gilles-Gaston Granger, *Formal Thought and the Sciences of Man*. 1983.
76. R. S. Cohen and L. Laudan (eds.), *Physics, Philosophy and Psychoanalysis*. 1983.
77. G. Böhme et al., *Finalization in Science*, ed. by W. Schäfer. 1983.
78. D. Shapere, *Reason and the Search for Knowledge*. 1983.
79. G. Andersson, *Rationality in Science and Politics*. 1984.
80. P. T. Durbin and F. Rapp, *Philosophy and Technology*. 1984.
81. M. Marković, *Dialectical Theory of Meaning*. 1984.

82. R. S. Cohen and M. W. Wartofsky, *Physical Sciences and History of Physics.* 1984.
83. E. Meyerson, *The Relativistic Deduction.* 1985.
84. R. S. Cohen and M. W. Wartofsky, *Methodology, Metaphysics and the History of Sciences.* 1984.
85. György Tamás, *The Logic of Categories.* 1985.
86. Sergio L. de C. Fernandes, *Foundations of Objective Knowledge.* 1985.
87. Robert S. Cohen and Thomas Schnelle (eds.), *Cognition and Fact.* 1985.
88. Gideon Freudenthal, *Atom and Individual in the Age of Newton.* 1985.
89. A. Donagan, A. N. Perovich, Jr., and M. V. Wedin (eds.), *Human Nature and Natural Knowledge.* 1985.
90. C. Mitcham and A. Huning (eds.), *Philosophy and Technology II.* 1986.
91. M. Grene and D. Nails (eds.), *Spinoza and the Sciences.* 1986.
92. S. P. Turner, *The Search for a Methodology of Social Science.* 1986.
93. I. C. Jarvie, *Thinking about Society: Theory and Practice.* 1986.
94. Edna Ullmann-Margalit (ed.), *The Kaleidoscope of Science.* 1986.
95. Edna Ullmann-Margalit (ed.), *The Prism of Science.* 1986.
96. G. Markus, *Language and Production.* 1986.
97. F. Amrine, F. J. Zucker, and H. Wheeler (eds.), *Goethe and the Sciences: A Reappraisal.* 1987.
98. Joseph C. Pitt and Marcella Pera (eds.), *Rational Changes in Science.* 1987.
99. O. Costa de Beauregard, *Time, the Physical Magnitude.* 1987.
100. Abner Shimony and Debra Nails (eds.), *Naturalistic Epistemology: A Symposium of Two Decades.* 1987.
101. Nathan Rotenstreich, *Time and Meaning in History.* 1987.
102. David B. Zilberman (ed.), *The Birth of Meaning in Hindu Thought.* 1987.
103. Thomas F. Glick (ed.), *The Comparative Reception of Relativity.* 1987.
104. Zellig Harris *et al.*, *The Form of Information in Science.* 1987
105. Frederick Burwick, *Approaches to Organic Form: Permutations in Science and Culture.* 1987.
106. M. Almási, *Philosophy of Appearances.* Forthcoming.
107. S. Hook, W. L. O'Neill, and R. O'Toole, *Philosophy, History and Social Action. Essays in Honor of Lewis Feuer.* 1988.
108. I. Hronszky, M. Fehér, and B. Dajka (eds.), *Scientific Knowledge Socialized. Selected Proceedings of the Fifth Joint International Conference on History and Philosophy of Science Organized by the IUHPS, Veszprém, 1984.* Forthcoming.
109. P. Tillers and E. D. Green (eds.), *Probability and Inference in the Law of Evidence. The Uses and Limits of Bayesianism.* 1988.
110. E. Ullmann-Margalit (ed.), *Science in Reflection. The Israel Colloquium: Studies in History, Philosophy, and Sociology of Science.* 1988.
111. K. Gavroglu, Y. Goudaroulis, and P. Nicolacopoulos (eds.), *Imre Lakatos and Theories of Scientific Change.* 1989.
112. Barry Glassner and Jonathan D. Moreno (eds.), *The Qualitative-Quantitative Distinction in the Social Sciences.* 1989.
113. K. Arens, *Structures of Knowing: Psychologies of the Nineteenth Century.* 1989.

114. A. Janik, *Style, Politics and the Future of Philosophy*. 1989.
115. F. Amrine (ed.), *Literature and Science as Modes of Expression*. 1989.
116. James Robert Brown and Jürgen Mittelstrass (eds.), *An Intimate Relation: Studies in the History and Philosophy of Science Presented to Robert E. Butts on his 60th Birthday*. 1989.
117. F. D'Agostino and I. C. Jarvie (eds.), *Freedom and Rationality: Essays in Honor of John Watkins*. 1989.
118. D. Zolo, *Reflective Epistemology: The Philosophical Legacy of Otto Neurath*. 1989.
119. Michael Kearn, Bernard S. Phillips and Robert S. Cohen (eds.), *George Simmel and Contemporary Sociology*. 1989.
120. Trevor H. Levere and William R. Shea (eds.), *Nature, Experiment, and the Sciences: Essays on Galileo and the History of Science in Honour of Stillman Drake*. 1989.
121. P. Nicolacopoulos (ed.), *Greek Studies in the Philosophy and History of Science*. 1990.
122. R. Cooke and D. Constantini (eds.), *Statistics in Science. The Foundations of Statistical Methods in Biology, Physics and Economics*. 1990
123. P. Duhem (ed.), *The Origins of Statics*. 1991.
124. K. Gavroglu and Y. Goudaroulis (eds.), Heike Kamerlingh Onnes, *Through Measurement to Knowledge. The Selected Papers of Heike Kamerlingh Onnes 1853–1926*. 1991.
125. M. Čapek, *The New Aspects of Time. Its Continuity and Novelties. Selected Papers in the Philosophy of Science*. 1991.

GPSR Compliance

The European Union's (EU) General Product Safety Regulation (GPSR) is a set of rules that requires consumer products to be safe and our obligations to ensure this.

If you have any concerns about our products, you can contact us on ProductSafety@springernature.com

In case Publisher is established outside the EU, the EU authorized representative is:

Springer Nature Customer Service Center GmbH
Europaplatz 3
69115 Heidelberg, Germany

Batch number: 10761599

Printed by Printforce, the Netherlands